ANIMAL SPECIES FOR DEVELOPMENTAL STUDIES

Volume 2
Vertebrates

ANIMAL SPECIES FOR DEVELOPMENTAL STUDIES

Volume 2
Vertebrates

Edited by
T. A. Dettlaff and S. G. Vassetzky

N. K. Kol'tsov Institute of Developmental Biology
Academy of Sciences of the USSR
Moscow, USSR

Translated into English by
G. G. Gause, Jr., and S. G. Vassetzky

Technical Editors
Frank Billett
and
L. A. Winchester

CONSULTANTS BUREAU • NEW YORK AND LONDON

Library of Congress Cataloging in Publication Data

(Revised for volume 2)
Animal species for developmental studies.

Translation of Ob"ekty biologii razvitiiâ.
Includes bibliographical references and index.
Contents: v. 1. Invertebrates — v. 2. Vertebrates.
1. Embryology, Experimental. 2. Developmental biology. 3. Laboratory animals. I.
Detlaf, Tat'iâna Antonovna. II. Vassetzky, S. G.
QL961.02413   1989   591.3'0724   89-22389
ISBN 0-306-11032-6

This translation is published under an agreement with the Copyright
Agency of the USSR (VAAP)

© 1991 Consultants Bureau, New York
A Division of Plenum Publishing Corporation
233 Spring Street, New York, N.Y. 10013

PREFACE

This volume is a revised and augmented edition of part of the book *Ob"ekty Biologii Razvitiya* (*Animal Species for Developmental Studies*) published in Russian in 1975 in the series of monographs *Problemy Biologii Razvitiya* (*Problems of Developmental Biology*) by Nauka Publishers, Moscow. That book described the development of organisms most frequently used in developmental biology studies. Data were provided for 22 animal species, belonging to different taxa, from protists to mammals. For the English edition we decided to divide the original book into two parts dealing with vertebrates and invertebrates, respectively. This volume deals with vertebrate species. When choosing these species, their advantages for laboratory studies, information available, and availability for experimentation in the USSR and in Europe were taken into account. This geographical criterion explains the absence in the book of a number of species widely used in the laboratories of the USA, Japan, and other countries, such as *Rana pipiens, Cynops pyrrhogaster*, and others.

Besides the classical laboratory animals, some fish have been described since the study of the mechanisms of their development and attempts to control their ontogenesis are of immediate value and the results obtained can be tested on the mass material. A study of the development of laboratory mammals is of special interest since current problems of modern medicine and veterinary sciences are tackled using these animals.

The description of every species is preceded by an introduction in which the advantages of working with this animal are stated and the problems studied (with the main references) are outlined. Data are also provided on the taxonomic status and distribution of the animal. Conditions of keeping the adult animals in the laboratory, methods of obtaining gametes and of artificial insemination, rearing of embryos and larvae, and the tables of normal development are given. For a number of species these tables were elaborated especially for the Russian edition: tables of comparable developmental stages for four mammal species (mouse, rat, golden hamster, rabbit); tables of normal development of the axolotl and the Russian sturgeon larvae. For the other species new drawings were made and new developmental stages (more detailed) were added as, for example, for the common frog. In the other cases the tables of normal development and the descriptions of the structure of embryos (with more or less significant additions), which had been published elsewhere, were reproduced.

In this book an attempt is undertaken to facilitate the study of temporal patterns of animal development by the introduction (for some animals) of relative (dimensionless) characteristics of the timing of developmental stages comparable in different animal species and at different temperatures. In this respect we considered it necessary to include a special introductory chapter in which the problem of time as one of the parameters of the development is examined and the method of relative characteristics of the development duration proposed in 1961 by one of us (T. A. Dettlaff) is substantiated. This first chapter demonstrates how the above-mentioned method can be used and what prospects it opens.

As a whole, the book ought to facilitate the selection of experimental species for development studies and the work with the chosen species. One has to bear in mind that the use of the normal tables is necessary for standardizing experimental materials and obtaining comparable results.

In conclusion, we would like to thank all those who kindly provided their permission to reproduce the normal tables, drawings, etc., as well as those who gave valuable advice.

T. A. Dettlaff
S. G. Vassetzky

CONTRIBUTORS

V. S. Baranov, Institute of Experimental Medicine, Academy of Medical Sciences of the USSR, Leningrad.

N. P. Bordzilovskaya, Institute of Developmental Biology, Academy of Sciences of the USSR, Moscow.

N. A. Chebotar', Institute of Experimental Medicine, Academy of Medical Sciences of the USSR, Leningrad.

N. V. Dabagyan, Biological Faculty, Moscow State University, Moscow.

T. A. Dettlaff, Institute of Developmental Biology, Academy of Sciences of the USSR, Moscow.

A. P. Dyban, Institute of Experimental Medicine, Academy of Medical Sciences of the USSR, Leningrad.

A. S. Ginsburg, Institute of Developmental Biology, Academy of Sciences of the USSR, Moscow.

G. M. Ignatieva, Institute of Developmental Biology, Academy of Sciences of the USSR, Moscow.

L. I. Khozhai, Institute of Experimental Medicine, Academy of Medical Sciences of the USSR, Leningrad.

A. A. Kostomarova, Institute of Developmental Biology, Academy of Sciences of the USSR, Moscow.

L. D. Liozner, Institute of Human Morphology, Academy of Medical Sciences of the USSR, Moscow.

V. F. Puchkov, Institute of Experimental Medicine, Academy of Medical Sciences of the USSR, Leningrad.

M. N. Ragozina, Institute of Evolutionary Morphology and Animal Ecology, Academy of Sciences of the USSR, Moscow.

T. B. Rudneva, Institute of Immunology, Academy of Medical Sciences of the USSR, Moscow.

N. A. Samoshkina, Institute of Experimental Medicine, Academy of Medical Sciences of the USSR, Leningrad.

O. I. Schmalhausen, Institute of Developmental Biology, Academy of Sciences of the USSR, Moscow.

L. A. Sleptsova, Biological Faculty, Moscow State University, Moscow.

S. G. Vassetzky, Institute of Developmental Biology, Academy of Sciences of the USSR, Moscow.

CONTENTS

Chapter 1
Introduction: Temperature and Timing
in Developmental Biology   1
T. A. Dettlaff

Chapter 2
The Russian Sturgeon *Acipenser güldenstädti*
Part I. Gametes and Early Development
up to Time of Hatching   15
A. S. Ginsburg and T. A. Dettlaff

Chapter 12
Laboratory Mammals: Mouse (*Mus musculus*),
Rat (*Rattus norvegicus*), Rabbit (*Oryctolagus cuniculus*),
and Golden Hamster (*Cricetus auratus*)   351
A. P. Dyban, V. F. Puchkov, N. A. Samoshkina,
L. I. Khozhai, N. A. Chebotar', and V. S. Baranov

Chapter 1

INTRODUCTION: TEMPERATURE AND TIMING IN DEVELOPMENTAL BIOLOGY

T. A. Dettlaff

Strictly ordered development and precisely coordinated timing of the appearance of new structures and functions is characteristic of animal development, but the mechanisms underlying such orderliness remain unclear [43, 51]. Investigations of temporal patterns in development have been limited until recently by the fact that the duration of any developmental period is a variable depending on a species-specific rate of development and, in poikilothermic animals, also on external conditions (temperature in particular), as well as on species-specific limits of the range of optimal temperatures which correspond to those of spawning.

In some poikilothermic animals there are no common spawning temperatures and the durations of their development cannot be compared at all. In most others the limits of the range of optimal temperatures are displaced with respect to each other, in more cold-loving species toward lower temperatures and in more warmth-loving species toward higher ones. Therefore, in the range of temperatures common for the species under comparison, the rates of their development change with temperature in different ways: at lower temperatures the embryos of cold-loving animals develop more rapidly, whereas at higher temperatures those of warmth-loving species develop more rapidly and, at some intermediate temperatures, they develop at the same rate (Fig. 1.1). This relationship must be taken into account, in particular, when using data on the timing of developmental stages in the normal tables obtained for different species, not infrequently at the same arbitrarily chosen temperature. For amphibian species it is usually 18°C, but the ratio of the rates of development determined at 18°C will be valid only for the given temperature and will prove to be different at other temperatures (see [19, 24]).

Thus, the duration of different developmental periods expressed in units of time (minutes, hours, days) is a variable which can be compared neither in different animals nor in the same animal under different conditions.

The application of relative, or dimensionless, criteria of developmental duration makes it possible to overcome these difficulties to a great extent, and to introduce timing into embryological investigations. With this aim, the durations of developmental stages or periods are expressed as a portion of the period taken as a unit of

1

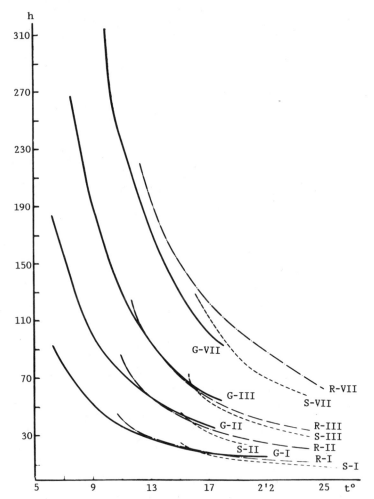

Fig. 1.1. Duration of same developmental periods in *Huso huso* (G), *Acipenser güldenstädti* (R), and *A. stellatus* (S) at different temperatures. Duration from fertilization to onset of gastrulation (I), end of gastrulation (II), closure of neural plate (III) [30], and onset of hatching (VII).

total developmental duration, rather than in units of absolute time. A few cases have been described in which relative characteristics of developmental duration were used for unravelling particular problems [2, 4, 21, 22, 36, 41]. In its more general form, the possibility of using relative criteria of developmental duration for studying the temporal patterns of animal development was considered by Dettlaff and Dettlaff [14, 15].

First, it was established that the duration of different developmental periods varies directly with temperature (within the limits of the range of optimal temperatures for a given species) and, hence, for every animal species the duration of any developmental period can be taken as a unit of total developmental duration. However, in the comparison of timing in different animals it is important that the period taken as a unit of developmental duration be biologically equivalent for these animals. In this connection it should be noted that, due to the wide distribution of heterochronies, most periods taken by other authors as units of developmental duration do not conform to this requirement. Thus, the period from fertilization to the onset of cleavage [22] is not equivalent in different animals, since their eggs are fertilized at different meiotic stages; the periods of cleavage, blastulation, and gastrulation [4] are not equivalent either, since gastrulation begins at different cleavage stages in different animals; the whole period of embryogenesis [21] cannot be compared since in different animals the embryos are liberated from their embryonic membranes at different developmental stages. As will be shown below, the early intercleavage interval [2], used for the comparison of the duration of the period from fertilization to the onset of gastrulation in different animals, is also unsuitable for valid comparison.

We proposed that the duration of one mitotic cycle, that is, the interval between the same phases of two successive divisions during synchronous cleavage divisions as a measure of time (τ_0), is comparable for a large number of animals [13]. For all animals which have a period of synchronous cleavage divisions in their development, τ_0 is a minimal cell cycle which has similar features and a similar structure. In this cycle the G_1 and G_2 phases are practically absent, and the M phase takes up half of the whole cycle. At interphase the nucleus has a karyomere structure, with no nucleoli or only their fibrillar component; the cytoplasm contains a large pool of DNA precursors, DNA synthesis enzymes, protein precursors of the mitotic apparatus, and other substances stored during oogenesis which provide for a cell cycle of minimal duration. As a result, τ_0 can serve as a comparable measure of developmental duration in all animals which have a period of synchronous cleavage divisions, including annelids, molluscs, crustaceans, insects, echinoderms, acipenserid and teleostean fish, amphibians, and birds. As there is no period of synchronous cleavage divisions in mammals or in some animals characterized by the mosaic type of development (such as *Tubifex*), we cannot propose τ_0 as a unit of developmental duration comparable with τ_0 in the other animal groups mentioned.

Thus, the use of τ_0 as a unit of developmental duration is theoretically possible in all animals which have a period of synchronous cleavage divisions. The problem is how to determine the value of τ_0 at different temperatures and for different animals. In many animals τ_0 can be determined by observing the interval between the appearance of cleavage furrows on the egg surface [9]. One has to bear in mind, however, that the ratios between karyo- and cytokinesis can differ in different species and they have to be determined specially. In acipenserid fish and amphibians with holoblastic eggs, τ_0 is taken as the interval between the appearance of the 1st and 2nd cleavage furrows on the egg surface [25, 39, 40]. In teleostean fish τ_0 is taken as half of the interval between the appearance of the 2nd and 4th cleavage furrows [29, 31; see also 26].

In animals with centrolecithal eggs in which the first cleavage divisions are not accompanied by the formation of cell boundaries, the duration of the mitotic cycle during the period of synchronous cleavage divisions can be determined by the interval between the same phases of successive cleavage divisions or by noting the degree of impairment of the nuclei at different mitotic phases by x rays and other factors, the interval between successive maxima of impairment being used to deter-

mine the cell cycle. There are such data for the mulberry silkworm [34, 37, 38].
For the honey bee in a hive in which a constant temperature of $35 \pm 0.5°C$ is maintained, τ_0 is (at this temperature) 30–35 min [20, 47].

In animals for which it is difficult to determine the values of τ_0 at different
temperatures, it is important to determine the value of τ_0 and the duration of different developmental periods of at least one temperature in the middle of the optimal
range. Thus, the temporal parameters of their development can be compared using
relative characteristics of developmental duration. Graphs expressing the dependence of τ_0 on temperature are available for four species of acipenserid fish, nine
species of teleostean fish, and seven species of amphibians (see [12]). In this book
they are given for *Asterina pectinifera*, *Acipenser güldenstädti*, *Salmo gairdneri*,
Misgurnus fossilis, *Triturus vulgaris*, *Ambystoma mexicanum*, *Xenopus laevis*,
Rana temporaria, and *Rana ridibunda*. Additionally, the value of τ_0 at one temperature is provided for *Pleurodeles waltlii* and at several temperatures for *Lymnaea
stagnalis*, *Strongylocentrotus dröbachiensis*, and *S. intermedius*.

To obtain the relative characteristics of the period (τ_n) from fertilization (or
from the appearance of the 1st cleavage furrow on the egg surface) to the onset of
any developmental stage, the value of τ_n in absolute time units is determined at a
temperature within the range of optimal temperatures under all other favorable conditions. In addition, to unify the method of determination of the value of τ_n and to
lessen the influence of natural variability in the timing of the same stages in different
embryos, τ_n is determined by the transition to the following stage of the first 5–
10% of embryos.

The values of τ_n determined in such a way, at well-defined developmental
stages on a large number of embryos from different females and at different
temperatures, fall regularly on the same curve, thus confirming the validity of this
method. The variability in timing of the same stages in different embryos, from
both the same and different females, does not usually exceed 10%. The time of
hatching, that is, of liberation of the embryos from their embryonic membranes, can
vary to a greater extent, since it depends not only on the time of onset of the definite
stage, but also on the conditions influencing the secretion and action of the hatching
enzyme.

The use of τ_0 units for the quantitative estimation of the variation of timing of
developmental stages in different animals allows us to compare its variation under
different conditions and in embryos from different females, and thus to study the
possible causes of variation.

Having determined the value of τ_n at an optimal temperature, one translates it
into minutes and divides by the value of τ_0 in minutes at the same temperature. In
such a way the same units of absolute time, temperature, and species-specific dependence of development rate on temperature are eliminated in both the denominator
and numerator. The resulting ratio τ_n/τ_0 is a relative (dimensionless) characteristic
of biological time, free both from temperature and from the species-specific
dependence of development rate on the temperature of development. The ratio does
not depend on the variation in the limits of the range of spawning temperatures in
different species. The use of τ_n/τ_0 as a criterion of time in development represents a
particular case of that in which several variables (in this case temperature, the
species-specific dependence of development rate on temperature, and the limits of
optimal conditions) are replaced by one generalized variable (in this case the

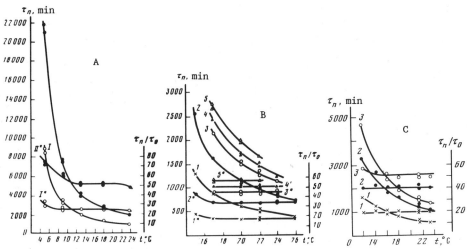

Fig. 1.2. Absolute (τ_n, in min) and relative (as τ_n/τ_0) durations of selected developmental periods in the embryos of: A, *Rana temporaria* [5]; B, *Xenopus laevis* [17, 35]; C, *Misgurnis fossilis* [11, 28]. A) Durations from fertilization to the onset (I) and end of gastrulation (II). B) Durations from the appearance of the first cleavage furrow on the egg surface to the following stages: 10, onset of gastrulation (1); 13, slitlike blastopore (2); 19, fusion of neural folds (3); 22, eye vesicles first discernible (4); 24, first muscle reaction to external stimuli (5). Stage numbers according to Nieuwkoop and Faber [35]. C) Durations from fertilization to the following stages: 10, onset of gastrulation (1); 20, end of epiboly (2); 26, stage of 10 pairs of somites (3). Stage numbers according to Chapter 5 of this volume. Numerals with asterisks refer to relative durations.

duration of one mitotic cycle during synchronous cleavage divisions, τ_0) which, in turn, is a function of all these factors [15].

The applicability of this method has proved to be very wide, since in a great number of studies [7, 8, 10–15, 26, 28, 52–55], as well as in studies considered in this review and in some chapters of this book, it was shown that proportional changes in the duration of τ_0, at different developmental periods at the same temperature, are observed not only during embryogenesis (Fig. 1.2) but also during oocyte maturation and after hatching. Such proportionality is also apparent at different levels of organization, from the molecular, via the subcellular and cellular, to the organic. Thus the ratio τ_n/τ_0 describes equally well the time of onset of synthesis of various RNAs, and the durations of meiotic phases of oocyte maturation, of mitotic phases of cleavage divisions, of gastrulation, and of embryonic and prelarval development.

The proportional changes in the duration of such different processes with temperature can be explained only by the hypothesis that some, as yet unknown, mechanisms working at the molecular level are involved.

At suboptimal temperatures this proportionality is disturbed and τ_0 cannot then serve as a measure of developmental duration. The change in the ratio τ_n/τ_0 (its decrease in the range of higher temperatures or increase in the range of lower temperatures) can serve as a criterion of the limits of the optimal temperature range for the development of the given species. Data on the limits of this range can be obtained by analyzing the curve expressing the dependence of the value of τ_0 on temperature. A temperature exceeding the lower limit by 1°C, and not resulting in a further decrease of the value of τ_0 or even inducing its increase, is an upper limit of the range of optimal temperatures. The sloping part of the curve coincides with spawning temperatures in species which spawn in spring and summer. The lower limit lies in the range of temperatures at which development is markedly slowed down. It is revealed more precisely, according to Masin et al. [32], if the dependence of τ_0 on temperature is presented in semilogarithmic coordinates (Fig. 1.3).

The application of dimensionless criteria of developmental duration makes it possible to predict the time of onset of certain developmental stages at various temperatures within the limits of the optimal temperature range without additional observations. For this purpose the value of τ_0 at a required temperature should be

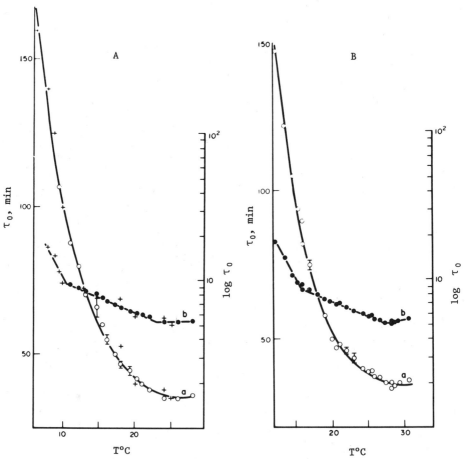

Fig. 1.3. Temperature dependence of the duration of the first mitotic cycle (τ_0). A) In *Rana temporaria*, + after Chulitskaya [5],○) after Mazin et al. [32]. B) In *R. ridibunda*, after Mazin et al. [32]. The same data are also presented on a semilogarithmic scale (b).

taken from the curve of τ_0 versus temperature and multiplied by the value of τ_n/τ_0 for the given developmental stage taken from the normal tables.

The time of onset of any developmental stage (τ_n) at different temperatures can be calculated in another way: The value of τ_{0_1} is taken at the temperature for which the time is known in hours (N_1), and the value of τ_{0_2} is taken at the temperature for which the time (N_2) should be determined as $N_2 = N_1(\tau_{0_2}/\tau_{0_1})$.

Thus, one can determine the times of successive developmental stages at different temperatures for meiosis in oocytes maturing *in vivo* and *in vitro*, since the rate of oocyte maturation is the same in both cases and depends on temperature as well as τ_0 [13, 54].

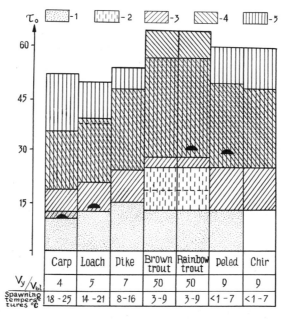

Fig. 1.4. Relative durations (expressed as number of τ_0 units) of corresponding periods of early development in different species of teleosts. 1) From fertilization to the prolongation of interphase (fall of mitotic interphase); 2) from the fall of the mitotic index to the onset of morphogenetic nuclear function; 3) from the onset of morphogenetic nuclear function to the onset of gastrulation; 4) from the onset of gastrulation to the end of epiboly; 5) from the onset of gastrulation to the formation of ten pairs of somites. Dashed lines indicate the onset of RNA synthesis in the loach and trout. Crescents show the age at which isolated blastoderms acquire the capacity to differentiate *in vitro*. V_y/V_{bl} is the ratio of yolk volume to the volume of the blastodisc; these volumes were calculated on the basis of egg diameter and blastodisc height at the two-blastomere stage, using the formula for the volume of a spherical segment. Data on spawning temperatures were taken from [28].

TABLE 1.1. Relative Durations (in τ_0 units) of the Same Developmental Periods in Different Amphibian Species (after [15])

Family	Species	Period*						Authors
		I	II	III	IV	V	I–V	
Pipidae	Xenopus laevis	3	11 / 21	7	16	14	51	Dettlaff and Rudneva [17]
Ranidae	Rana temporaria	3	14 / 26	9	25	14	65	Chulitskaya [5]
	Rana pipiens	3	26		24	16	66	Moore [33]†
	Rana palustris	3	26.6		19	21	67–68	Moore [33]†
Ambystomidae	Ambystoma mexicanum	3	10.5	7	19	10.5	50	Skoblina [49]; Dettlaff and Rudneva (unpubl.); Bordzilovskaya and Dettlaff [3]
Salamandridae	Triturus helveticus		19.3		12	15.3	46.6	Gallien and Bidaud [23]†
	Pleurodeles waltlii		18.7		16	16	50.7	Gallien and Durocher [24]†

*Periods: I, from fertilization to the appearance of the 1st cleavage furrow; II, from the end of I to the fall of mitotic index (synchronous divisions and desynchronization); III, from the end of II to the onset of gastrulation (blastulation); IV, from the end of IV to the formation of neural tube (neurulation). I–V, total period.
†Recalculation in τ_0 units (see [8]).

The possibility of predicting the timing of different stages, and the duration of different developmental periods, is of great value both for laboratory studies and in the artificial breeding of fish [16, 18]. The ratio τ_n/τ_0 also allows the determination of the number of nuclear divisions during the period of cleavage (from the first cleavage division metaphase to the stage when the mitotic index falls and the cell cycle increases).

On the other hand, the possibility of calculating biologically equivalent time intervals with the help of τ_0 values permits us to study more accurately the effect of various factors on development. For this purpose the time of effect (exposure) at different temperatures should amount to the same number of τ_0 units.

The method of using the relative characteristics of developmental duration allows a comparative study of temporal patterns of embryogenesis in various species. Thus, in closely related but geographically isolated *Rana* species the duration of the same early developmental periods is practically equal, but differs from that in *Urodeles* and *Xenopus laevis* (Table 1.1). Constant relative duration of similar developmental periods was established for different geographically isolated species of sea urchins (see Vol. I, Table 10.3), two species each of *Acipenser, Coregonus*, and *Salmo*, as well as for the genera *Acipenser* and *Polyodon,* which belong to different families of the Acipenseriformes [53]. Heterochronies of various changes during early embryogenesis in teleosteans were characterized quantitatively by means of dimensionless criteria (Fig. 1.4).

This method allows the determination of the stages from which the development of *Huso huso* is accelerated as compared with that of *Acipenser güldenstädti*, and of the latter as compared with *A. stellatus*: the same developmental periods having a smaller number of τ_0 units (see [12, 16, 30]).

Close similarity was found to exist between the relative durations of successive mitotic phases of the first synchronous divisions in all those species studied which show the regulative type of development (Table 1.2). This fact suggests marked evolutionary conservatism, not only of morphology but of temporal parameters of mitosis as well. Using this criterion, Selman [48] established the similarity of the time of cleavage furrow determination in amphibians and acipenserid fish and its difference from that in sea urchins.

The regular character of all the data obtained confirms the expediency of using relative (dimensionless) criteria of developmental duration for comparative investigations.

One of the aspects of the term "timing" considered above, that is, the time of onset of various stages and duration of developmental periods, is also expressed by the word "schedule." Meanwhile, "timing" also means coordinated action. Coordination of various changes within embryos in time under optimal conditions is achieved, as we have already seen, by their triggering at the same rate, which changes in the same way in different parts of the embryo and at different levels of its organization under changing temperature conditions. Therefore, coordination is achieved of independent processes and of those interrelated by morphogenetic and other correlations. "Reserves" of regulative abilities of the embryo (duration of inducing activity of the rudiments, reserves of competent material, and others) [11, 46] provide the integrity of the developing organism when the processes are somewhat desynchronized under suboptimal conditions. However, with greater desynchronization of changes in the embryo, for example, morphogenetic movements with respect to induction effects of chordamesoderm rudiments on ectoderm during gastrulation, some abnormalities appear and the embryos die.

TABLE 1.2. Relative Durations (in fractions of τ_0) of the Phases of the First Cleavage Divisions in the Eggs of Sea Urchin, Russian Sturgeon, Lake and Rainbow Trout, Pike, Carp, Loach, Axolotl, and South African Clawed Toad

Species	Interval between fixations in fractions of τ_0	Temperature (°C)	τ_0, min	Prometaphase	Metaphase	Anaphase	Telophase + interphase	Authors
Psammechinus miliaris		11.0 15.0 21.0	67 44 30	— — —	0.20 0.19 0.22	0.14 0.13 0.15	— — —	Recalculation of data of Agrell [1] by Dettlaff [7]
Acipenser güldenstädti Colch.	0.03	16.2 ± 0.1 23.0	60 40	0.1 0.1	0.20 0.20	0.14 0.15	0.56 0.55	Dettlaff [7]
Salmo trutto	0.04	1.8 ± 0.3	840	0.07	0.21	0.14	0.58	Dettlaff [7]
Salmo gairdneri	0.04	4.8	405	0.07	0.19	0.14	0.60	Ignatieva [28]
Esox lucius*	0.03	10.0	90	0.07; 0.10*	0.17; 0.17	0.10; 0.10	0.66; 0.63	Ignatieva [28]
	0.035	18.0	42.5	0.07; 0.08	0.18; 0.14	0.11; 0.11	0.64; 0.67	
Cyprinus carpio*	0.05	18.0	39	0.05; 0.05	0.15; 0.10	0.15; 0.18	0.65; 0.67	Ignatieva [28]
Misgurnus fossilis	0.03	19.0	36	0.08	0.19	0.15	0.58	Ignatieva and Kostomarova [29]
Ambystoma mexicanum	0.03	18.0	104	0.05	0.21	0.15	0.59	Skoblina [50]
Xenopus laevis	0.03	20	34	0.09	0.20	0.18	0.53	Rudneva [40]

*In these species the duration of various mitotic phases was determined during the 1st and 2nd cleavage divisions.

Finally, "timing" also means estimation of the time of onset of some process or of the settling of a definite rhythm. This aspect of timing has been discussed at length in relation to rhythmic and differentiating events proceeding in various tissues and organs [6, 42, 44, 51]. The solution of these problems involves data on ooplasmic segregation, cycles of DNA replication, the onset of differential activity of genes, inductive interactions, and determination of the fate of cellular material and its differentiation, realized in succession but at different times in various parts of the embryo. The relative characteristics of developmental duration (in τ_0 units) are also useful in dealing with this third aspect of timing: it can be used to obtain comparable time parameters for all the above-mentioned events, as applied to the formation of different organs and for various animals.

REFERENCES

1. Y. Agrell, "The thermal dependence of the mitotic stages during the early development of the sea urchin," *Ark. Zool.* **11**, 382–392 (1958).
2. N. J. Berill, "Cell division and differentiation in asexual and sexual development," *J. Morphol.* **57**, 353–427 (1935).
3. N. P. Bordzilovskaya and T. A. Dettlaff, "Table of stages of the normal development of axolotl embryos," *Axolotl Newsletter* **7**, 2–22 (1979).
4. G. ten Cate, *Intrinsic Embryonic Development*, North Holland Publ. Co., Amsterdam (1956).
5. E. V. Chulitskaya, "Relative duration of cleavage and gastrulation and latent differentiation of rudimentary labyrinth in *Rana temporaria* embryos at different temperatures," *Dokl. Akad. Nauk SSSR* **160**, 489–492 (1965).
6. J. Cooke, "The control of somite number during amphibian development: Models and experiment," in *Vertebrate Limb and Somite Morphogenesis*, D. A. Ede et al., eds., Cambridge University Press (1977), pp. 443–448.
7. T. A. Dettlaff, "Mitotic dynamics of the first cleavage divisions in the eggs of sturgeons (at various temperatures) and of trout," *Exp. Cell Res.* **29**, 490–503 (1963).
8. T. A. Dettlaff, "Cell divisions, duration of interkinetic states and differentiation in early stages of embryonic development," *Adv. Morphogen.* **3**, 323–363 (1964).
9. T. A. Dettlaff, "Determination of the duration of mitotic cycle during synchronous cleavage divisions," in *Methods in Developmental Biology*, T. A. Dettlaff, V. Ya. Brodsky, and G. G. Gause, eds. [in Russian], Nauka, Moscow (1974), pp. 136–139.
10. T. A. Dettlaff, "Some temporal and thermal patterns of embryogenesis in poikilothermic animals," in *Problems of Experimental Biology* [in Russian], Nauka, Moscow (1977), pp. 269–287.
11. T. A. Dettlaff, "Adaptation of poikilothermic animals to development under varying temperatures and problem of integrity of developing organism," *Ontogenez* **12**, 227–242 (1981).
12. T. A. Dettlaff, "The rate of development in poikilothermic animals calculated in absolute and relative time units," *J. Therm. Biol.* **11**, 1–7 (1986).
13. T. A. Dettlaff, "Development of the mature egg organization in amphibians, fish, and starfish during the concluding stages of oogenesis, in the period of maturation," in *Oocyte Growth and Maturation*, T. A. Dettlaff and S. G. Vassetzky, eds., Consultants Bureau, New York (1988).

14. T. A. Dettlaff and A. A. Dettlaff, "On relative dimensionless characteristics of development duration in embryology," *Arch. Biol.* **72**, 1–16 (1961).

15. T. A. Dettlaff and A. A. Dettlaff, "Dimensionless criteria as a method of quantitative characterization of animal development," in *Mathematical Biology of Development*, A. I. Zotin, ed. [in Russian], Nauka, Moscow (1982), pp. 25–39.

16. T. A. Dettlaff, A. S. Ginsburg, and O. I. Schmalhausen, *Development of Acipenserid Fish. Maturation of Oocytes, Fertilization, Development of Embryos and Prelarvae* [in Russian], Nauka, Moscow (1981).

17. T. A. Dettlaff and T. B. Rudneva, "Dimensionless characteristics of the duration of embryonic development of the spur-toed frog," *Sov. J. Dev. Biol.* **4**, 423–432 (1973).

18. T. A. Dettlaff, S. G. Vassetzky, and S. I. Davydova, "Maturation of sturgeon oocytes under different temperatures and time of obtaining eggs after the hypophysial injection," in *6th Congrès Intern. Reprod. Amm. Insem. Artif. Paris I* (1968), pp. 129–131.

19. M. A. DiBerardino, "Frogs," in *Methods in Developmental Biology*, F. H. Wilt and N. K. Wessels, eds., T. Y. Crowell Co., New York (1967), pp. 53–74.

20. E. Y. DuPraw, "The honeybee embryo," in *Methods in Developmental Biology*, F. H. Wilt and N. K. Wessels, eds., T. Y. Crowell Co., New York (1967), pp. 183–217.

21. G. P. Eremeyev, "On synchronism in avian embryogenesis," *Arkh. Anat., Gistol. Embriol.* **37**, 67–70 (1959).

22. H. J. Fry, "Studies on the mitotic figure. V. The schedule of mitotic changes in developing *Arbacia* eggs," *Biol. Bull.* **70**, 89–99 (1936).

23. L. Gallien and O. Bidaud, "Table chronologique du développement chez *Triturus helveticus*," *Bull. Soc. Zool. Fr.* **84**, 22–32 (1959).

24. L. Gallien and M. Durocher, "Table chronologique du développement chez *Pleurodeles waltlii* Michah," *Bull. Biol. Fr. Belg.* **91**, 97–114 (1957).

25. A. S. Ginsburg, "Fertilization in Acipenserid fish. 1. The fusion of gametes," *Tsitologiya* **1**, 510–526 (1959).

26. G. M. Ignatieva, "Regularities of early embryogenesis in teleosts as revealed by studies of the temporal pattern of development. 1. The duration of the mitotic cycle and its phases during synchronous cleavage divisions," *Wilhelm Roux's Arch. Dev. Biol.* **179**, 301–312 (1976).

27. G. M. Ignatieva, "Regularities of early embryogenesis in teleosts as revealed by studies of the temporal pattern of development. 2. Relative duration of corresponding periods of development in different species," *Wilhelm Roux's Arch. Dev. Biol.* **179**, 313–325 (1976).

28. G. M. Ignatieva, *Early Embryogenesis of Fish and Amphibians* [in Russian], Nauka, Moscow (1979).

29. G. M. Ignatieva and A. A. Kostomarova, "Duration of mitotic cycle during synchronous cleavage divisions (τ_0) and its dependence on temperature in loach embryos," *Dokl. Akad. Nauk SSSR* **168**, 221–224 (1966).

30. E. V. Igumnova, "Chronological patterns of embryonic development of the beluga," *Sov. J. Dev. Biol.* **6**, 38–43 (1975).

31. A. A. Kostomarova and G. M. Ignatieva, "Ratio of the processes of karyo- and cytotomy during the period of synchronous cleavage divisions in the loach (*Misgurnus fossilis* L.), *Dokl. Akad. Nauk SSSR* **183**, 490–493 (1968).

32. A. L. Mazin, V. N. Vitvitzky, and V. Ya. Aleksandrov, "Temperature dependence of oocyte cleavage in three thermophilically different frogs of the genus *Rana*," *J. Therm. Biol.* **4**, 57–61 (1979).

33. J. A. Moore, "Temperature tolerance and rates of development in the eggs of Amphibia," *Ecology* **20**, 459–478 (1939).

34. A. Murakami, "Comparison of radiosensitivity among different silkworm strains with respect to the killing effect on the embryos," *Mutat. Res.* **8**, 343–352 (1969).

35. P. D. Nieuwkoop and J. Faber, *Normal Table of Xenopus laevis (Daudin)*, North Holland Publ. Co., Amsterdam (1956).

36. L. du Nouy, *Biological Time*, Methuen, London (1936).

37. V. P. Ostryakova-Varshaver, "Cytology of fertilization in the mulberry silkworm with reference to the differences in the sensitivity of successive phases of the process to high temperature," *Dokl. Akad. Nauk SSSR* **83**, 921–924 (1952).

38. V. P. Ostryakova-Varshaver, "Effect of high temperature in embryogenesis of the mulberry silkworm (*Bombyx mori* L.)," in *Tr. Inst. Morfol. Zhivotn. Akad. Nauk SSSR* **21**, 81–103 (1958).

39. N. N. Rott, "Correlation between karyo- and cytokinesis during the first cell divisions in the axolotl (*Ambystoma mexicanum* Cope)," *Sov. J. Dev. Biol.* **4**, 175–177 (1973).

40. T. B. Rudneva, "Duration of karyomitosis and cell division in cleavage divisions II–IV in the clawed frog, *Xenopus laevis*," *Sov. J. Dev. Biol.* **3**, 526–530 (1972).

41. D. A. Sabinin, *Developmental Physiology of Plants* [in Russian], Izd. Akad. Nauk SSSR, Moscow (1936).

42. N. Satoh, "On the 'clock' mechanism determining the time of tissue-specific enzyme developing during ascidian embryogenesis," *J. Embryol. Exp. Morphol.* **54**, 131–139 (1979).

43. N. Satoh, "Timing mechanism in early embryonic development," *Differentiation* **22**, 156–163 (1982).

44. N. Satoh and S. Ikegami, "A definite number of aphidicolin-sensitive cell cycle events are required for acetylcholinesterase development in the presumptive muscle cells of ascidian embryos," *J. Embryol. Exp. Morphol.* **61**, 1–13 (1981).

45. N. Satoh and S. Ikegami, "On the 'clock' mechanism determining the time of tissue-specific enzyme development during ascidian embryogenesis. 2. Evidence for association of the 'clock' with the cycle of DNA replication," *J. Embryol. Exp. Morphol.* **64**, 61–71 (1981).

46. I. I. Schmalhausen, *Factors of Evolution: The Theory of Stabilizing Selection*, Blackiston Co., New York (1949).

47. M. Schnetter, "Morphologische Untersuchungen über das Differenzierungszentrum in der Embryonalentwicklung der Honigbiene," *Z. Morphol. Oekol. Tiere* **29**, 114 (1935) (cited in [16]).

48. G. G. Selman, "Determination of the first two cleavage furrows in developing eggs of *Triturus alpestris* compared with other forms," *Dev., Growth Differ.* **24**, 1–6 (1982).

49. M. N. Skoblina, "Characteristics of duration of main stages of embryogenesis in *Ambystoma mexicanum*," in *4th Embryological Conference*. Abstracts [in Russian], Leningrad Univ. Press (1963), pp. 172–173.

50. M. N. Skoblina, "Dimensionless characteristics of duration of mitotic phases during synchronous cleavage divisions in the axolotl," *Dokl. Akad. Nauk SSSR* **160**, 700–703 (1965).
51. M. H. L. Snow and P. P. Tam, "Timing in embryological development," *Nature (London)* **286**, 107 (1980).
52. P. Valouch, J. Melichna, and F. Sládecek, "The number of cells at the beginning of gastrulation depending on the temperature in different species of amphibians," *Acta Univ. Carol., Biol.* 195–205 (1971).
53. S. G. Vassetzky, "Fish of the family Polyodontidae," *Vopr. Ikhtiol.* **11**, 26–42 (1971).
54. S. G. Vassetzky, "Meiotic divisions," in *Oocyte Growth and Maturation,* T. A. Dettlaff and S. G. Vassetzky, eds., Consultants Bureau, New York (1988).
55. T. A. Dettlaff, G. M. Ignatieva, and S. G. Vassetsky, "The problem of time in developmental biology: its study by the use of relative characteristics of development duration," in *Sov. Sci. Rev. I Physiol. Gen. Biol.* **1**, Part A, 1–88 (1987).

Chapter 2

THE RUSSIAN STURGEON *Acipenser güldenstädti.* Part I. GAMETES AND EARLY DEVELOPMENT UP TO TIME OF HATCHING

A. S. Ginsburg and T. A. Dettlaff

The embryos and larvae of sturgeons are of great interest for comparative embryological studies as they are considered to be the most primitive among the higher fishes. They are also good for experimental embryological studies since they are highly resistant to unfavorable conditions; in this respect the Russian sturgeon is of special interest, as its viability markedly exceeds that of not only teleosteans but also other sturgeons studied, namely, the giant and stellate sturgeons. In sturgeon eggs, although the yolk is distributed unequally, it is not separated from the cytoplasm. In this respect they are similar to amphibian eggs and, therefore, for such experiments as explantation, isolation, and grafting of rudiments sturgeons are more suitable than teleosts. The follicle-enclosed sturgeon oocytes are very suitable for studying mechanisms of maturation, since they can mature *in vitro* under the influence of gonadotropic hormones and progesterone and subsequently can be fertilized. Like the embryos, the oocytes tolerate microsurgery quite well. Large numbers of synchronously developing sturgeon embryos can be produced easily, thus allowing diverse experimental and biochemical studies. Different species of sturgeons spawn at about the same time and easily cross with each other; thus studies on hybrid forms and, in principle, heteroplastic transplantations are possible. Since sturgeons are used for artificial breeding, a study of their development is of practical value (see [34]).

Data on the Russian sturgeon given in this chapter are also important with respect to recent attempts at artificial propagation of sturgeons in the USA [38], France [124], and other countries, since the development of different representatives of the genus *Acipenser* is very similar.

The development of sturgeons was first outlined for the sterlet (*Acipenser ruthenus*) by Kowalewsky et al. [80]; the embryogenesis of this species was then described in detail by Zalensky [112, 113]. An intensive study of the development of sturgeons began after the method of pituitary injections had been introduced in pisciculture [42, 43], thus allowing the artificial reproduction of the main species of sturgeons.

15

Later the normal development of the stellate, Russian, and giant sturgeons was studied, as well as their developmental defects [32, 34, 57, 58]. The period of oogenesis has been actively studied in different species of sturgeons. A number of papers have been devoted to studies of previtellogenesis and vitellogenesis [13, 41, 47, 91–96, 99]. The period of oocyte maturation has received most attention, especially in the following areas: dynamics of meiosis [22, 75–77, 89, 107, 108]; the role of nucleus and cytoplasm in the development of egg organization and properties [2, 23, 35, 83, 87, 100–102]; changes in the fine structure of the cytoplasm [1] and follicle wall [97]; the development of cortical reaction [60]; changes in thermosensitivity of the oocyte and follicular epithelium [28, 106]; and the effect of gonadotropic pituitary hormones and progesterone on the oocyte and follicular cells *in vitro* [20, 27, 28, 36, 62–66]. Other investigations include the structure and properties of the mature gametes and their fusion, the cortical reaction of the egg upon fertilization and artificial activation; acrosomal reaction of the spermatozoon, and mechanism of the block to polyspermy [9–12, 21, 24, 33, 45, 49, 51, 53–55, 59]. The processes of formation and fusion of the pronuclei have been described [48, 50, 90], and the development of bilateral symmetry of the embryo studied [44]. Changes in the egg membranes during development and peculiarities of water metabolism in the embryos have also been investigated [114, 115, 118]. Detailed studies also include the period of cleavage: the duration of mitotic phases [25], mechanism of cytokinesis [120, 121, 123], variations in the pattern of cleavage [46], desynchronization of nuclear divisions and changes in the structure of the cell cycle upon transition to the blastula stage, and the influence of the cytoplasm on nuclear divisions [17, 81, 82]. There are descriptions of spontaneous abortive parthenogenesis [4, 32]. A fate map for the late blastula stage has been constructed, and the morphogenetic movements of cells during gastrulation described [7, 32, 72, 74]. Data have been obtained on primary embryonic induction [56] and the regional pattern of the inductive action of the chordamesoderm [70, 71], on the time of onset of morphogenetic nuclear function [3], and on the mechanism of the archenteron cavity formation [119, 122]. Regularities of eye development have been studied [5, 6, 18, 19], as well as those of labyrinth development [14–16] and the development of the hatching gland, including conditions of hatching enzyme secretion [69, 73, 116, 117]. The intensity of respiration and aerobic glycolysis [79, 103], changes in heat production of embryos during development [40], and the effect of oxygen content and salinity on development [39, 110, 111] have also been studied. Finally, sturgeons were used for the elaboration of dimensionless (relative) criteria of developmental duration [26, 30, 31].

2.1. TAXONOMY, DISTRIBUTION, AND REPRODUCTION

The Russian sturgeon *Acipenser güldenstädti* Brandt belongs to the family Acipenseridae, order Acipenseriformes, class Pisces.

The genus *Acipenser* includes 16 further species, among these the stellate sturgeon *A. stellatus* Pallas and the sterlet *A. ruthenus* L. The giant sturgeon *Huso huso* (L.) belongs to another genus of the same family.

Russian sturgeons are found in the basins of the Caspian Sea (*A. güldenstädti* Brandt and *A. güldenstädti persicus* Borodin) and of the Black and Azov Seas (*A. güldenstädti colchicus* V. Marti).

The diploid number of chromosomes in the Russian sturgeon *A. güldenstädti* from the Azov Sea is 250 ± 8 and from the Caspian Sea 247 ± 8. It belongs, thus, to the multichromosomal species of sturgeons, whereas the giant sturgeon, stellate sturgeon, sterlet, and ship [*Acipenser nudiventris* (Lov.)] have many fewer chromosomes ($2n \sim 120$). It is suggested that the multichromosomal species originated in the course of evolution from ancestors with small numbers of chromosomes via tetraploidization [105].

Like most sturgeons, the Russian sturgeon is an anadromous fish: it lives in the sea and enters the rivers for spawning only. In fish caught in the river mouths the gonads do not always have the capacity to react to gonadotropic hormones. The completion of maturation of the oocytes and spermatozoa takes place while the fish travel upstream and on the spawning grounds. In their search for conditions suitable for spawning (sturgeons spawn in places with a strong current and a dense, usually pebbly, river bed), the fish can travel upstream for many hundreds of kilometers [8]. At present, however, in most of the rivers in the USSR where sturgeons are found, the ways to the main spawning grounds are blocked by dams, and this has led to a sharp fall in their natural reproduction. The North Caspian sturgeon has recently started to assimilate the lower Volga flow spawning grounds [78, 109].

During spawning, ovulation proceeds gradually, from the caudal part of the ovary to the cranial part. The ovulated eggs pass into the body cavity, where a large amount of body-cavity fluid is accumulated by the beginning of ovulation. The eggs are not retained in the body cavity for long. As ovulation proceeds they pass through short paired oviducts and exit into water via the genital opening. Simultaneously, the male releases portions of spermatozoa which are dispersed by the current and fertilize the eggs. The spawning of each female lasts several hours.

The Russian sturgeon spawns in the Volga River in the spring at temperatures of 9–16°C [34].

2.2. HORMONAL STIMULATION

Nowadays the reproduction of sturgeons occurs largely in hatcheries, where the maturation of gametes is induced by hormonal stimulation. Corresponding to the conditions of natural spawning, sturgeon eggs are obtained at the hatcheries in April–May at water temperatures of 10–16°C.

The maturation and ovulation of the eggs is induced by a single intramuscular injection into the female of a pituitary preparation derived from any sturgeon species. Depending on the female's size, between 40 and 70 mg of acetonized pituitary powder in 1 ml of Ringer solution, for poikilothermic animals, or distilled water are injected. In addition, a synthetic analog of luliberin, (des-Gly-NH_2^{10})-D-Ala^6-Pro-NH-Et^9-LH-RH, is also used to stimulate maturation. It is injected intramuscularly in a dose of 1 µg/kg female weight or usually 20 µg per female [67, 68]. The latent period of egg maturation after stimulation with the luliberin analog exceeds by 15–25% that after the injection of pituitary suspension. In all other respects (effectivity and quality of gametes) these stimuli are identical. After a certain time, depending on temperature [34, 37], a female showing signs of maturation is killed, the body wall is cut through along the mid-abdominal line, and all eggs ovulated or readily leaving the ovary are taken out.

To obtain eggs of good quality, it is important to keep the female in running, well-aerated water until ovulation is completed under optimal temperature conditions. Of no less importance is the timely sacrifice of the female, since the retention of ovulated eggs in the body cavity results in a rapid fall in their fertilizability and deterioration of their quality (increase in the percentage of abnormal embryos after fertilization). A single female of *A. güldenstädti colchicus* can produce from 167 to 610 thousand eggs (data for 26 females).

After the eggs have been collected, the spermatozoa are taken from a male injected with 20–35 mg of pituitary powder or 20 μg of the synthetic luliberin analog per individual. The male is stunned, and spermatozoa stripped into a dry vessel. Spermatozoa can be taken 2–3 times from the same male. When necessary, the males can be used for 3–4 days. The volume of a single ejaculate is 25–500 cm^3.

Eggs and spermatozoa for experimental purposes are collected at the time when they are obtained at the hatcheries for breeding. The eggs are taken in dry wide test tubes (30–35 mm wide, with flat bottoms) up to half the height of the tube and with a layer of body-cavity fluid on top. The tubes are closed with corks, placed in a cool place, and screened from direct sunlight. It is desirable that the temperature of preservation corresponds to that during the period of maturation of females in fishponds, or is somewhat lower.

The eggs retain their fertilizability in body-cavity fluid at the spawning temperatures for 6–8 h or more. The preservation of eggs at 5 and 25°C for 60–100 min did not affect their fertilizability (there are no comparable data for longer periods of preservation), but the preservation of eggs for the same period at 0 and 30°C induced their activation and markedly decreased their fertilizability, usually down to zero [54, pp. 85, 86].

Spermatozoa are put in dry test tubes (those of each male in a separate tube), and the tubes are closed tightly with corks and preserved at a low temperature, preferably on ice. "Dry" semen kept at spawning temperatures preserves its fertilizing capacity for 12–36 h and at 0–4°C usually not less than 2—3 days (sometimes up to 8 days).

2.3. OOCYTE MATURATION *IN VIVO* AND *IN VITRO*

Pituitary gonadotropic hormones induce the maturation of the older generation of follicles, i.e., of the oocytes covered with a three-layered follicle envelope consisting of follicular epithelium, connective tissue layer, and mesothelium [97]. They contain a large germinal vesicle displaced toward the animal region and are characterized by a distinct polarity. The vegetal part of the oocyte is filled with large yolk granules and lipid droplets; the main mass of cytoplasm containing small yolk granules and lipid inclusions is concentrated in the animal region, which constitutes the upper third of the oocyte (Fig. 2.1). The sturgeon oocyte is thus telolecithal. The oocyte is covered by a two-layered vitelline membrane (zona radiata interna and zona radiata externa) and an outer jelly coat which adjoins the follicle epithelium. In the center of the animal region of the oocyte envelopes, there are channels which prior to ovulation are occupied by outgrowths of large follicle cells; these become the micropylar channels.

Several stages can be distinguished in the process of oocyte maturation according to the oocyte's properties (absence or presence of maturation inertia, contractility of the cortical cytoplasm, capacity for cortical reaction, etc.), the position

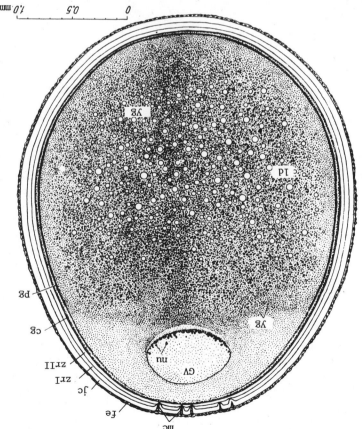

Fig. 2.1. Fully grown oocyte of the Russian sturgeon (cut along animal–vegetal axis). *yg* Yolk granules; *ld* lipid droplets; *GV* germinal vesicle; *cg* cortical granules; *mc* micropylar channels; *pg* pigment granules; *jc* jelly coat; *fe* follicle epithelium; *nu* nucleoli; *zrI* zona radiata externa; *zrII* zona radiata interna.

of the germinal vesicle, and the progress of maturation (meiotic) divisions. Oocytes at different stages of maturation do not differ in their external appearance. Their structure in sections is shown in Fig. 2.2. To determine rapidly the stage of germinal vesicle breakdown, the oocytes can be fixed in boiling water or alcohol:acetic acid mixture (9:1) and cut manually with a safety razor blade. The tentative time of the onset of successive stages of oocyte maturation in the ovary can be determined by the curve of t_0 versus temperature (Fig. 2.6); the duration of the interval from the moment of pituitary injection to every stage in t_0 is indicated in Fig. 2.2 (derived from one female). One should bear in mind, however, that the duration of the period from the moment of pituitary injection to that of germinal vesicle break-down depends not only on temperature but also on the initial condition of the fe-male, which can differ markedly in different females. In the female body the oocyte reaches the metaphase stage of the second maturation division (metaphase II; Fig. 2.2) and ovulates at this stage.

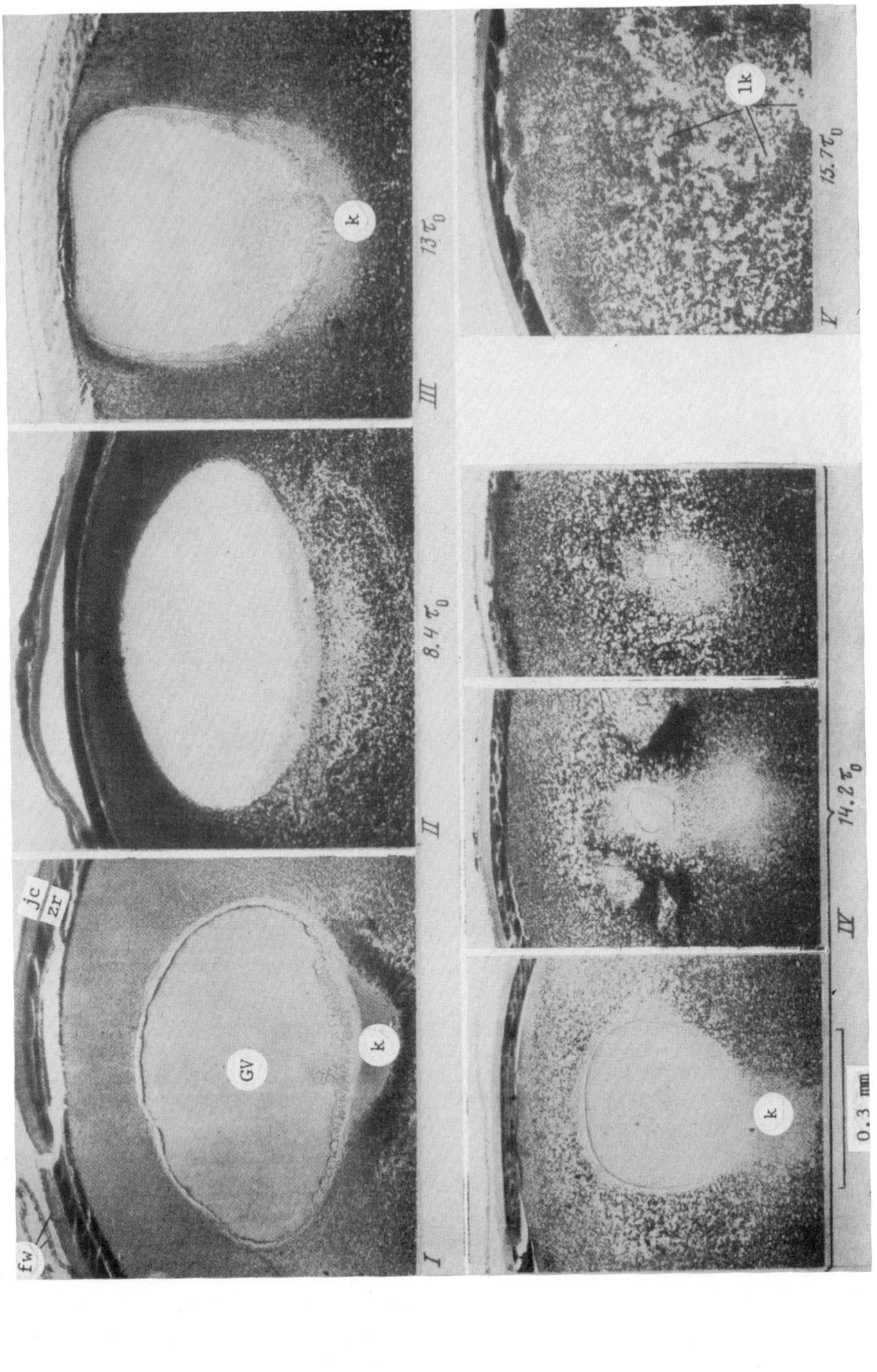
I
II
III
IV
V
fw
jc
zr
GV
k
1k
8.4τ0
13τ0
14.2τ0
15.7τ0
0.3 mm

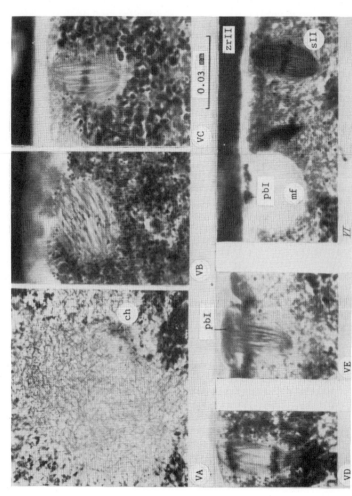

Fig. 2.2. Structure of the animal region of the Russian sturgeon oocyte at successive stages of maturation. I) Initial stage; full-grown oocyte at the moment of pituitary injection, no maturation inertia; II) stage of maturation inertia; III) stage of maximal approach of GV to the oocyte surface, corresponding to the stage of "white spot"; IV) stage of GV breakdown; V) 1st maturation division, karyoplasm is distributed in the animal cytoplasm (A, B, early and late prometaphase; C, metaphase; D, anaphase; E, late telophase); VI) oocyte at metaphase II, mature egg. sII) Spindle of the 2nd maturation division; mf) maturation funnel; GV) germinal vesicle; k) karyoplasm distributed in cytoplasm; lk) lacunae of karyoplasm in cytoplasm; pbI) 1st polar body; fw) follicle wall; ch) chromosomes; zr) zona radiata; zrII) bilaminar zona radiata; zrII) zona radiata interna. Relative time is indicated in τ_0 units (observations on one female).

In Russian and stellate sturgeons oocyte maturation can be induced *in vitro* in Ringer solution with an increased sodium bicarbonate content (1.5–2 g/liter) and with acetonized pituitary powder added (4 μg/ml) [65] . It is necessary that the temperature be maintained within the limits of spawning temperatures. Even short-term cooling or overheating of oocytes covered with a follicle envelope prevents their *in vitro* maturation under the influence of gonadotropic pituitary hormones [20, 28, 29]. The oocytes should be kept for some time in the solution with pituitary hormones, after which they acquire maturation inertia and are capable of completing their maturation in the Ringer solution without hormones. The duration of the hormone-dependent period depends on the concentration of hormones in the medium, the temperature, and the condition of the oocytes [62].

The *in vitro* oocyte maturation can also be induced by progesterone, the percentage of maturation being, as a rule, higher. The follicle-enclosed oocytes are placed in Ringer solution to which progesterone is added (1–5 μg/ml). The progesterone is previously dissolved in alcohol, the necessary volume of solution being taken by a micropipette into a Petri dish and dried in air to evaporate the alcohol; then the necessary volume of Ringer solution is added.

Under the effect of progesterone oocytes mature over a wider range of temperatures and tonicity values than under the effect of pituitary hormones. They do this not only in the presence of the follicle envelope but after its removal as well [20, 28, 35, 36]. To obtain a higher ovulation rate, they are best incubated in a solution containing both progesterone and pituitary suspension. If the oocytes are being prepared for operative procedures, the maturation medium should be sterilized by boiling, and antibiotics (penicillin, 500,000 U/liter and streptomycin, 0.25 g/liter) should be added. Oocytes matured and ovulated *in vitro* are capable of fertilization and normal development.

For experiments it is better to take oocytes from females within 24 h after capture. Oocytes taken from females held in captivity may react to gonadotropic pituitary hormones *in vitro* much less well or not at all. The oocytes are taken with the help of a special device ("hand spike"), a sharpened steel rod with a recess having cutting edges. The body wall in the region of the ovary is pierced with this hand spike, which is turned around several times, taken out, and placed in a test tube with Ringer solution. The oocytes are then washed out from the recess.

2.4. STRUCTURE OF MATURE GAMETES

The data given below were obtained for *Acipenser güldenstädti colchicus* at the Don River and have already been published elsewhere [32, 34, 54, 57, 58].

2.4.1. Ovulated Egg

On the average, the size of the ovulated egg is 3.1 × 3.5 mm. The color is brownish-grey. The animal region is characterized by a distinct pattern with a light polar spot in the center surrounded by a ringlike accumulation of pigment granules. There is another pigment ring at the border with the vegetal region (Fig. 2.7, 1 and 1an). In some eggs there is an intermediate ring between these two. The diameter of the spot and the width and intensity of pigmentation of the concentric dark and light rings vary within a wide range, not only in the eggs from different females but also in eggs from the same female. The animal region is usually lighter than the uniformly colored vegetal region, but sometimes it is pigmented so intensely that it becomes darker than the vegetal region; in such eggs the pigment rings become indiscernible and a light spot only is preserved in the center. Individual rearing of eggs with different pigment patterns has shown that eggs with one, two, and three pigment rings, as well as with a uniform dark color of the animal region, are capable of fertilization and normal development.

The eggs of *A. güldenstädti colchicus* are, as a rule, more intensely pigmented than those of *A. güldenstädti* and *A. güldenstädti persicus*. Sometimes, but very rarely, the eggs are fully devoid of dark pigment and have a pale-yellow color (such eggs were found in albino sturgeons).

In its internal structure, the mature egg has much in common with the primary oocyte. It is characterized by the same distribution of yolk inclusions, a layer of cortical granules is at the surface, and there is a deeper layer of smaller pigment granules, their greater or lesser concentration providing the characteristic pigmentation of the egg (Figs. 2.1 and 2.3). The germinal vesicle membrane is disintegrated and the karyoplasm is now distributed in the cytoplasm of the animal region as a branched net of lacunae (Fig. 2.2, V). The nucleus at metaphase II is near the egg surface in the vicinity of the maturation funnel and 1st polar body (Fig. 2.2, VI).

The egg envelopes are 110–150 μm thick. They are somewhat thinner in the animal pole region (80–110 μm) where there is a group of micropylar channels, usually about a dozen (but the number varies from 1 or 2 to 52). The openings of the micropylar channels are distributed without any order at a distance of 40–80 μm (more rarely 100 μm) from each other (Fig. 2.4) and together they occupy a small area: 120–320 μm in diameter in the case of 5–10 micropyles, and up to 1000–1100 μm in the case of over 30 micropyles [54]. The membranes of the unfertilized eggs are not sticky and are rather strong; the egg can support a load of 30–40 g [115].

2.4.2. Spermatozoon

The spermatozoon of *A. güldenstädti* is about 60 μm long. Its head is rodlike, 1.5–2 μm wide, and about 9 μm long (Fig. 2.5). As in the spermatozoon of the stellate sturgeon [55], there is a caplike acrosome at the anterior, somewhat tapering, head end. The nucleus constitutes the main bulk of the head; it is penetrated along its length by three spirally convoluted channels. These channels contain filaments which are thrown out when the sperm meets the egg and they form the core of the acrosomal process. The middle part of the sperm contains two centrioles and a body of lamellar structure similar to the appendage of the centriolar complex in the spermatids of sharks and some teleostean fish. The proximal centriole adjoins the nucleus posteriorly and is oriented transversely to the head's longitudinal axis. The distal centriole adjoins the proximal one at a right angle and constitutes the flagellum basal body. The centrioles are surrounded by the mitochondrial body which arises by fusion of individual mitochondria. This body envelops the flagellum base and is separated from the latter by a deep fold of plasma membrane, forming a sleeve around it. The length of the tail is 6–7 times that of the rodlike head. The axial filament complex is of the usual (9 + 2) type. Two marginal ridges run along the flagellum on each side of this complex in a plane passing through both the central fibrils and peripheral doublets 3 and 8. Each ridge represents a fold of plasma membrane filled by fine granular material without any supporting structures. These

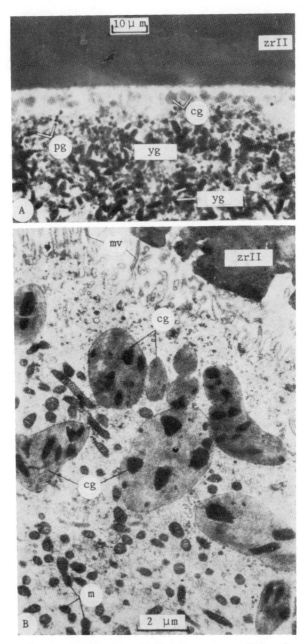

Fig. 2.3. Structure of mature egg cortex. A) Russian sturgeon, light microscope; B) stellate sturgeon, electron micrograph; yg) yolk granules; cg) cortical granules; m) mitochondria; mv) microvilli; pg) pigment granules; zrII) zona radiata interna.

ridges appear to increase the efficiency of the swimming movements of the flagellum. They are absent only in the terminal tail region (Fig. 2.5), 2.1–3.0 μm long.

2.5. ARTIFICIAL INSEMINATION

It is desirable to use for insemination milt of whole milk color, with a yellowish shade; 1 ml of such a sample contains over 2 billion spermatozoa. In the absence of dense milt, one can use more liquid milt of skim-milk color, with between 1 and 2 billion spermatozoa per ml. To inseminate small amounts of eggs, even watery semen of whey color with less than 1 billion spermatozoa per ml can be used, since the spermatozoa in such milt are often well activated.

Sturgeon spermatozoa, immobile in the seminal fluid ("dry" semen), begin an active forward movement after the addition of water. Within 1 min some spermatozoa begin oscillating movements and, within 3 min, usually less than half the spermatozoa swim. Within 5–10 min only a few single spermatozoa are moving forward; such "long-lived" spermatozoa preserve the capacity of active movement for a long time, up to 60 min or more, but they constitute an insignificant portion of the total number of spermatozoa. The fertilizing capacity of semen diluted with water is therefore rather rapidly reduced (usually within 10 min after dilution).

The quality of semen should be checked before use. In semen of good quality all or almost all spermatozoa swim actively in water. To estimate the quality of semen, the following five-grade scale can be used [88].

Grade 5. Rapid progressive movement of all spermatozoa.
Grade 4. Rapid progressive movement of most spermatozoa, but in the field of vision of the microscope there also occur spermatozoa with slow, zigzag-like and oscillatory movement.
Grade 3. Rapid progressive movement of a proportion of spermatozoa, but spermatozoa with zigzag-like and oscillatory movement predominate; there are also immobile spermatozoa.
Grade 2. Spermatozoa with progressive movement rare, a proportion of spermatozoa with oscillatory movement; about 75% of immobile spermatozoa.
Grade 1. All spermatozoa are immobile.

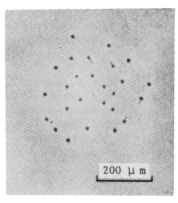

Fig. 2.4. Entries of micropylar channels.

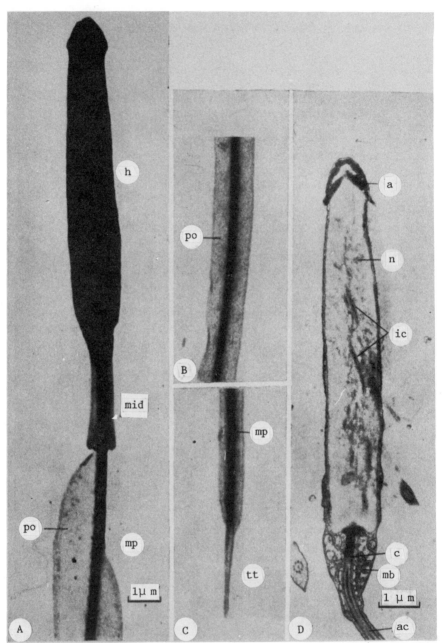

Fig. 2.5. Structure of spermatozoon of the Russian sturgeon (electron micrographs). A) Head and middle part; B) main part; C) tail tip; D) longitudinal section of head; a) acrosome; ic) intranuclear channels; po) plasma membrane outgrowth; h) head; mp) main part; tt) tip of tail; ac) axial complex of fibrils; mid) middle part; c) distal centriole; n) nucleus; mb) mitochondrial body.

Such a scale may be useful in research where it may be necessary to compare semen from different males.

Unlike spermatozoa, mature eggs are fertilizable in water at spawning temperatures for a comparatively long time – for 2 h or more. However, some eggs lose this capacity within a much shorter period, since water can induce their activation.

For artificial insemination of a small number of eggs, they are placed in a Petri dish and the excess of body-cavity fluid is pipetted off, since not only the spermatozoa remain inactive, but also fertilization is hampered in the body-cavity fluid [54, p. 287]. The necessary amount of "dry" semen is poured into a small dry flask and water is added just before insemination. To obtain the highest percentage of fertilization and normal embryonic development, dense semen of good quality is diluted with water in a ratio of 1:200 [52, 61]; if liquid semen is used, the quantity of water should be decreased.

The semen is rapidly mixed with water by circular motion of the flask, and poured onto the eggs. The Petri dish with the fertilized eggs is shaken gently several times and then the eggs are distributed evenly over the bottom using a feather. The dish is left until the eggs adhere to the glass. Thereafter the suspension of spermatozoa is replaced by filtered river or dechlorinated tap water. The layer of water in the dish should be just sufficient to cover the eggs, thus providing good conditions of gas exchange. If the eggs are incubated in the laboratory for a long time, the water should be changed not less than twice a day.

In those cases where a large number of eggs is fertilized, the process is carried out as at the hatchery. The eggs are placed in a dry enamel basin and the excess of body-cavity fluid is poured off. The necessary amount of "dry" semen is taken (if possible, a mixture of semen from 3–4 males is used) and diluted with water. A dilution of 1:200 is optimal. To inseminate 1 kg of eggs, 10 ml of semen is taken, poured into a vessel with 2 liters of water, mixed, and rapidly added to the basin containing eggs. The eggs are thoroughly mixed with the sperm by slow circular hand movements for 3–5 min. The water with sperm is then poured off, and the eggs are "degummed" by a suspension of infusorian powder and placed in an incubator.

2.6. FERTILIZATION

In sturgeons fertilization is normally monospermic, but with highly concentrated sperm suspensions, or in the case of poor physiological conditions of the eggs themselves, several (usually not more than four) spermatozoa can penetrate via different micropylar channels. All these spermatozoa are involved in development, thus leading to the formation of supernumerary cleavage furrows (see Section 2.8.2). The polyspermic eggs develop abnormally and, as a rule, perish before hatching; a few may develop to abnormal larvae which die [45].

With correct sperm dosage and good quality eggs, the percentage of polyspermic eggs in *A. güldenstädti colchicus* does not exceed 4–6%. When eggs of good quality are fertilized by concentrated sperm suspensions, the incidence of polyspermic eggs amounts to 20%, and when eggs of poor quality are inseminated, it is still higher (the maximal incidence of polyspermy attained under experimental conditions was 74%).

2.7. NORMAL DEVELOPMENT UP TO TIME OF HATCHING

The whole period of embryonic development of the sturgeon from fertilization to hatching comprises 35 stages (stage 36 – embryos and prelarvae at the period of mass hatching). The embryos of the species studied, Russian, stellate, and giant sturgeons, are very similar. The photographs of the stellate sturgeon embryos were published earlier [32], as well as drawings of the Russian sturgeon (*A. güldenstädti colchicus*) embryos at most of these stages [34, 57, 58]. The full set of drawings of the Russian sturgeon embryos is published in this volume (Figs. 2.7 to 2.14; the numbers of the drawings correspond to the numbers of the stages). The drawings were prepared by A. S. Ginsburg (one of the authors) and E. N. Smirnova, using a drawing device.

To determine the time of onset of all developmental stages, in addition to the available data [30, 32], Russian sturgeon embryos were fixed round the clock with formaldehyde (1:9) at intervals equal to the duration of one cleavage division (τ_0), and more often in the periods of transition from one stage to another.

The time of fixation of a sample where the embryos at the given stage were first found was considered as the time of onset of this stage. Since the temperature varied somewhat during the incubation (17.9–20.0°C), the mean temperature was calculated for every period between the onset of two successive stages. The value of τ_0 for the given mean temperature was then found from the curve of τ_0 versus temperature (Fig. 2.6), the time interval between two successive stages was divided by the value of τ_0, and the duration of this interval in τ_0 was thus determined.

The data obtained are presented in Table 2.1: numbers of stages, their diagnostic features, and time from insemination to the onset of every stage τ_n both in τ_0 units and in hours and minutes at 18°C (the absolute time was determined by multiplying the number of τ_0 units by the value of τ_0 at 18°C).

A more detailed description of the structure of sturgeon embryos at successive developmental stages is given below (including external features, as well as some data on changes in internal structure).

Stage 1. Egg within a few minutes after fertilization; it does not differ from the unfertilized egg (Fig. 2.7). There is a light spot in the center of the animal region surrounded by a ringlike accumulation of pigment (for variations of pigment pattern see Section 2.4.1). Envelopes adhere tightly to the egg; they have not yet begun to swell. The egg usually lies on its side (animal–vegetal axis directed horizontally).

Stage 2. Egg after secretion of hydrophilic colloid and rotation (Fig. 2.7). The animal region is flattened and a large perivitelline space forms between it and the egg envelopes. The pigment pattern of the animal region changes: the pigment concentrates to the center while the light polar spot disappears. The envelopes surrounding the egg are swollen and their layers become discernible. The jelly coat becomes sticky. The egg rotates inside its envelopes and becomes oriented with the animal pole upward.

Stage 3. Light crescent stage (Fig. 2.7). The displacement of the superficial cytoplasm layer is complete; the pigment accumulation which was in the center of the animal region at stage 2 becomes displaced. At the margin of the animal region a light, sometimes white, crescentlike area is often seen. The intensity of the light crescent and the precision of its boundaries vary greatly; sometimes it is not discernible. The egg has acquired a bilaterally symmetrical structure; the plane passing through the middle of the crescent and the animal and vegetal poles is the plane of bilateral symmetry of the embryo.

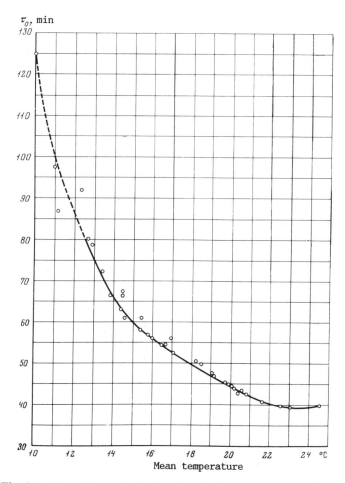

Fig. 2.6. Dependence of the duration of one mitotic cycle during synchronous cleavage divisions (τ_0) on temperature in the Russian sturgeon. At 12°C and lower temperatures the curve calls for precision.

Stage 4. First cleavage stage (Fig. 2.8). The first cleavage furrow divides the animal region and progresses to the darker vegetal material which is richer in yolk.

Stage 5. Second cleavage stage (Fig. 2.8). The animal region is divided by meridional furrows into four approximately equal parts. The first cleavage furrow descends below the equator and the second cleavage furrow progresses to the darker vegetal material.

Stage 6. Third cleavage stage (Fig. 2.8). The third cleavage furrows divide the animal region into eight blastomeres. The first furrow is completed in the area of the vegetal pole but does not penetrate deeply into the vegetal cytoplasm which is rich in yolk inclusions. The cleavage pattern at this stage can vary. In elongated eggs the third cleavage furrows can run almost parallel or at a small angle to the first furrow, forming a characteristic figure (as in Fig. 2.8, 6an); in spherical eggs the

TABLE 2.1. Chronology of the Embryonic Development of the Russian Sturgeon, *Acipenser güldenstädti colchicus*

Stage number	Time from fertilization		Diagnostic features of the stages
	h. min at 18°C	τ_n/τ_0	
1	0	0	Egg at the moment of fertilization
2	~00.50	~1	Light polar spot disappeared; perivitelline space formed
3	~1.40	~2	Pigment accumulation in the animal region is eccentric; a light crescent formed
4	2.55	3.5	Animal region is divided by the first cleavage furrow
5	3.45	4.5	Animal region is divided into four blastomeres
6	4.35	5.5	Animal region is divided into eight blastomeres
7	5.25	6.5	Fourth cleavage furrows laid in the animal region
8	6.15	7.5	Fifth cleavage furrows laid in the animal region
9	7.30	9.0	Seventh cleavage division; the furrows divide completely the vegetal part of the egg
10	8.20	10.0	Cleavage cavity (blastocoele) is forming; nuclear divisions in the animal region are still synchronous
11	10.00	12.00	Early blastula; blastomeres in the animal region clearly discernible under low magnification; desynchronization of nuclear divisions in the animal region
12	12.30	15.0	Late blastula; individual cells in the animal region not discernible under low magnification
12+	14.10	17.0	Changes in the cell cycle structure in the animal blastomeres; fall of mitotic index
13	16.15	19.5	Onset of gastrulation; an intensely pigmented band formed above the equator in the region of future dorsal blastopore lip
14	17.05	20.5	Early gastrula; dorsal blastopore lip appears as a short slit
15	22.55	27.5	Middle gastrula; animal material covers 2/3 of the embryo's surface; blastopore closed in a ring
16	25.00	30.0	Large yolk plug
17	27.50	32.5	Small yolk plug
18	31.40	38.0	Gastrulation completed, slitlike blastopore
19	32.30	39.0	Early neurula; neural folds are delineated around the neural plate head end
20	33.45	40.5	Wide neural plate; neural folds clearly seen around the neural plate head end

TABLE 2.1 (continued)

Stage number	h. min at 18°C	τ_n/τ_0	Diagnostic features of the stages
	Time from fertilization		
21	34.45	41.7	Onset of rapprochement of neural folds; rudiments of excretory system are first indicated
22	36.40	44.0	Late neurula; neural folds almost touching in the trunk region; rudiments of excretory system elongated
23	37.30	45.0	Neural tube closed; suture in the region of fusion of neural folds clearly seen
24	–	–	Eye protrusions formed in prosencephalon; a thickening arises in the anterior part of excretory system rudiments
25	–	–	Lateral plates reach the anterior head end, their tapering ends approaching one another in front of hatching gland rudiment; a thickening arises in the region of tail rudiment
26	50.00	60.0	Lateral plates fuse and at the site of their fusion heart rudiment is forming; tail rudiment is being separated.
27	53.20	64.0	Heart rudiment present as a short tube
28	57.30	69.0	Heart rudiment present as a straight elongated tube; trunk muscles do not respond to external stimulation
29	60.00	72.0	Heart tube S-shaped, onset of heartbeat; trunk muscles respond to external stimulation by twitchings
30	62.05	74.5	In the embryo within the envelopes tail end approaches the head; tail begins to straighten
31	–	–	In the embryo within the envelopes tail end approaches the heart; embryo can move its head and tail
32	78.00	93.5	In the embryo within the envelopes tail end reaches the head
33	–	–	Tail straightens fully after the removal of the envelopes
34	–	–	Embryo taken out of the envelopes is capable of slow progressive movement
35	96–100 h	116.0–120.0	Hatching of advanced embryos; after the removal of envelopes embryo is capable of rapid progressive movement
36	–	–	Mass hatching; branchial clefts and mouth not yet perforated; rudiments of pectoral fins formed; a distinct pigment spot in the eye; blood pink

furrows usually run radially. The cleavage pattern is not, as a rule, geometrically regular, the blastomeres having different form and size.

Stage 7. Fourth cleavage stage (Fig. 2.8). In the center of the animal region the latitudinal furrows separate small blastomeres; in those blastomeres which were wide enough at the previous stage, the furrows run radially.

Stage 8. Fifth cleavage stage (Fig. 2.8). There are over a dozen small blastomeres in the center of the animal region. The process of separation of blastomeres from each other continues in the vegetal region.

Stage 9. Seventh cleavage stage (Fig. 2.8). The separation of the blastomeres progresses. Furrows completely divide the vegetal region rich in yolk inclusions. Small slits between the blastomeres form inside the embryo.

Stage 10. Late cleavage stage (Fig. 2.8). Successive cleavage divisions result in a decrease of blastomere size. In the intermediate zone (at the boundary between the animal and vegetal regions) relatively small blastomeres (but larger than micromeres) have been separated from the macromeres by horizontal division. Inside the embryo, slits between the small animal blastomeres at the boundary with the blastomeres rich in yolk increase by accumulation of fluid. Thus the formation of the cleavage cavity (blastocoele) begins. From this stage on nuclear divisions in the animal blastomeres begin to desynchronize.

Stage 11. Early (blastomere) blastula (Fig. 2.8). In the animal region individual blastomeres are discernible under low magnification. They are spherical and adjoin each other loosely. A cleavage cavity (blastocoele) is seen through the blastula roof. In sections, it has irregular form: individual blastomeres cut into it. Blastomeres of the intermediate zone are distinct (see Fig. 2.8, 11).

Stage 12. Late (epithelial) blastula (Fig. 2.9). In the animal region individual cells are not discernible under low magnification. They adjoin closely and acquire an epithelial character. The blastocoele is larger than at stage 11, its walls becoming smoother in outline and the roof more compact and thinner. Within $2\tau_0$, at stage 12+, the cell cycle in the animal blastomeres becomes more extended and the mitotic index falls. Figure 2.15 shows a fate map at this stage, constructed from the results of vital staining [7]. The cells of the marginal zone lying in a band (occupying 30–40°) above the equator contain the material of the notochord, somites, lateral plates, and the most anterior region of the archenteron roof (prechordal plate, parachordal endoderm). Later on, together with the cells of the marginal zone of the embryo ventral half, they form the skeleton, muscles, coelom lining, and circulatory, excretory, and genital systems. Above the marginal zone lies the territory of future ectoderm, including the presumptive neural plate, which later forms the nervous system, and the presumptive epidermis. The cells of the vegetal hemisphere represent the material of the endoderm; they form the digestive system and its derivatives.

Stage 13. Onset of gastrulation (Fig. 2.9). A narrow intensely pigmented band with a vague outline is formed in the intermediate zone, slightly above the equator.

Figs. 2.7–2.14. Stages of normal development of the Russian sturgeon *Acipenser güldenstädti colchicus* V. Marti (drawings by A. S. Ginsburg and E. N. Smirnova). Drawing numbers correspond to those of stages. an) Animal view; veg) vegetal view; d) dorsal view; v) ventral view; h) view from head region; t) view from tail region. Drawings without letter designations are lateral views (lateral views from stage 25 on are drawn from embryos taken from the envelopes prior to fixation).

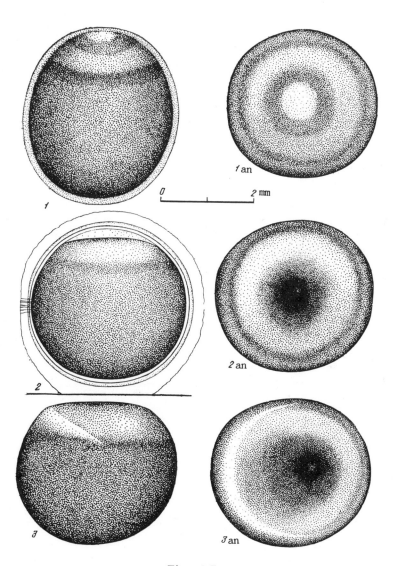

Fig. 2.7.

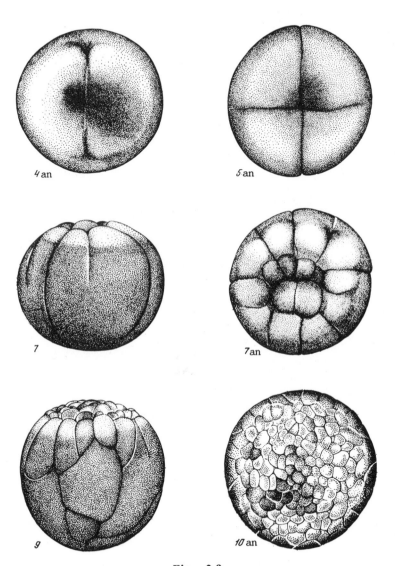

Fig. 2.8.

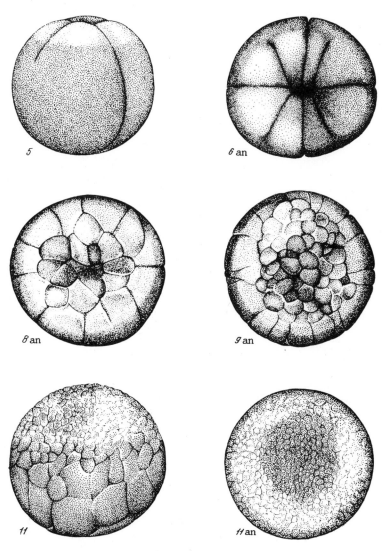

Fig. 2.8, continued.

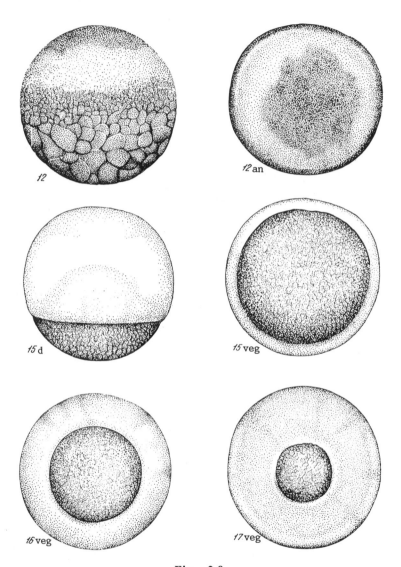

Fig. 2.9.

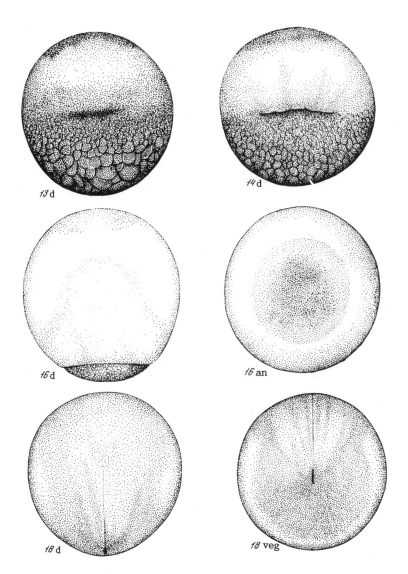

Fig. 2.9, continued.

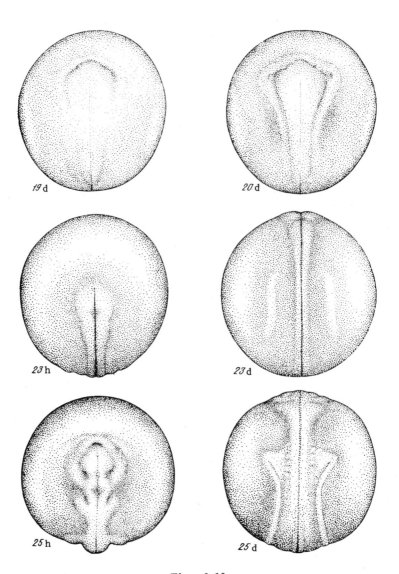

Fig. 2.10.

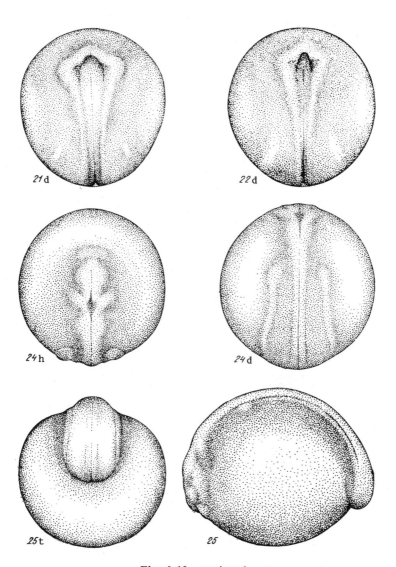

Fig. 2.10, continued.

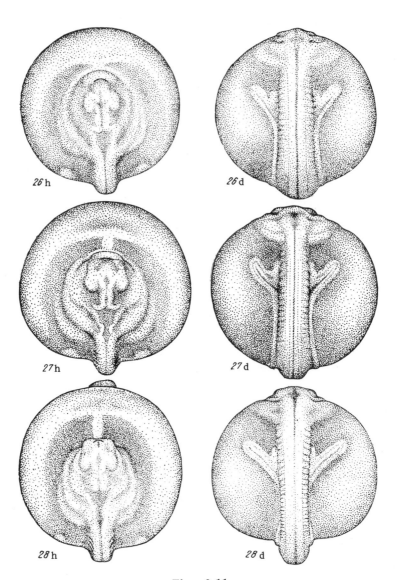

26 h
26 d
27 h
27 d
28 h
28 d

Fig. 2.11.

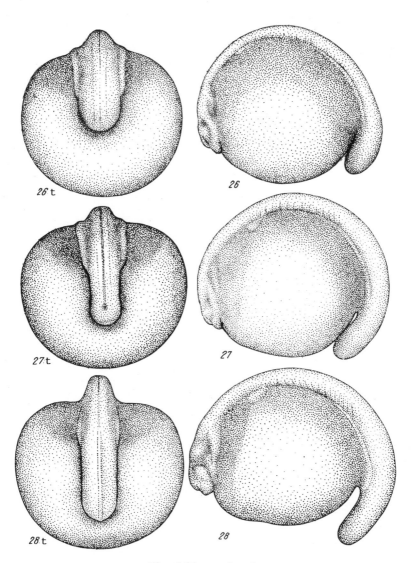

Fig. 2.11, continued.

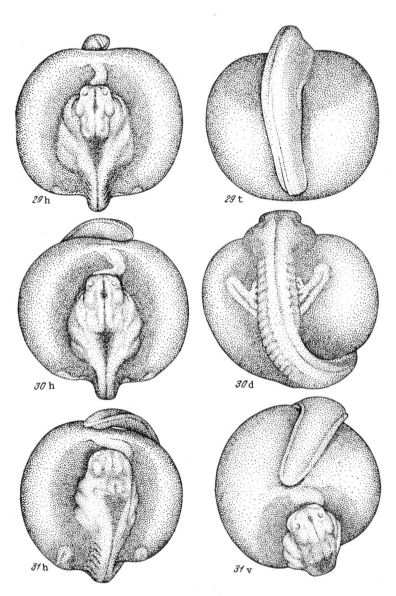

Fig.   2.12.

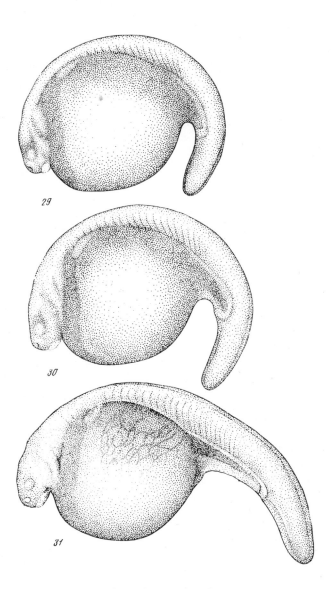

Fig. 2.12, continued.

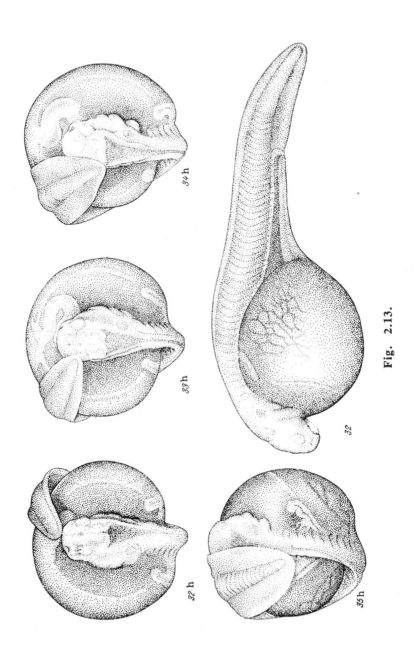

34 h

33 h

32

32 h

35 h

Fig. 2.13.

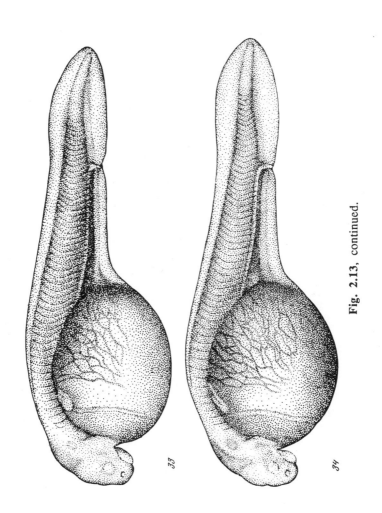

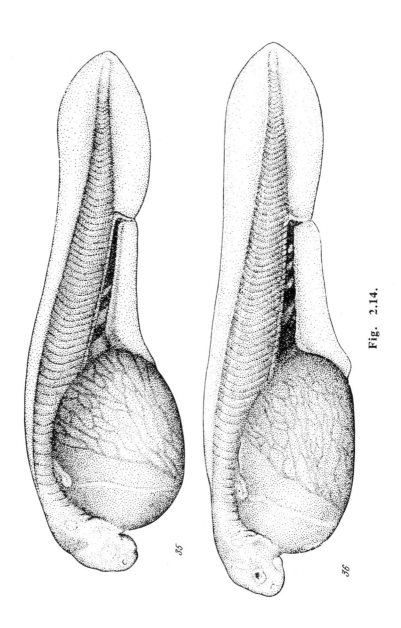

35

36

Fig. 2.14.

Stage 14. Early gastrula (Figs. 2.9 and 2.16). The dorsal blastopore lip forms in the position of the pigment band. It appears as a short shallow slit through which the process of invagination has begun. Later the slit progressively widens and curves, forming the blastopore's lateral lips.

Stage 15. Middle gastrula (Fig. 2.9). Animal material covers two-thirds of the embryo's surface. The ventral blastopore lip forms, and the blastopore closes to form a ring. On the dorsal side the darker material of the gastral cavity (archenteron) floor is visible through the overlying cell layers.

Stage 16. Large yolk plug stage (Figs. 2.9 and 2.16). The margin of the ring-shaped blastopore contracts to a smaller diameter, but the yolk plug is still large (16veg). The blastocoele is visible from above as a dark round spot (16an).

Stage 17. Small yolk plug stage (Fig. 2.9). The surface of the embryo, except the small yolk plug, is covered with light-colored animal material. The embryo maintains the same position within the envelopes, yolk plug downward. Between stages 17 and 18, the embryo turns by 90° due to the displacement of the center of gravity. As a result of this rotation, the dorsal side of the embryo is now turned upward.

Stage 18. Slitlike blastopore (Figs. 2.9 and 2.16). The process of gastrulation is completed; the blastopore margins close to form a narrow slit through which the archenteron cavity communicates with the environment. A shallow neural groove appears on the dorsal surface of the embryo.

Stage 19. Early neurula (Fig. 2.10). Neural folds are discernible around the head end of the neural plate; they are slightly raised. The neural groove is distinct.

Stage 20. Broad neural plate (Fig. 2.10). Neural folds around the head end of the neural plate are distinct. At the level of the future mesencephalon where the neural plate is widest, the folds are markedly thickened and divided into outer and inner parts.

Stage 21. Onset of approach of neural folds (Fig. 2.10). Rudiments of the excretory system are first seen as short light cords through the epidermis on each side of the neural tube in the trunk region.

Stage 22. Late neurula (Fig. 2.10). Onset of neural plate closure in its anterior part in the region of future prosencephalon. Neural folds in the trunk region are close to each other. Rudiments of the excretory system are elongated and can be clearly seen.

Stage 23. Closed neural tube (Figs. 2.10 and 2.17). The suture in the region of fusion of the neural folds is clearly seen as a shallow groove (23h). The head region of the neural tube, which is elongating, has begun to divide into three brain vesicles. Rudiments of the excretory organs are markedly elongated, slightly bent, compact cords, still without thickening in the anterior part (23d).

Stage 24. Appearance of eye protrusions and thickening of the anterior end of the rudiments of the excretory system (Figs. 2.10 and 2.18). The suture in the region of fusion of the neural folds is less distinct (24h). Eye protrusions form in the posterior part of the prosencephalon. On each side of the rhombencephalon, thickenings of the epithelium internal layer – ear placodes, rudiments of labyrinths – have arisen. In front of and adjoining the brain, a light crescent-shaped plate is delineated, the rudiment of the hatching gland. On each side of the mesencephalon two light winglike structures run forward around the prosencephalon. These are rudiments of the first pair of visceral arches, later forming the mandibular arch. Behind them, at the boundary between the prosencephalon and the mesencephalon, the second pair of arches is already delineated, not yet in winglike form. In the anterior part of the rudiments of the excretory system, a thickening forms which is a rudiment of the pronephros, collecting duct, and upper part of the excretory ducts. The rudiments of the excretory system are somewhat elongated caudally.

Stage 25. Rapprochement of lateral plates and formation of thickening in the region of tail rudiment (Fig. 2.10). Two parts are discernible in the prosencephalon, the future telencephalon and diencephalon. Eye protrusions approach the epidermis and are clearly seen from outside; their bases have become narrower (formation of the eye stalks). The cavity of the rhombencephalon begins to increase. On each side of the brain's anterior end olfactory placodes are formed (rudiments of olfactory sacs). In front of the brain the hatching gland rudiment is more distinct. Rudiments of the first pair of visceral arches reach the level of the anterior margin of the eye vesicles; rudiments of the second pair of arches become winglike. Behind them, rudiments of branchial arches are delineated. Endodermal pharyngeal pouches are discernible in the throat region. Lateral plates reach the anterior head end; their tapering ends approach one another in front of the hatching gland rudiment. In the pronephros area rudiments of pronephric tubules are delineated; a flexure is formed at the position of transition of the rudiment of the collecting duct into that of the excretory duct (25d). The excretory ducts of the pronephros almost reach the level of the neural tube's caudal end. In the rudiments of the excretory system a lumen has appeared, transforming them from solid cords into hollow tubes. An elevation has appeared in the posterior end of the embryo, forming a not-yet-separated tail rudiment (25), the end of which approaches the dark field margin, that is, the area where the gut floor is visible (25t).

Stage 26. Fusion of lateral plates and onset of separation of the embryo tail region (Figs. 2.11 and 2.19). On each side of the prosencephalon, olfactory pits are discernible. Eye vesicles are approaching the epidermis. The diencephalon cavity is markedly increased (26h). As a result of invagination of the ear placodes and contraction of their margins, the ear vesicles are formed, preserving their connection with the epidermis. Lateral plates fuse and at the site of fusion the heart rudiment forms. The first two pairs of endodermal gill pouches approach the epidermis. A loop formed by the collecting and excretory ducts of the pronephros has markedly elongated (26d). The tail rudiment (26d) is separated as a short, wide lobe (26t) and the tail end extends beyond the dark area. The head is not yet distinct.

Stage 27. Short heart tube (Fig. 2.11). Head separation begins. The head rudiments spread on the ventral body region are drawn to the mid-dorsal line and the head is elevated, only slightly as yet. The wall of the eye vesicle thickens in the area of contact with the epidermis, forming the retinal rudiment. The heart rudiment is present as a short tube (27h). The pericardial cavity is formed; its boundaries are better seen laterally. The tail rudiment is elongated and constricted (27t, 27).

Stage 28. Straight elongated heart tube (Fig. 2.11). The embryo is still immobile; the trunk muscles do not yet react to external stimulation. The head is markedly elevated, separated by a skin fold (28). As a result of these changes, the hatching gland rudiment is being displaced to the lower head surface. The retinal rudiment is markedly thickened and begins to invaginate: this marks the onset of transformation of eye vesicles into eye cups. The internal layer of the epidermis is compacted and thickened in the region of contact with retinal rudiment (formation of lens rudiment). On the ventral tail surface, at its base, a pit (cloaca rudiment) appears into which the pronephric ducts open. The tail rudiment is somewhat elongated and rod-shaped (28t, 28). The epidermis thickens at the position of finfold formation on the tail.

Stage 29. Formation of S-shaped heart and beginning of heartbeat (Fig. 2.12). Head separation continues (29). The heart tube is elongated and bent into an S shape (29h). Heart muscle contractions have just begun but are rare and irregular. Trunk muscles are also capable of contraction: weak muscle twitchings are observed in response to external stimulation. Blood vessels in the yolk sac wall are more numerous. Active hemopoiesis proceeds. The lens rudiment is markedly increased. The tail rudiment is somewhat elongated, and the fin-fold rudiment has appeared on it (29t, 29).

Stage 30. Approach of tail end to heart (Fig. 2.12). In the embryo within the envelopes, the tail and a part of the dorsal region as far forward as the pronephros are bent to one side (30d). The embryo can move its tail and head. After the removal of the envelopes it is seen that the tail has begun to straighten (30); it is compressed and surrounded by a narrow fin fold. There is a dense network of blood vessels in the body wall on the dorsal side of the yolk sac (not shown in 30).

Stage 31. Tail end reaches heart (Fig. 2.12). In the embryo within the envelopes, the dorsal region, from the head backward, is bent sideways (31h). If the envelopes are removed, the embryo turns its head to the right and left, makes swinging tail movements, and rolls from one side to the other. The tail is markedly straightened (31). The fin fold is clearly discernible.

Stage 32. Tail end touches head (Fig. 2.13). The head begins to bend sideways. The embryo can move actively within the envelopes; the dorsal region bends from side to side along the vertical axis. If the envelopes are removed, the embryo makes swimming movements on the spot, but is not yet capable of progressive movement. The tail is straightened to a greater extent than at stage 31 and the fin fold is wider (32).

Stage 33. Tail end somewhat overlaps the head (Fig. 2.13). The head is bent sideways. If the envelopes are removed, the tail is completely straightened (33); the fin fold continues to widen.

Stage 34. Tail end reaches diencephalon (Fig. 2.13). The embryo moves actively within the envelopes. If the envelopes are removed, the embryo is capable of slow progressive movement. The fin fold is markedly widened.

Stage 35. Onset of hatching (Figs. 2.13 and 2.14). The tail end reaches the pronephros, but only if the envelopes remain unstretched. Prior to hatching, however, the envelopes are often markedly stretched and the embryo within is somewhat straightened; in such cases the position of the tail end cannot serve as a diagnostic feature of the stage. If the envelopes are removed, the embryo is capable of rapid progressive movement. The yolk sac is spherical. A pigment spot appears in the eye. The mouth invagination is outlined. The pectoral fin rudiments may be indicated as hardly discernible folds. The blood is colorless or has a weak yellowish-pink shade.

Stage 36. Prelarva soon after hatching in the period of mass hatching (Figs. 2.14 and 2.20). The yolk sac is elongated and egg-shaped. The fin fold is wider than at stage 35; it is widest in the posterior third of the tail and ventrally beyond the tail base, where it forms a keel and extends downward along the ventral wall of the yolk sac. There is a distinct pigment spot in the eye (pigmented region of fundus oculi which can already perceive light). In the gill area two folds are delineated at the position of the first pharyngeal pouches; branchial clefts are not yet perforated. Foldlike elevations of epidermis are seen behind the pronephroses, the rudiments of pectoral fins. The blood is pink. A small triangular mouth invagination is visible on the lower head surface, but no mouth opening has formed as yet. The most

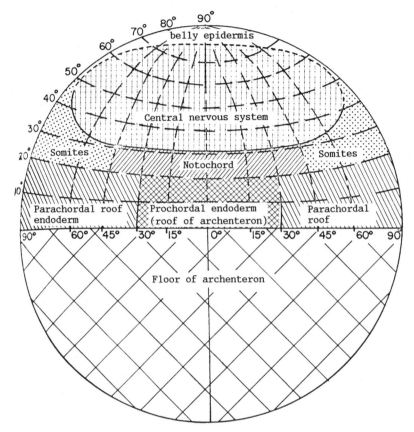

Fig. 2.15. Fate map of the Russian and stellate sturgeons at the late blastula stage (7). Dorsal blastopore lip will appear along 0° latitude and from 15° left to 15° right longitude.

posterior region of the intestine, the cloaca, communicates with the environment. The excretory ducts of the pronephros open into the cloaca, but the latter does not yet communicate with the intestinal lumen since as yet the posterior part of the intestine has no lumen. At the stage of hatching, therefore, the digestive system is still closed.

2.8. DEVELOPMENTAL DEFECTS

2.8.1. Parthenogenetic Cleavage

Mature eggs can be activated if put in water, and develop parthenogenetically for some time; some eggs behave in the same way if they are not fertilized, due to low concentration of sperm or poor quality of the semen. A few eggs are sometimes activated in the female body.

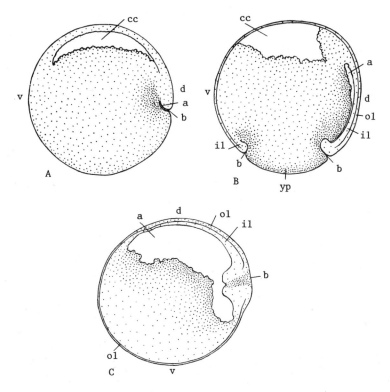

Fig. 2.16. Sections of embryos at stages 14 (early gastrula, A), 16 (large yolk plug, B), and 18 (slitlike blastopore, C). b) Blastopore; v) ventral side of embryo; il) invaginated cell layer; yp) yolk plug; ol) outer gastrula layer; a) archenteron; cc) cleavage cavity (blastocoele); d) dorsal side of embryo.

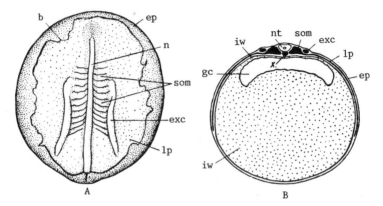

Fig. 2.17. Structure of the Russian sturgeon embryo at stage 23 (closed neural tube). A) Dorsal view after removal of covering epithelium and neural tube; B) cross section (scheme). lp) Lateral plate; exc) rudiment of excretory system; b) boundary of denuded area (after removal of epithelium and neural tube); nt) neural tube; gc) gut cavity; iw) intestine wall; som) somites; n) notochord; ep) covering epithelium.

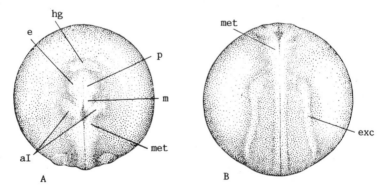

Fig. 2.18. Stage 24. Appearance of eye protrusions and thickening of anterior end of excretory system rudiments. A) Head view; B) dorsal view. exc) Rudiment of excretory system; e) eye rudiment; aI) rudiments of first pair of visceral arches; hg) hatching gland rudiment; met) metencephalon; p) prosencephalon; m) mesencephalon.

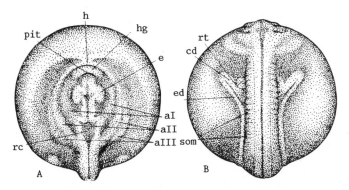

Fig. 2.19. Stage 26. Fusion of lateral plates and onset of separation of embryo's tail region. A) Head view; B) dorsal view. ed) Pronephric excretory duct; pit) invagination in the place of pituitary rudiment formation; e) eye rudiment; aI, aII) first and second pairs of visceral arches; aIII) rudiments of branchial arches; hg) hatching gland rudiment; rt) renal tubules; rc) rhombencephalon cavity; h) heart rudiment; cd) pronephric collecting duct; som) somites.

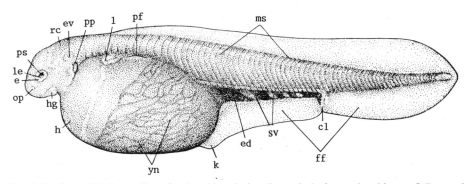

Fig. 2.20. Stage 36. Larva just after hatching during the period of mass hatching. pf) Pectoral fin rudiment; hg) hatching gland; pp) pharyngeal pouches; k) keel of fin fold; cl) cloaca; ms) muscle segments; op) olfactory pit; l) loop formed by pronephric collecting and excretory ducts; ps) pigment spot in eye; ff) fin fold; h) heart; yn) net of blood vessels in yolk sac wall; ev) ear vesicle; sv) spiral valve; le) lens; all other designations as in Fig. 2.19.

Eggs activated without fertilization frequently cleave. This cleavage is greatly delayed compared with that of fertilized eggs and proceeds at a slower rate. In those cases, however, where egg activation takes place in the female body, the cleavage of the activated eggs may begin earlier than that of the fertilized ones. Sometimes this cleavage even occurs in the ovary or body cavity.

Parthenogenetic cleavage is irregular (Fig. 2.21), a significant part of the egg often not cleaving at all. In some cases cleavage is arrested quite early, after a few cleavage furrows have formed, while in other cases it continues until the stage of many blastomeres (Fig. 2.21C). The latter is the limiting stage; parthenogenetic development never goes beyond it, the embryos never start gastrulation, and slowly die. The boundaries between the cells disappear gradually and the embryo acquires a whitish marmoreal coloring (Fig. 2.21D). The degeneration of these eggs coincides in time with the gastrulation of fertilized eggs in the same batch.

2.8.2. Defects of Cleavage of Fertilized Eggs

True defects of cleavage, among which defects due to polyspermic fertilization are most frequent, should be distinguished from variations in the form and position of cleavage furrows in the normal embryo.

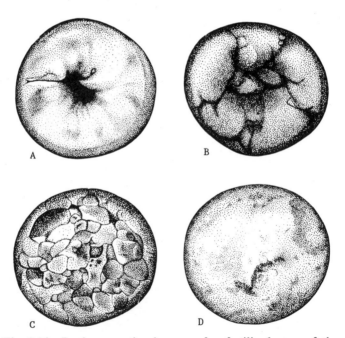

Fig. 2.21. Parthenogenetic cleavage of unfertilized eggs of the Russian sturgeon. A) Beginning of cleavage; B) advanced cleavage; C) limiting stage of development of the unfertilized eggs; D) moribund unfertilized egg.

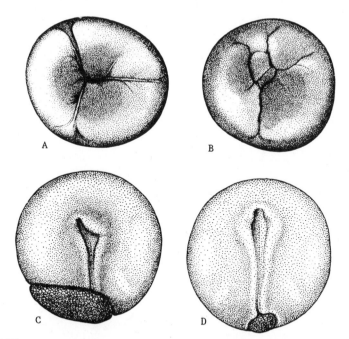

Fig. 2.22. Developmental defects in the Russian sturgeon embryos during cleavage and neurulation. A, B) Polyspermic eggs at stage 4 (first cleavage); C, D) formation of neural plate in embryos with a remaining yolk plug (C, large yolk plug, neural plate shortened and bent; D, small yolk plug, neural plate almost normal).

In polyspermic eggs, which are found in every batch in greater or lesser quantity, the supernumerary cleavage furrows become apparent during the first cleavage division, the animal region dividing at once into three, four, or more blastomeres, depending on the intensity of polyspermy (Fig. 2.22). At that time only dispermic eggs do not differ, as a rule, from normal monospermic ones, one cleavage furrow being laid in the animal region. However, during the second cleavage of dispermic eggs, two furrows are laid in each of the first two blastomeres and the animal region is divided into six blastomeres. From this stage on (stage 5), all polyspermic eggs can be easily distinguished from the normal ones by the presence of supernumerary blastomeres.

Marked defects of cleavage are also caused by temperature damage of the oocyte cytoplasm. These include a great diversity in the time of appearance of individual furrows and distortion of the cleavage pattern. A greater or lesser part of the egg sometimes does not cleave at all (mosaic cleavage). The incubation of such eggs results in a high percentage of markedly abnormal embryos.

2.8.3. Defects of Gastrulation

In some embryos epiboly is delayed and neurulation begins in the presence of a more or less large yolk plug (Fig. 2.22C, D). If both epiboly and invagination are

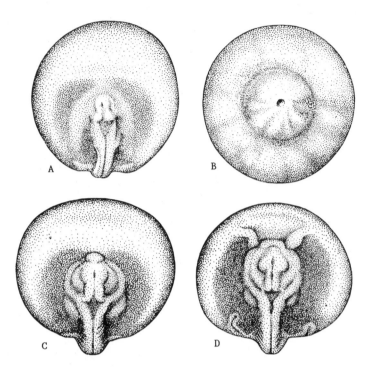

Fig. 2.23. Developmental defects in the Russian sturgeon embryos at stages 26 and 28. A, B) Grossly abnormal embryos at stage 26 with underdeveloped anterior body regions (A, prosencephalon absent; B, head and trunk regions at axial organs absent, only tail rudiment developed); C, D) abnormal forms at stage 28 with defects of heart development (C, heart rudiment absent; D, two heart rudiments developed).

affected, grossly abnormal embryos arise with different degrees of underdevelopment of the anterior body regions: embryos with forebrains of decreased size; embryos without telencephalon and diencephalon (Figs. 2.23A and 2.24A); embryos in which the head begins with rhombencephalon and acephalous ones (Fig. 2.24B); finally, there are some embryos devoid not only of a head but also of a dorsal trunk region, with a tailbud on the yolk sac (Fig. 2.23B) which later differentiates into a more or less developed tail (Fig. 2.24C).

2.8.4.   Defects of Subsequent Development

Defects of heart development may lead to the appearance of embryos fully devoid of a heart tube (see Fig. 2.23C) or, in cases where the lateral plates do not fuse in front of the head, of embryos with two heart tubes (see Fig. 2.23D).

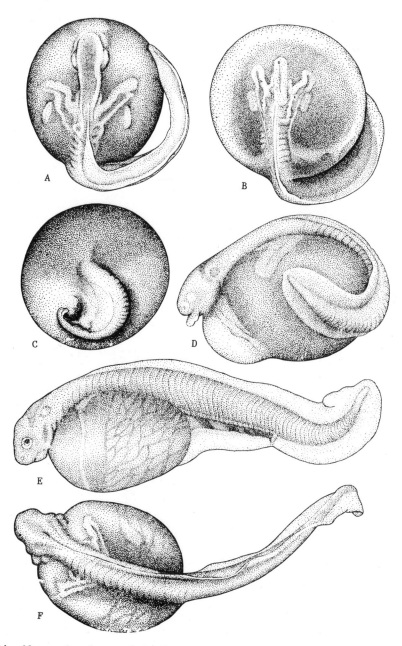

Fig. 2.24. Abnormal embryos of the Russian sturgeon at stage 35. A) Embryo without telencephalon and diencephalon; B) acephalous embryo; C) only abnormal tail developed; D) grossly abnormal embryo without a fully separated head and with a bent, shortened tail and hydropericardium; E, F) shortened embryos with bent tails.

At later developmental stages, besides abnormal forms with underdeveloped anterior body regions due to gastrulation defects (see Section 2.8.3), embryos occur with shortened and curved axial structures, hydropericardium (see Fig. 2.24D–F), and some other less significant defects.

Some abnormal embryos die after the formation of the heart rudiment, but many of them remain alive at the time of hatching of the normal embryos. The vast majority of grossly abnormal embryos cannot liberate themselves from the envelopes, but if the envelopes are removed, these abnormal forms can still live for a long time until the yolk reserves in the intestine are exhausted.

REFERENCES

1. T. B. Aizenshtadt and T. A. Dettlaff, "Ultrastructure of stellate sturgeon oocytes during maturation. 1. Annulate lamellae and Golgi complex," *Sov. J. Dev. Biol.* **3**, 220–229 (1972).
2. T. B. Aizenshtadt and M. N. Skoblina, "Fine structure of frog oocytes maturing *in vitro* under the influence of hypophyseal hormones after the removal of their germinal vesicles," *Tsitologiya* **12**, 713–717 (1970).
3. I. P. Arman and A. A. Neyfakh, "Morphogenetic function of nuclei at early developmental stages of Acipenseridae, studied with the aid of the radiation method," *Dokl. Akad. Nauk SSSR* **137**, 745–748 (1961).
4. B. L. Astaurov, "Abortive parthenogenesis in the Acipenserid fish (*Acipenser stellatus, A. güldenstädti, Huso huso*)," *Dokl. Akad. Nauk SSSR* **78**, 173–176 (1951).
5. E. A. Baburina, "Development of eyes and their function in *Acipenser güldenstädti* and *A. stellatus*," *Tr. Inst. Morfol. Zhivotn. Akad. Nauk SSSR* **20**, 148–186 (1957).
6. E. A. Baburina, *Development of Eyes in Cyclostomata and Pisces with Respect to Their Ecology* [in Russian], Nauka, Moscow (1972).
7. W. W. Ballard and A. S. Ginsburg, "Morphogenetic movements in acipenserid embryos," *J. Exp. Zool.* **213**, 69–103 (1980).
8. L. S. Berg, *Fish of Fresh Waters of the USSR and Adjacent Countries, Pt. I*, 4th edn. [in Russian], Izd. Akad. Nauk SSSR, Moscow (1948).
9. G. N. Cherr and W. H. Clark, Jr., "Fine structure of the envelope and micropyles in the eggs of the white sturgeon, *Acipenser transmontanus* Richardson," *Dev., Growth Differ.* **24**, 341–352 (1982).
10. G. N. Cherr and W. H. Clark, Jr., "An acrosome reaction in sperm from the white sturgeon, *Acipenser transmontanus*," *J. Exp. Zool.* **232**, 129–139 (1984).
11. G. N. Cherr and W. H. Clark, Jr., "Gamete interaction in the white sturgeon *Acipenser transmontanus* : A morphological and physiological review," *Environ. Biol. Fish.* **14**, 11–22 (1985).
12. G. N. Cherr and W. H. Clark, Jr., "An egg envelope component induces the acrosome reaction in sturgeon sperm," *J. Exp. Zool.* **234**, 75–85 (1985).
13. D. A. Chmilevsky and E. V. Raikova, "Incorporation of ^{3}H-thymidine into early meiotic prophase oocytes of *Acipenser ruthenus* L.," *Tsitologiya* **18**, 689–692 (1976).

14. E. V. Chulitskaya, "Latent differentiation of the auditory vesicle material in *A. güldenstädti* and *A. stellatus*," *Dokl. Akad. Nauk SSSR* **138**, 718–721 (1961).
15. E. V. Chulitskaya, "Correlation between the latent differentiation of the ear vesicle material and the cell generations at the same developmental stages in embryos of Acipenseridae (*Huso huso, A. güldenstädti, A. stellatus*)," *Dokl. Akad. Nauk SSSR* **139**, 506–509 (1961).
16. E. V. Chulitskaya, "A study of inductive effect of the mesoderm on the material of the auditory vesicle in acipenserid fish and amphibians," *Dokl. Akad. Nauk SSSR* **144**, 245–247 (1962).
17. E. V. Chulitskaya, "Desynchronization of cell divisions in the course of egg cleavage and an attempt at experimental shift of its onset," *J. Embryol. Exp. Morphol.* **23**, 359–374 (1970).
18. N. V. Dabagyan, "The role of the mesenchyme in the development of pigment epithelium of the eye in *A. güldenstädti*," *Dokl. Akad. Nauk SSSR* **119**, 391–394 (1958).
19. N. V. Dabagyan, "Regulatory properties of the eye in the embryos of Acipenseridae," *Dokl. Akad. Nauk SSSR* **125**, 938–940 (1959).
20. S. I. Davydova, "Effect of temperature and time of keeping female sturgeons in captivity on maturation of oocytes under the influence of hormones *in vitro*," *Sov. J. Dev. Biol.* **4**, 415–420 (1972).
21. T. A. Dettlaff, "Cortical granules and substances secreted from the animal portion of the egg at the period of activation in Acipenseridae," *Dokl. Akad. Nauk SSSR* **116**, 341–344 (1957).
22. T. A. Dettlaff, "Rate of propagation of fertilization impulse and the dynamics of the completion of the second maturation division in the eggs of Acipenserid fish," *Dokl. Akad. Nauk SSSR* **140**, 967–969 (1961).
23. T. A. Dettlaff, "The ovulation and activation of oocytes of Acipenserid fish *in vitro*," in *Sympos. on Germ Cells and Earliest Stages of Development,* Intern. Inst. Embryol. and Fond. A. Baselli, Milan (1961), pp. 141–144.
24. T. A. Dettlaff, "Cortical changes in acipenserid eggs during fertilization and artificial activation," *J. Embryol. Exp. Morphol.* **10**, 1–26 (1962).
25. T. A. Dettlaff, "Mitotic dynamics of the first cleavage divisions in the eggs of sturgeon (at various temperatures) and of trout," *Exp. Cell Res.* **29**, 490–503 (1963).
26. T. A. Dettlaff, "Some thermal temporal patterns of embryogenesis," in *Problems of Experimental Biology* [in Russian], Nauka, Moscow (1977), pp. 269–287.
27. T. A. Dettlaff and S. I. Davydova, "Effect of triiodothyronine on ripening of oocytes in stellate sturgeon after the action of low temperatures and keeping females in captivity," *Sov. J. Dev. Biol.* **5**, 454–462 (1974).
28. T. A. Dettlaff and S. I. Davydova, "Differential sensitivity of cells of follicular epithelium and oocytes in the stellate sturgeon to unfavorable conditions and correcting influence of triiodothyronine," *J. Gen. Comp. Endocrinol.* **39**, 236–243 (1979).
29. T. A. Dettlaff and S. I. Davydova, "Differential sensitivity of follicle epithelium cells and oocytes in the stellate sturgeon to unfavorable conditions and the correcting effect of triiodothyronine," in *Mechanisms of Hormonal Regulations and Role of Feedback Relations in Phenomena of Development and Homeostasis* [in Russian], Nauka, Moscow (1981), pp. 322–331.

30. T. A. Dettlaff and A. A. Dettlaff, "On relative dimensionless characteristics of the development duration in embryology," *Arch. Biol.* **72**, 1–16 (1961).

31. T. A. Dettlaff and A. A. Dettlaff, "Dimensionless criteria as a method of quantitative characterization of animal development," in *Mathematical Biology of Development* [in Russian], Nauka, Moscow (1982), pp. 25–39.

32. T. A. Dettlaff and A. S. Ginsburg, *The Embryonic Development of Acipenserid Fish (Stellate, Russian, and Giant Sturgeons) with Reference to the Problems of Their Breeding* [in Russian], Izd. Akad. Nauk SSSR, Moscow (1954).

33. T. A. Dettlaff and A. S. Ginsburg, "Acrosomal reaction in acipenserid fish and the role of calcium ions in the coupling of gametes," *Dokl. Akad. Nauk SSSR* **153**, 1461–1464 (1963).

34. T. A. Dettlaff, A. S. Ginsburg, and O. I. Schmalhausen, *Development of Acipenserid Fish. Maturation of Oocytes, Fertilization, Development of Embryos and Prelarvae* [in Russian], Nauka, Moscow (1981).

35. T. A. Dettlaff and M. N. Skoblina, "The role of germinal vesicle in the process of oocyte maturation in Anura and Acipenseridae," *Ann. Embryol. Morphogen., Suppl. I*, 133–151 (1969).

36. T. A. Dettlaff, M. N. Skoblina, and S. I. Davydova, "Intercellular influences in the process of oocyte maturation in acipenserid fish," in *Abstr. of Symp., Intercellular Interactions in the Processes of Differentiation and Growth* [in Russian], Tbilisi (1968), pp. 5–6.

37. T. A. Dettlaff, S. G. Vassetzky, and S. I. Davydova, "Maturation of sturgeon oocytes under different temperatures and time of obtaining eggs after the hypophyseal injection," in *6th Congrès Intern. Reprod. Anim. Insemin. Artif. Paris I.* (1968), pp. 129–131.

38. S. I. Doroshov, W. H. Clark, Jr., P. B. Lutes, R. L. Swallow, K. E. Beer, A. B. McGuire, and M. D. Cochran, "Artificial propagation of the white sturgeon, *Acipenser transmontanus*," *Aquaculture* **32**, 93–104 (1983).

39. B. M. Drabkina, "The effect of different water salinity on the survival of sperm, eggs, and larvae of *A. güldenstädti*," *Dokl. Akad. Nauk SSSR* **138**, 492–495 (1961).

40. V. S. Faustov and A. I. Zotin, "Variations in the heat emitted by eggs of fish and amphibians during their development," *Dokl. Akad. Nauk SSSR* **162**, 965–968 (1965).

41. N. S. Gabaeva, "Development of the follicle and egg-shell formation in the course of oogenesis of sturgeon," *Nauchn. Dokl. Vyssh. Shk. Biol. Nauki* **12**, 15-21 (1974).

42. N. L. Gerbilsky, "Method of pituitary injections and its role in pisciculture," in *Method of Pituitary Injections and Its Role in Reproduction of Fish Reserves* [in Russian], Izd. LGU, Leningrad (1941), pp. 5–35.

43. N. L. Gerbilsky, "Experimental and methodological bases of sturgeon pisciculture in the lower Kura flow," in *Tr. Lab. Osn. Rybovodstva* **2**, 5–28 (1949).

44. A. S. Ginsburg, "Establishment of bilateral symmetry in the eggs of acipenserid fish," *Dokl. Akad. Nauk SSSR* **90**, 477–480 (1953a).

45. A. S. Ginsburg, "Developmental defects in acipenserid fish related to insemination conditions," *Dokl. Akad. Nauk SSSR* **92**, 1097–1100 (1953b).

46. A. S. Ginsburg, "Variations in cleavage pattern in acipenserid fish," *Dokl. Akad. Nauk SSSR* **95**, 1117–1120 (1954).

47. A. S. Ginsburg, "Lipids in the oocytes and in the eggs of *Acipenser stellatus*," *Dokl. Akad. Nauk SSSR* **111**, 236–239 (1956).
48. A. S. Ginsburg, "Monospermy in acipenserid fish upon normal fertilization and the consequences of penetration into the egg of supernumerary spermatozoa," *Dokl. Akad. Nauk SSSR* **114**, 445–447 (1957).
49. A. S. Ginsburg, "The time of contact establishment between the egg and spermatozoon in acipenserid fish," *Dokl. Akad. Nauk SSSR* **115**, 845–848 (1957).
50. A. S. Ginsburg, "Fertilization in acipenserid fish. 1. The fusion of gametes," *Tsitologiya* **1**, 510–526 (1959).
51. A. S. Ginsburg, "The block to polyspermy in sturgeon and trout with special reference to the role of cortical granules (alveoli)," *J. Embryol. Exp. Morphol.* **9**, 173–190 (1961).
52. A. S. Ginsburg, *Instruction for Artificial Insemination of the Eggs of Acipenserid Fish* [in Russian], Glavrybvod Gos. Komiteta po Rybnomu Khosyaistvu pri SNKH SSSR, Moscow (1963).
53. A. S. Ginsburg, "Differences in morphology and properties of gametes in fish and their importance for the elaboration of adequate methods of artificial insemination," in *VIe Congrès Intern. Reprod. Anim. Insem. Artif., Paris*, Vol. 2 (1968), pp. 1037–1040.
54. A. S. Ginsburg, *Fertilization in Fish and the Problem of Polyspermy*, Israel Program for Scientific Translations, Ltd., IPST Cat. No. 600418 (1972).
55. A. S. Ginsburg, "Fine structure of the spermatozoon and acrosome reaction in *Acipenser stellatus*," in *Problems of Experimental Biology* [in Russian], Nauka, Moscow (1977), pp. 246–256.
56. A. S. Ginsburg and T. A. Dettlaff, "Experiments on transplantation and removal of organ rudiments in embryos of *Acipenser stellatus* in early developmental stages," *C. R. (Doklady) Acad. Sci. USSR* **44**, 209–212 (1944).
57. A. S. Ginsburg and T. A. Dettlaff, *Embryonic Development of Acipenserid Fish* [in Russian], Izd. Akad. Nauk SSSR, Moscow (1955).
58. A. S. Ginsburg and T. A. Dettlaff, *Development of Acipenserid Fish. Maturation of Oocytes, Fertilization, and Embryogenesis* [in Russian], Nauka, Moscow (1969).
59. A. S. Ginsburg and G. P. Nikiforova, "Cortical reaction in the stellate sturgeon egg during fertilization and artificial activation," *Sov. J. Dev. Biol.* **9**, 189–199 (1978).
60. A. S. Ginsburg and G. P. Nikiforova, "Development of cortical reaction during oocyte maturation in *Acipenser stellatus*," *Sov. J. Dev. Biol.* **10**, 385–392 (1979).
61. A. S. Ginsburg, S. E. Zubova, and L. A. Filatova, "An efficient method of artificial insemination of acipenserid eggs," in *Sturgeon Fishery in Water Bodies of the USSR* [in Russian], Izd. Akad. Nauk SSSR, Moscow (1963), pp. 47–55.
62. B. F. Goncharov, "A study of the transition of amphibian and acipenserid oocytes from growth to maturation," Author's Abstract of Candidate's Dissertation, Moscow (1971).
63. B. F. Goncharov, "Determination of the acipenserid pituitary gonadotropic activity by *in vitro* oocyte maturation," in *Sturgeons and Problems of Sturgeon Fishery* [in Russian], Pishchevaya Promyshlennost, Moscow (1972), pp. 257–262.

64. B. F. Goncharov, "The physiological state of the elder generation of stellate sturgeon follicles during their spawn migration," in *Ecological Physiology of Fish, Abstracts*, Part 2 [in Russian], Naukova Dumka, Kiev (1976), pp. 139–140.

65. B. F. Goncharov, "Effect of the cultivation medium composition on the capacity of acipenserid follicles to mature in response to the action of gonadotropic hormones," in *2nd All-Union Conference on Problems of Early Ontogenesis of Fish, Abstracts* [in Russian], Naukova Dumka, Kiev (1978), pp. 77–78.

66. B. F. Goncharov, "Use of temporal parameters of the reaction of follicles to hormones *in vitro* for the estimation of breeding quality of acipenserid fish," in *Sturgeon Fishery of Inland Water Bodies of the USSR, Abstracts* [in Russian], Astrakhan (1979), p. 60.

67. B. F. Goncharov, "A synthetic analog of luteinizing hormone-releasing hormone as a new promising stimulator of gamete maturation in Acipenseridae," *Dokl. Akad. Nauk SSSR* **276**, 1002–1006 (1984).

68. B. F. Goncharov, "Prospects of using synthetic luliberin analogs for obtaining mature gametes of sturgeons," in *Ecological Physiology and Biochemistry of Fish* [in Russian], Vilnius (1985), pp. 384–385.

69. G. M. Ignatieva, "The hatching enzyme secretion in explants of the hatching gland of acipenserid embryos," *Dokl. Akad. Nauk SSSR* **128**, 212–215 (1959).

70. G. M. Ignatieva, "The regional nature of the inductive action of chordamesoderm in embryos of acipenserid fish," *Dokl. Akad. Nauk SSSR* **134**, 233–236 (1960).

71. G. M. Ignatieva, "Inductive properties of the chordamesodermal rudiment prior to invagination and the regulation of its defects in sturgeon embryos," *Dokl. Akad. Nauk SSSR* **139**, 503–505 (1961).

72. G. M. Ignatieva, "A comparison of the dynamics of invagination of chordamesoderm material in embryos of *Acipenser stellatus, A. güldenstädti*, and axolotl," *Dokl. Akad. Nauk SSSR* **151**, 1466–1469 (1963).

73. G. M. Ignatieva, "A morphological study of the hatching gland in the acipenserid embryos under different conditions of development," in *Problems of Modern Embryology* [in Russian], Izd. MGU, Moscow (1964), pp. 274–279.

74. G. M. Ignatieva, "Correlation of epiboly and invagination processes at the period of gastrulation in *Acipenser stellatus* embryos," *Dokl. Akad. Nauk SSSR* **165**, 970–973 (1965).

75. B. N. Kazansky, "On maturation and fertilization of sturgeon eggs," *Dokl. Akad. Nauk SSSR* **89**, 757–760 (1953).

76. B. N. Kazansky, "Nuclear changes in sturgeon oocytes during the transition of the organism to the spawning state after hypophyseal injection," *Dokl. Akad. Nauk SSSR* **98**, 1045–1048 (1954).

77. B. N. Kazansky, "Analysis of phenomena occurring in the acipenserid oocytes upon application of hypophyseal injections," in *Proceedings of the Conference on Pisciculture* [in Russian], Izd. Akad. Nauk SSSR, Moscow (1957), pp. 130–138.

78. P. N. Khoroshko, "Ecology and efficiency of reproduction of the acipenserid fish in the lower Volga flow," Author's Abstract of Candidate's Dissertation, Astrakhan (1968).

79. P. A. Korzhuev, I. S. Nikolskaya, and L. I. Radzinskaya, "Respiration of acipenserid eggs during the period of incubation," *Vopr. Ikhtiol.*, **14**, 113–118 (1960).

80. A. Kowalewsky, P. Owsjannikow, and N. Wagner, "Die Entwicklungsgeschichte der Störe. Vorläufige Mitteilung," *Bull. Acad. Sci. St. Petersburg*, **14**, 317–325 (1870).

81. E. V. Krenig (Chulitskaya), "On the correlation between the processes of cleavage and gastrulation in *Acipenser güldenstädti* and *A. stellatus*," *Dokl. Akad. Nauk SSSR* **134**, 984–986 (1960).

82. A. A. Neyfakh and N. N. Rott, "A study of the ways of realization of radiation injuries in early fish development," *Dokl. Akad. Nauk SSSR* **119**, 261–264 (1958).

83. L. A. Nikitina, "Transplantation of germinal vesicle material in enucleated oocytes of *Acipenser stellatus*," *Dokl. Akad. Nauk SSSR* **205**, 1487–1489 (1972).

84. L. A. Nikitina, "Transplantation of nuclei from growing oocytes into fully grown enucleated sturgeon oocytes," *Sov. J. Dev. Biol.* **5**, 251–254 (1974).

85. L. A. Nikitina, "Presence of a cytotomy factor in the karyoplasm of *Acipenser* oocytes at the onset of the period of rapid growth," *Wilhelm Roux's Arch. Dev. Biol.* **184**, 109–113 (1978).

86. L. A. Nikitina, "A study of properties of nuclei of growing acipenserid oocytes by means of serial cloning," *Ontogenez* **13**, 40–45 (1982).

87. L. A. Nikitina, "Changes in the nuclei of the stellate sturgeon and axolotl oocytes produced by transplantation into the cytoplasm of mature activated eggs," *Ontogenez* **14**, 261–265 (1983).

88. G. M. Persov, "Account of investigation in sturgeon fishery with respect to application of the method of hypophyseal injections," in *Method of Pituitary Injections and Its Role in Reproduction of Fish Reserves* [in Russian], Izd. LGU, Leningrad (1941), pp. 42–50.

89. G. M. Persov, "On maturation divisions in oocytes and early stages of male pronucleus formation in acipenserid fish (sterlet and Russian sturgeon)," *Dokl. Akad. Nauk SSSR* **98**, 681–683 (1954).

90. G. M. Person, "Formation of pronuclei and their movement and fusion in the sterlet (*Acipenser ruthenus* L.)," *Dokl. Akad. Nauk SSSR* **103**, 737–740 (1955).

91. E. V. Raikova, "Ultrastructure of the oocytes of sturgeons at the end of pre-vitellogenesis. 1. Ultrastructure of the nucleus," *Sov. J. Dev. Biol.* **3**, 58–66 (1972).

92. E. V. Raikova, "Ultrastructure of sturgeon oocytes at the end of previtellogenesis. 2. Fine structure of the cytoplasm," *Tsitologiya* **15**, 1352–1361 (1973).

93. E. V. Raikova, "Ultrastructure of the sterlet oocytes during early vitellogenesis. 1. Nuclear ultrastructure," *Tsitologiya* **16**, 679–684 (1974).

94. E. V. Raikova, "Ultrastructure of the sterlet oocytes during early vitellogenesis. 2. Cytoplasmic fine structure," *Tsitologiya* **16**, 1345–1351 (1974).

95. E. V. Raikova, "Evolution of the nucleolar apparatus during oogenesis in Acipenseridae," *J. Embryol. Exp. Morphol.* **35**, 667–687 (1976).

96. E. V. Raikova, G. Steinert, and C. Thomas, "Localization of amplified ribosomal DNA in meiotic prophase oocytes of Acipenseridae by *in situ* hybridization with ^{125}I ribosomal RNA," *Tsitologiya* **19**, 338–341 (1977).

97. O. B. Trubnikova and L. V. Ryabova, "Preovulatory changes of the ovarian follicle of *Acipenser stellatus* Pall.," *Ontogenez* **20**, 532–542 (1989).

98. E. V. Serebryakova, "Some data on chromosome complexes in Acipenseridae," in *Genetics, Selection, and Hybridization of Fish* [in Russian], Nauka, Moscow (1969), pp. 105–113.

99. I. A. Shmanzar', "A cytoplasmic study of DNA synthesis during early oogenesis in acipenserids," *Tsitologiya* **18**, 27–30 (1976).

100. M. N. Skoblina, "Ripening of cortex of anuclear oocytes in frog and stellate sturgeon under the influence of gonadotrophic hypophyseal hormones," *Dokl. Akad. Nauk SSSR* **183**, 982–984 (1968).

101. M. N. Skoblina, "Independence of cortex maturation from germinal vesicle material during the maturation of amphibian and sturgeon oocytes," *Exp. Cell Res.* **55**, 142–144 (1969).

102. M. N. Skoblina, "An experimental study of the role of the nucleus in the process of oocyte maturation in amphibians and acipenserid fish," Author's Abstract of Candidate's Dissertation, Moscow (1970).

103. R. I. Tatarskaya, K. A. Kafiani, and S. I. Kanopkaite, "Some enzymes of phosphorus metabolism and the rate of respiration and anaerobic glycolysis in the embryonic development of sturgeons," *Biokhimiya* **23**, 527–539 (1958).

104. V. P. Vasiliev, "Chromosome numbers in Cyclostomata and Pisces," *Vopr. Ikhtiol.* **20**, 387–422 (1980).

105. V. P. Vasiliev, *Evolutionary Karyology of Fish* [in Russian], Nauka, Moscow (1985).

106. S. G. Vassetzky, "The thermosensitivity of sturgeon eggs at the stages of maturation and cleavage," *Zh. Obshch. Biol.* **27**, 583–595 (1966).

107. S. G. Vassetzky, "Studies of oocyte maturation and early embryogenesis in acipenserid fish with respect to possible application of the method of thermal sex control in fish," Author's Abstract of Candidate's Dissertation, Moscow (1966).

108. S. G. Vassetzky, "Dynamics of the first maturation division in sturgeon oocytes," *Zh. Obshch. Biol.* **31**, 84–93 (1970).

109. A. D. Vlasenko, "Biological foundations of natural reproduction of the acipenserid fish in the regulated Volga and Kuban Rivers," Author's Abstract of Candidate's Dissertation, Moscow (1982).

110. Yu. G. Yurovitsky, "Morphological characteristics of sturgeon embryo (*Acipenser güldenstädti* Brandt) under various oxygen conditions," *Vopr. Ikhtiol.* **4**, 315–329 (1964).

111. Yu. G. Yurovitsky and P. N. Reznichenko, "Morphophysiological peculiarities of sturgeon embryos upon incubation at different oxygen regimes," in *Sturgeon Fishery in Water Bodies of the USSR* [in Russian], Izd. Akad. Nauk SSSR, Moscow (1963), pp. 77–82.

112. V. V. Zalensky, "The life history of the sterlet (*Acipenser ruthenus*). Part 1. Embryogenesis," in *Tr. Obshch. Estestv. Ispytat. Kazan. Univ.* **7**, 1–226 (1878).

113. V. V. Zalensky (W. Salensky), "Zur Embryologie der Ganoiden," *Zool. Anz.* **1**, 243–245, 266–269, 288–291 (1878).

114. A. I. Zotin, "Water consumption from environment by developing acipenserid eggs," *Dokl. Akad. Nauk SSSR* **89**, 377–380 (1953).

115. A. I. Zotin, "Changes in the egg membranes' strength during embryogenesis of acipenserid fish," *Dokl. Akad. Nauk SSSR* **92**, 443–446 (1953).

116. A. I. Zotin, "Hatching enzyme in acipenserid embryos," *Dokl. Akad. Nauk SSSR* **92**, 685–687 (1953).
117. A. I. Zotin, "Peculiarities of hatching enzyme secretion in the acipenserid and salmonid embryos," *Dokl. Akad. Nauk SSSR* **95**, 1121–1124 (1954).
118. A. I. Zotin, *Physiology of Water Metabolism in the Embryos of Fish and Cyclostomata* [in Russian], Izd. Akad. Nauk SSSR, Moscow (1961).
119. A. I. Zotin, "The mechanism of transition of blastocoele liquor into the archenteron cavity in embryos of *Acipenser güldenstädti*," *Dokl. Akad. Nauk SSSR* **142**, 968–971 (1962).
120. A. I. Zotin, "The mechanism of cleavage in amphibian and sturgeon eggs," *J. Embryol. Exp. Morphol.* **12**, 247–262 (1964).
121. A. I. Zotin, "The mechanism of cytokinesis," *Usp. Sovrem. Biol.* **71**, 66–84 (1971).
122. A. I. Zotin and A. Y. Krumin', "Formation of archenteron cavity in acipenserid embryos," *Zh. Obshch. Biol.* **20**, 313–321 (1959).
123. A. I. Zotin and R. V. Pagnaeva, "The time of determination of the cleavage furrow position in eggs of sturgeon and axolotl," *Dokl. Akad. Nauk SSSR* **152**, 765–768 (1963).
124. E. Zuccelli, "L'esturgeon des Landes," *Sci. Avenir*, No. 467, 62–67 (1986).

Chapter 3

THE RUSSIAN STURGEON *Acipenser Güldenstädti*
Part II. LATER PRELARVAL DEVELOPMENT

O. I. Schmalhausen

A prelarval period can be distinguished in the development of sturgeons as a distinct and separate phase; this is also the case for many teleosteans [1, 38, 40, 41, 51, 70, 71]. The period begins with the liberation of the embryo from the envelopes and terminates with its transition to an active feeding form. In this period the embryo is called a prelarva, sometimes referred to as a free embryo or eleutheroembryo [6]. Prelarval development continues at the expense of yolk reserves, but the prelarva acquires relationships with the environment which are not achieved by the embryo.

The general morphoecological characteristics of the prelarval development of sturgeons have been provided by Zalensky [69], Strelkovsky [59], Dragomirov [25], and others. By using morphological, physiological, and ecological criteria some authors distinguished a number of phases in prelarval development (see [56]).

The description of successive prelarval developmental stages of the Russian (*Acipenser güldenstädti*) and stellate (*A. stellatus*) sturgeons obtained at sturgeon hatcheries is given by Dragomirov [19, 22], Zaryanova [71], and Dettlaff et al. [16], and of the giant sturgeon (*Huso huso*) by Gordienko [32] and Schmalhausen [51]. In addition, there are descriptions of prelarvae of the Russian and stellate sturgeons caught in nature [1], and also for "kaluga" (*H. dauricus*) and Amur (*A. schrencki*) sturgeons [58], shovelnose (*Pseudoscaphyrhynchus*) sturgeon [29], and black (*A. oxyrhynchus*) and short-nosed (*A. brevirostris*) sturgeons [7]. These data are, however, not comparable in most cases due to selection of different stages and different diagnostic features.

Meanwhile, a number of stages can be distinguished in the prelarval development of sturgeons which are not only characteristic for individual species but also common for different species (and genera), at least for the well-studied ones (Russian, stellate, and giant sturgeons), since the differences between the prelarvae of these species can be reduced to those in skin pigmentation, absolute size, proportion of some body parts, and behavior. These differences are most pronounced at the end of the prelarval period although some of them can already be found at the stage of hatching [1, 20].

The prelarval development of sturgeons is naturally divided into two periods: from hatching to the onset of active respiratory movements and from onset of these movements to the transition to active (exogenous) feeding [48]. Ten stages are distinguished in prelarval development [51, 56], and some of these stages coincide with those described by Dragomirov [22].

The description is given below for the prelarvae of *Acipenser güldenstädti colchicus* V. Marti at the same developmental stages which were earlier described for the giant sturgeon [51].

Prelarvae were obtained from eggs incubated at the hatcheries. The observations were carried out both at the hatcheries and in the laboratory. The structure of prelarvae was studied on live and fixed (Bouin's fluid) materials under a binocular microscope. In some cases fixed prelarvae were prepared (skin removed, individual branchial arches isolated, etc.). The structure of prelarvae at successive developmental stages is described by external diagnostic features.

The internal structure of prelarvae, as well as the development of some organ systems, has been described by Ostroumov [39], Kryzhanovsky [36, 37], Ryndzyunsky [42], Severtzov [57], Disler [17, 18], Schmalhausen [44–49, 51], Yakovleva [67, 68], Dragomirov [21, 23, 24, 26–28], Baburina [2–4], Gerbilsky [30, 31], Titova [62], Vinnikov and Titova [64], Krayushkina [35], Caloianu-Iordachel [13], Chusovitina [14], Vernidub et al. [63], and Timofeev [61].

3.1. STAGES OF NORMAL DEVELOPMENT FROM HATCHING TO THE END OF THE PRELARVAL PERIOD

The numbers of the stages given below (see also [51]) continue those embryonic stages for sturgeons suggested by Dettlaff and Ginsburg [15] (see Chapter 2). Figures 3.1 to 3.3 show the sturgeon prelarvae at successive developmental stages. The numbers of the drawings correspond to those of the stages. The drawings are made of larvae developing at a mean temperature of 18.6°C. The duration of prelarval development at this temperature is 9 days. In Table 3.1 the main diagnostic features of all stages are described briefly.

To provide dimensionless (relative) characteristics of the time of onset of successive developmental stages, the mean temperature was determined for every interval between fixations (about 24 h), the value of τ_0 for this temperature was determined by the curve on Fig. 2.6, and the absolute duration of development was divided by the value of τ_0. In addition, the duration of each interval between the fixation at 18°C was calculated. The data obtained are presented in Table 3.1.

Stage 36 (after Dettlaff and Ginsburg [15]). Prelarvae during the period of mass hatching; mouth opening and gill clefts not yet formed; hatching gland still discernible (Figs. 3.1, 36 and 3.3, 36). Prelarvae 9–10.5 mm long (prelarvae fixed in Bouin's fluid were measured). They are rather dark since the cells of the epidermis still contain a large amount of melanin. The head is small with respect to the trunk and bent toward the massive ventral region (the so-called yolk sac) of elongated egg form. The mouth opening is absent. The hatching gland is still seen on the lower head surface in front of the mouth invagination. There are no barbel rudiments. The trunk and tail are bordered by a fin fold. The upper and lower lobes of the tail fin fold are equally wide, that is, the tail at this stage is still protocercal. Between the pre- and postanal parts of the fin fold, separated by a small groove, there is a rudiment of the cloaca into which pronephric excretory ducts open. The alimentary canal is still closed. A pre-anal fin fold begins at the posterior surface of the yolk sac; this anterior region of the fold is markedly widened but does not yet form a process characteristic of subsequent stages (keel). The dorsal fin fold is constricted cranially and ends at the level of anterior muscle segments. Rudiments of pectoral fins are either absent or appear as hardly discernible skin thickenings just behind the pronephros and the upper part of Cuvier's ducts. There are no ventral fin rudiments.

The segmentation of trunk muscles is completed by this stage (~40 muscle segments to cloaca). The anterior segments form ventral processes; the primary-symmetrical form of the myotomes [42] is thereby already disturbed at the stage of hatching. The segmentation of tail muscles is incomplete (the posterior part of the mesoderm is not yet segmented). The posterior end of the notochord is bent slightly upward. There are paired rudiments of the maxillary, hyoid, and first branchial arches. Gill clefts are still absent. Grooves are seen at the sites of the first two pairs of gill clefts. Respiration is effected via the body surface and a network of blood vessels in the posterior part of the yolk sac. These vessels obtain blood from the caudal and subintestinal veins, from the segmental trunk vessels, and from develop-ing parts of the posterior cardinal veins. In front they fuse in paired vitelline veins opening in the heart alongside the Cuvier's ducts. The blood is yellowish.

The sensory organs are weakly developed. The rudiments of the olfactory organs are spheroid, with one external opening. The eyes are not pigmented, except at small areas of high light sensitivity in their bottom part [4]. According to Baburina [4], just-hatched prelarvae are indifferent to light; this reaction is preserved during the whole prelarval period. On each side of the diencephalon, clearly discernible due to the transparency of the covering tissues, there are ear vesicles (rudiments of labyrinths); the latter have no signs of subdivisions. The seismosensory (lateral line) system of the head is represented by paired rudiments of supraoptic, suboptic, and temporal lines, as well as by paired rudiments of the hyomandibular complex of organs [17, 25]. The suboptic rudiments begin below the ear vesicles, pass around the eyes from behind, and terminate at the level of the lower eye margin. The shorter supraoptic rudiments are above the eyes.

Stage 37. Characterized by the appearance of the mouth opening and the first pair of gill clefts and distinct rudiments of pectoral fins (Fig. 3.1, 37). Prelarvae 10.5–11.5 mm long. Embryonic pigment is still present in some regions of the covering epithelium. The head begins to straighten. The mouth is broken through in the middle part. Four round tubercles are outlined on the lower head side in front of the mouth; these are the rudiments of barbels situated at the anterior mouth boundary. The form of the yolk sac changes: it elongates and its ventral outline becomes less concave. On both sides of the ventral regions short folds are visible: they are formed in the wall of the intestinal tract rudiment and subdivide its wide part into two. These folds are directed obliquely (down and forward) and coincide with the anterior boundary of the yolk sac vascular network. The distance between this network and Cuvier's ducts has increased and also increases further during the whole period of yolk respiration. Rudiments of pectoral fins appear as distinct folds. The tail grows faster than the posterior trunk part and becomes relatively longer, and this continues further at subsequent developmental stages. The anterior end of the pre-anal part of the fin fold begins to form a keel. In the posterior trunk part muscle segments begin to grow forward and downward; this growth can clearly be seen in the anterior segments. The first gill cleft is open. Divisions are

TABLE 3.1. Chronology of Prelarval Development of the Russian Sturgeon *Acipenser güldenstädti colchicus* V. Marti

Stage No.	Time from hatching days, hours at 18°C	τ_n/τ_0	Diagnostic features of stages
36	0	0	Prelarva at the stage of mass hatching; mouth opening and gill clefts absent; hatching gland still seen
37	23 h	28	Rudiments of barbels appear; mouth opening has broken through; onset of yolk sac division into stomach and intestinal parts; rudiments of pectoral fins clearly seen as small skin folds; rudiments of gill filaments absent; a rudiment of seismosensory system lateral line appears
38	2 days, 3 h	63	First melanocytes discernible; endodermal fold separating stomach and intestine incomplete; first muscle buds develop in the region of dorsal and anal fins; gill filament rudiments appear on operculum and first branchial arch; lateral line of seismosensory system reaches the level of Cuvier's duct
39	3 days, 3 h	93	Stomach separated from intestine; dorsal and anal fins separated; lateral line of seismosensory system reaches the level of pectoral fin posterior margin; an accessory row of seismosensory system appears
40	4 days, 4 h	123	Ventral fin rudiment discernible as a narrow skin fold; ventral processes of muscle segments in the region of pectoral fin descend below its base; first irregular mandibular movements are observed; lateral line of seismosensory system does not reach the level of stomach end; accessory row terminates above pectoral fin
41	5 days, 4 h	153	Margins of olfactory lobes close but do not yet fuse; frequent mandibular movements; lateral line of seismosensory system terminates above spiral gut; its accessory row extends beyond the posterior boundary of pectoral fin; a short rudiment of dorsal row appears
42	6 days, 3 h	181	A rudiment of pyloric appendage appears; lobes of olfactory organ fused; lateral line of seismosensory system reaches the level of ventral fin anterior margin; dorsal row begins to bend
43	7 days, 5 h	213	Rostrum assumes horizontal position; ventral fin reaches the margin of pre-anal fold; rudiments of secondary filaments in the first gill appear; lateral line of seismosensory system reaches the ventral fin posterior margin; accessory row terminates above spiral gut; dorsal row bent and begins to grow in parallel with lateral line

TABLE 3.1 (continued)

Stage No.	Time from hatching		Diagnostic features of stages
	days, hours at 18°C	τ_n/τ_0	
44	8 days, 2 h	238	Barbel bases displaced forward and their tips do not reach the anterior mouth boundary; mass discarding of pigment plugs; lobes of ventral fins descend below the pre-anal fold margin; a mesenchyme band (common rudiment of dorsal scutes) appears in the dorsal fin fold; anterior transverse commissure of seismosensory system displaced dorsally and not seen from the ventral side; lateral line extends beyond the level of ventral fin posterior margin; accessory row does not reach the ventral fin anterior boundary
45	9 days, 1 h	266	Stage of transition to active feeding; individual rudiments of scutes in the dorsal fin fold are seen; lateral line of seismosensory system extends beyond the level of dorsal fin middle part; accessory row almost reaches the level of ventral fin anterior margin; dorsal row extends beyond the level of pectoral fin posterior margin

delineated in the ear vesicle. Short rudiments of the lateral line of the trunk seis-mosensory system are discernible.

Stage 38. Characterized by the appearance of rudiments of filaments on oper-cular fold and the first branchial arch (Figs. 3.1, 38 and 3.3, 38). Prelarvae 11.5–12 mm long. Single, small, still rather pale melanocytes are visible through the covering epithelium on the upper head surface. Such melanocytes are also found on the dorsal side of the first trunk muscle segments, as well as on the muscle seg-ments and fin fold in the region of the future dorsal fin. The ventral body region is flattened laterally and ventrally. A fold separating the digestive system rudiment into stomach and intestine reaches the middle of the lateral surface of the prelarva. The tail is slightly bent upward. The height of the muscle segments is increased, and their dorsal processes in the region of the dorsal fin form discernible rudiments of muscle buds. The ventral processes of the anterior trunk segments do not yet reach the lower margin of the opercular fold; in the region of the pectoral fin they run to its base. At the level of the future anal fin hardly discernible rudiments of muscle buds are seen on the ventral processes of the muscle segments. The oper-cular fold and first branchial arch each have a row of still short rudiments of gill fil-aments. As a result of a change in the direction of blood flow in the gill vessels, branchial respiration begins in the prelarvae. The olfactory pits are slightly elon-gated. Pigmentation of the iris begins. The supraoptic line of the seismosensory system elongates and terminates above the olfactory pit. The suboptic line runs outside the lateral barbel and terminates at the anterior margin of its base. The lateral trunk line reaches the 4th to 6th muscle segments.

Stage 39. Characterized by the division of the intestine rudiments into two parts, stomach and intestine, by a fold of the rudiment wall (Fig. 3.1, 39). Prelarvae 12.0–13.2 mm long. Their pigmentation is markedly intensified. Melanocytes are darker and distributed all over the surface of the trunk and tail muscle segments. Folds are delineated in the mouth angles, marking the onset of differentiation of lips. The rudiment of the upper lip consists of two parts. Teeth rudiments appear. The widened part of the digestive system rudiment is divided in two regions, stomach and intestine. The yolk sac vascular network is reduced, and the left vitelline vein degenerates. On the right side a short connective tissue cord is directed upward from the intestine and the anterior end of the dorsal pancreas rudiment; it intersects the most posterior part of the stomach surface and runs under the ventral processes of muscle segments to the dorsal mesenterium. The ventral processes of the anterior muscle segments almost reach the lower operculum margin. Enlarged pectoral fins are somewhat displaced downward. The fin fold is already widened where the dorsal fin is forming, and begins to widen in the region of the anal fin and the lower lobe of the tail fin. The tail fin is separated from the anal and dorsal fins by shallow grooves. Skin thickenings appear in the positions of the future ventral fins.

The second gill cleft has opened. Rudiments of the filaments of the opercular half-gill (situated on the opercular fold) and of the first row of the first gill are slightly elongated, but in the second row of the first gill they are still short. The first row of filaments of the second gill is delineated. The olfactory pits are trapeziform. The pigmentation of the iris is more marked. The anterior ends of the suboptic lines of the seismosensory system move forward to the level of the olfactory organs and bend toward each other, forming rudiments of the premaxillary lines. Rudiments of the trunk lateral line reach the level of the posterior margin of the pectoral fins, and rudiments of the accessory row of the lateral line appear.

Stage 40. Characterized by the appearance of rudiments of the ventral fins and the onset of mandibular movements (Figs. 3.1, 40 and 3.3, 40). Prelarvae ~13.00 mm long. Their pigmentation is intensified. Melanocytes form loose accumulations on each side of the mesencephalon and the anterior part of the medulla oblongata, on the dorsal side of the anterior muscle segments, on the boundary between the axial and ventral muscles (alongside pectoral fins), and on the dorsal fin and all over the lateral side of the posterior trunk part and tail. The barbel rudiments are elongated. The median recess of the upper lip has begun to smooth over. A connective tissue cord supporting the intestine moves forward but still lies on the posterior part of the stomach. The dorsal rudiment of the pancreas lying in parallel with the intestine stretches to this level as well.

The bases of the pectoral fins do not yet reach the middle of the lateral surface of the stomach. Muscle buds differentiate in them. The first signs of a reduction of the anterior keel-shaped process of the pre-anal fin fold appear, but it is still clearly seen. Rudiments of ventral fins appear as narrow longitudinal folds, and the lower

Fig. 3.1 and 3.2. Prelarvae of the Russian sturgeon *Acipenser güldenstädti colchicus* V. Marti at successive developmental stages. Lateral view. Numbers of drawings correspond to those of stages.

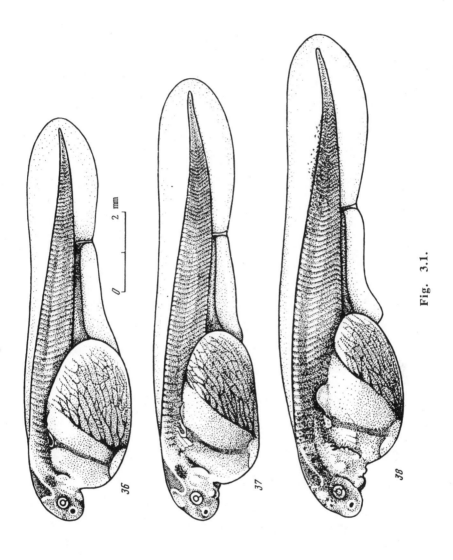

Fig. 3.1.

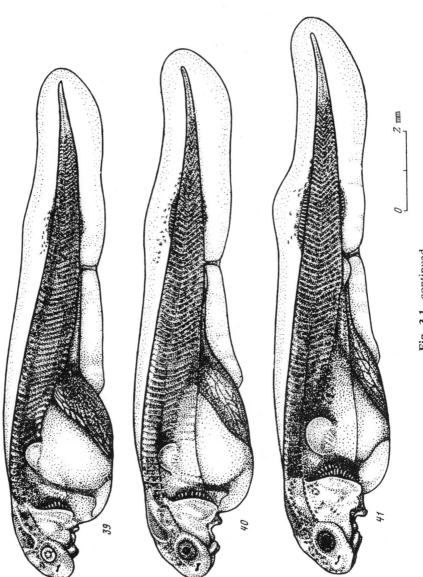

Fig. 3.1, continued.

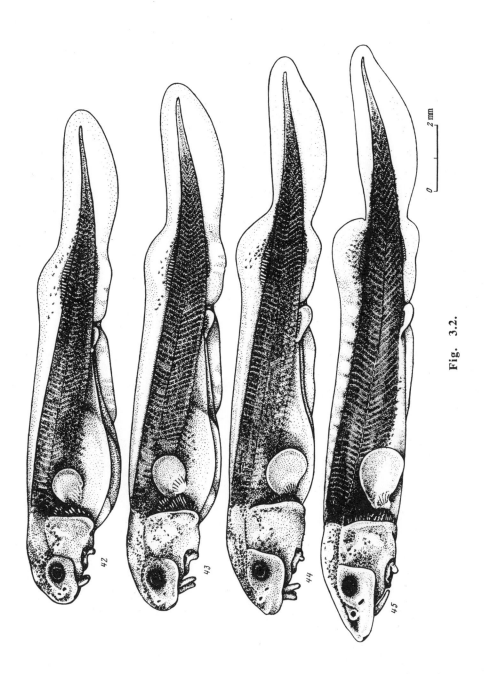

Fig. 3.2.

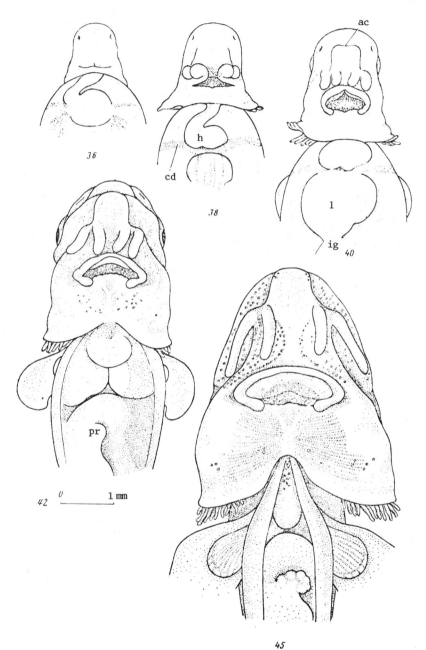

Fig. 3.3. Prelarvae of the Russian sturgeon at successive developmental stages. Ventral view. Numbers of drawings correspond to stages. cd) Cuvier's duct; ig) intermediate gut; pr) rudiment of pyloric appendage; ac) anterior transverse commissure; l) liver; h) heart.

lobe of the tail fin continues to widen. Ventral processes of muscle segments over-grow the lateral surface of the stomach halfway and descend below the bases of the pectoral fins.

The third gill cleft is still incomplete. The filaments of the opercular half-gill and the first row of the first gill reach the level of Cuvier's duct. The rudiments of filaments of the second row of the first gill have elongated, and the rudiments of filaments of the first row of the second gill are clearly discernible. The first respiratory mandibular movements are observed.

The olfactory pits elongate, and their margins give rise to two lobes partitioning off the olfactory opening; the upper lobe is more developed. The iris is distinguished by its light-brown color. Not yet fully separated semicircular channels of the labyrinths are visible through the transparent covering tissues. The anterior ends of the right and left premaxillary lines of the seismosensory system have fused on the lower head surface and formed an anterior transverse commissure at the level of the olfactory rudiments (Fig. 3.3, 40). The lateral lines of the trunk seismosensory system almost reach the posterior boundary of the stomach; paired rudiments of the accessory rows terminate above the pectoral fins.

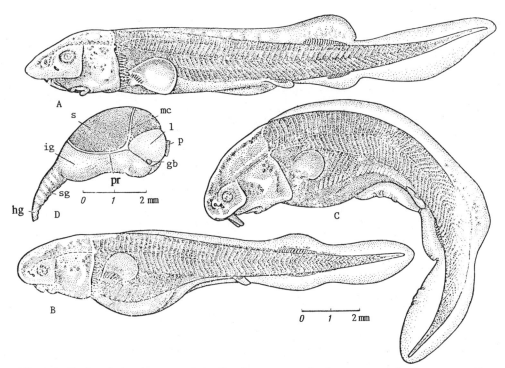

Fig. 3.4. Prelarval development of the Russian sturgeon in the presence of a brass net. A) Normal prelarva at stage 45; B) prelarva of the same age which developed from the stage of hatching in the presence of a brass net piece; C) prelarva of the same age which developed in a Chalikov's incubator in the river; D) digestive system of the prelarva shown in C; mc) mesenterial cord; gb) gall bladder; p) postpharynx or esophagus; hg) hindgut; pr) pancreas rudiment; s) stomach; sg) spiral gut; all other designations as in Fig. 3.3.

Stage 41. Characterized by the closure of lobes partitioning off the olfactory opening and onset of rhythmic respiratory movements (Fig. 3.1, 41). Prelarvae 13–14 mm long. The first taste buds appear on the tips of the barbels. The upper lip is constricted in the middle part, and short lateral lobes of the lower lip are clearly visible. The liver margin adjacent to the intestine is divided into two parts, and in the right part the gall bladder rudiment is visible.

The connective tissue cord running from the intestine to the dorsal mesenterium is displaced forward and now lies across the stomach, midway between its end and the posterior margin of the pectoral fin. The bases of the pectoral fins are oblique with respect to the longitudinal body axis and begin to constrict. The keel-shaped process of the pre-anal fin fold is markedly reduced. The posterior part of the ventral fin has widened. The tail is bent smoothly upward. The ventral processes of the muscle segments have overgrown more than half the lateral body surface. The gill cleft between the second and third gills has broken through. The second row of filament rudiments on the second gill have formed, and filament formation begins in the first row of the third gill. Movements of the visceral apparatus become stronger and more rhythmic. Lobes partitioning off the external olfactory opening now touch each other but are not yet fused. The iris has become darker. The anterior ends of the supraoptic rudiments of the head seismosensory system have extended over the upper olfactory lobe. Fusion of the premaxillary lines forming the anterior transverse commissure is clearly seen on the lower side of the rostrum. As the rostrum straightens, this commissure is displaced forward and almost reaches the level of the anterior margin of the olfactory organs. Developing pitlines (neuroepithelial follicles) of the opercular group are seen on the operculum. The lateral line reaches the level of the spiral intestine. The accessory row of lateral system organs extends beyond the vertical of the pectoral fin's posterior boundary. Paired rudiments of the dorsal row appear.

Stage 42. Characteristic features are the fusion of lobes partitioning off the olfactory opening and the appearance of a rudiment of the pyloric appendage (Figs. 3.2, 42 and 3.3, 42). Prelarvae 14–15 mm long. Melanocytes appear on the lower surface of the opercula and the anterior part of the ventral region. The rostrum continues to straighten. Taste buds appear on the lips. The liver is divided into two lobes. A small tubercle, the rudiment of pyloric appendage, is separated from the anterior end of the intermediate gut on its left side; it is clearly visible ventrally. The connective tissue cord lying above the stomach on the right side reaches the posterior margin of the pectoral fin. The anterior end of the dorsal pancreas rudiment is located here as well. The pectoral fin has descended to the middle of the lateral surface of the prelarva and is displaced forward. Its anterior end touches the filaments of the opercular half-gill and first gill. The keel-shaped process of the pre-anal fold is smoothing out. The ventral fins change their form; a lobe forms in their posterior part which does not yet reach the pre-anal fin fold margin. Muscle buds differentiate at the bases of the ventral fins. The anal fin's caudal part widens. Ventral processes of muscle segments overgrow the whole lateral surface of the body. Cuvier's ducts, located in front of the pectoral fins, are not visible (they are covered from above by a muscle layer and gill filaments). Rudiments of the first-row filaments are distinct on the third branchial arch. Rhythmic movements of the visceral apparatus occur.

The lobes partitioning off the external olfactory openings fuse. The anterior transverse commissure of the seismosensory system lies in front of the olfactory organs. Rudiments of neuroepithelial follicles are visible on the lower side of the rostrum. The lateral line of the trunk reaches the anterior level of the ventral fin. An accessory row of lateral line organs extends to the end of the stomach. The rudiment of the dorsal row of the lateral system intersects the anterior muscle segments obliquely downward and begins to bend backward. At this stage the prelarvae settle on the substratum.

Stage 43. Characteristic features are the appearance of rudiments of secondary filaments in the first gill; the ventral fins reach the margin of the pre-anal fin fold (Fig. 3.2, 43). Prelarvae 15–16 mm long. The rostrum assumes a horizontal position. The pyloric rudiment begins to subdivide into lobes. The anal opening has broken through. The connective tissue cord attaching the intermediate gut to the dorsal mesenterium is no longer visible, as it is obscured by a muscle layer and displaced under the pectoral fin. The lobes of the pectoral fins widen.

The ventral fins reach the margin of the pre-anal fin fold, and its keel is fully reduced. Rudiments of secondary filaments appear in the first gill; in the second row they are less pronounced. Rudiments of second-row filaments appear in the third gill. The fourth gill cleft has broken through. Opercula cover the anterior margin of the pectoral fins.

The iris is dark. The anterior transverse commissure of the seismosensory system is displaced to the anterior end of the rostrum, but is still visible from the ventral side. The lateral line of the trunk seismosensory system reaches the level of the ventral fin's posterior margin, the accessory row terminates above the spiral gut, and the dorsal row is already bent and its posterior part is parallel to the lateral line. The end of this row is situated above the pectoral fin.

Stage 44. Characteristically the margins of the ventral fins extend beyond the pre-anal fin fold; the common rudiment of the dorsal scutes appears (Fig. 3.2, 44). Prelarvae 16.2–17.0 mm long. The pigmentation of the prelarva is markedly intensified. Melanocytes migrate downward and cover almost the whole flank of the prelarva. Dorsally the melanocytes are situated in spots above the pectoral fins. Rudiments of bony scales appear on the upper head surface. The relative size of the head increases due to marked elongation of the rostrum. The barbel bases are displaced forward and the barbel tips do not reach the anterior mouth margin; the shorter middle barbels terminate at a greater distance from the mouth than the longer lateral ones.

The pyloric appendage continues to subdivide into lobes; three lobes are usually discernible. A mass discarding of pigment plugs takes place. A narrow mesenchyme cord, the common rudiment of the dorsal scutes, appears in the dorsal fin fold and is visible. The height of the dorsal fin is markedly increased. The anterior margin of the pre-anal fin fold is sharply constricted. Lobes of the ventral fins descend below the margin of the pre-anal fin fold. The tail end is attenuated and begins to bend downward. Segmental trunk muscles overgrow a part of the ventral surface of the prelarva.

In both rows of filaments of the second gill, rudiments of secondary filaments appear. The anterior transverse commissure of the seismosensory line is displaced dorsally and is not visible from the ventral side. The lateral line of the trunk seismosensory system extends beyond the ventral fin's posterior margin, and the accessory row does not reach its anterior boundary. The end of the dorsal row rudiment reaches the vertical of the pectoral fin's posterior margin.

Stage 45. This stage is marked by the transition to active feeding. The prelarva is capable of seizing movements with the mouth. There are separate rudiments of dorsal scutes (in the Russian and stellate sturgeons) (Figs. 3.2, 45 and

3.3, 45). Prelarvae 17–18 mm long. Transition to active (exogenous) feeding; after this stage the prelarvae should be called larvae. The prelarvae are intensively pigmented. The rostrum lengthens still more, and the middle barbels are now at a significant distance from the mouth. Lateral barbels almost reach the anterior mouth margin. Mandibular and subpharyngeal teeth begin to break through. The pectoral fins are displaced to the ventral side; the pre-anal fin fold is still present, but its width is less than that of the anal fin and of the fin fold's dorsal part. Mesenchyme rudiments of dorsal scutes are discernible in the latter. The dorsal and anal fins are not yet fully separated from the tail fin. The tail tip is attenuated and bent downward, and the ventral lobe of the tail fin, which began to widen at stage 39, is now much wider than its dorsal lobe. The discarding of pigment plugs ends. Segmental muscles of the right and left flanks of the prelarva are not yet fused on the ventral trunk surface.

The spiracle has broken through in some prelarvae. Spiracular gill filaments begin to appear. The first two gills are pectinate. Rudiments of secondary leaflets appear in the first row of the third gill. Rudiments of first-row filaments are formed in the fourth gill. The iris is fully pigmented. At this stage subject vision appears and the oculomotor muscles begin to function [4]. Most neuroepithelial follicles of the head are now open. The end of the lateral line extends beyond the level of the dorsal fin's middle region. The accessory row of the seismosensory system almost reaches the level of the ventral fin's anterior margin, and the dorsal row extends beyond the level of the pectoral fin's posterior margin.

3.2. EMBRYOGENESIS DURING THE TWO MAIN PHASES OF PRELARVAL LIFE

3.2.1. From Hatching to the Onset of Rhythmic Respiratory Movements

The stages of prelarval development have been characterized above, and the chronology of these stages has been provided (Table 3.1). Now we will consider changes in the prelarvae during two qualitatively different periods of their development as a whole on the basis of both their external and internal structure.

The first of these periods, from hatching to the onset of rhythmic respiratory movements, is characterized by preparation for the definitive way of respiration. Significant changes in the gill region take place at this time [36, 37, 46–48, 57, 61]. At the beginning of this period the gills are still absent and respiration is carried out, as before, via a network of yolk sac blood vessels, which develops and begins to function long before the hatching of the embryos. Just after hatching, the intensity of respiration increases steeply [34]. At the beginning of this period the major part of the venous blood from the caudal and subintestinal veins, segmental trunk vessels, and developing parts of the posterior cardinal veins goes into a network of yolk sac blood vessels where it is oxygenated. Venous blood from the head and anterior trunk region goes into Cuvier's ducts and from there into the heart.

After hatching, the cephalic blood vessels are rearranged. Within a few hours the aortic arches begin to differentiate into afferent and efferent vessels in the first two pairs of branchial septa, and loops of vessels form which grow into rudiments of the gill filaments. The direction of blood flow in a commissure connecting the mandibular and hyoid aortic arches changes and, as a result, the mandibular vessel below the commissure degenerates. The head, which obtained previously mixed blood from the heart via the ventral aorta, now obtains oxygenated blood from the vascular network made up by anastomosing efferent vessels of the opercular half-gill and from the first gill efferent vessel. The prelarvae now start passive gill respiration. The formation of gill vessels proceeds gradually in all the other branchial arches.

As the gill respiration develops, the importance of the yolk sac blood vessels for respiration falls sharply due to a change of blood flow direction in the posterior cardinal veins, so that an ever-increasing part of the venous blood goes directly to Cuvier's ducts and a correspondingly decreasing part into the yolk sac vascular network, which begins to be reduced. The change of blood flow direction in the posterior cardinal veins takes place gradually, so that the subintestinal vein collects blood from an ever-decreasing number of segmental vessels and finally degenerates. All blood from the caudal vein goes directly to the posterior cardinal veins. The major part of the venous blood therefore goes into the heart, and the gills obtain nonoxygenated blood.

At the beginning of this period the gill rudiments are small and immobile, and the mouth is open. By its end, at stage 40, the first rare mandibular movements are observed, and the mouth acquires the ability to close.

The visceral apparatus during the first period is still rudimentary. The cartilaginous elements of the maxillary arch appear at stage 38, and chondrification of the hyoid arch begins at stage 40, at the very end of this period. The visceral muscles begin to differentiate somewhat earlier.

All these changes create conditions for the transition to active gill respiration, which is realized at the beginning of the next developmental period.

At the stage of hatching, the digestive system consists of the alimentary canal and rudiments of the digestive glands, liver, and dorsal pancreas. The alimentary canal is divided into two parts, a widened anterior (yolk sac) part and a narrow posterior part. The digestive system is closed, the mouth is still absent, and the pharyngeal cavity is separated from the alimentary canal cavity by a massive septum so that, even after the formation of the mouth opening at stage 37, no communication with the environment is established; the narrow posterior part of the alimentary canal still lacks a lumen.

During this period the yolk sac is divided into two parts, an anterior stomach part and a posterior intestinal part. The division is formed by the ingrowth of an obliquely directed fold of the alimentary canal wall. The fold goes along the boundary between the first part of the yolk sac, not covered by a vascular network, and its second part, covered by a network of vessels which takes part in respiration at these stages. The anterior fold wall becomes the lower stomach wall, and its posterior fold wall becomes the upper wall of the anterior part of the intestine. In parallel with these morphogenetic processes, the yolk sac vascular network is reduced and loses its initial function. In the posterior part of the intestine a spiral gut with a spiral valve forms. The mucous membrane of the whole intestine becomes folded and differentiates. The liver becomes bilobate. The gall duct and gall bladder form and, in connection with the former, two ventral pancreas rudiments appear. By the end of this period, therefore, the pancreas consists of three independent rudiments.

The process of intracellular digestion begins before the end of embryogenesis: thus, at the late neurula stage the cells of the nervous system and sense organ rudiments are almost completely devoid of yolk.

Besides the intracellular digestion of yolk, which also continues after hatching, the yolk is also disintegrated in the alimentary canal cavity. This cavity is filled with endodermal cells rich in yolk which do not form part of the alimentary canal wall; this yolk is already utilized by the time of hatching. As the yolk is utilized, the granules of embryonic pigment are discarded from the cells and form a melanin plug in the alimentary canal cavity.

Sense organs in the just-hatched larva are also as yet weakly differentiated. At this stage the olfactory organ rudiment is a spheroid sac, the bottom part of which is made up by a thickened internal layer of epithelium (its external layer is already atrophied): both layers are present in the marginal zone. The bottom of the olfactory sac is smooth; sensory epithelium differentiates there. The olfactory sac is connected with the forebrain by a short branch of the olfactory nerve. The sac cavity communicates with the environment through a circular primary olfactory opening. As the head length increases, the olfactory rudiment elongates and its opening becomes oval. Two lobes appear in the marginal zone of the olfactory organ opening on its dorsal and ventral sides and grow toward each other: this is the onset of partitioning off the primary olfactory opening. A short rudiment of the first longitudinal fold partitions off the olfactory sac bottom [49].

Definitive receptor cells differentiate and retinal layers form in the eye; at the same time, transitory photoreceptor cells degenerate in the area of high light sensitivity. At the end of this period all retinal layers are already discernible and pigmentation of the iris begins [4].

The seismosensory system lines are represented by thickenings of the covering epithelium internal layers which grow with the rudiments of corresponding nerves. During this period rudiments of all the head canals differentiate; by its end their neuromasts communicate with the environment. Rudiments of the trunk seismosensory system line and accessory row appear and grow caudally. Rudiments of neuroepithelial follicles are still not visible.

Ear vesicles differentiate, and by the end of this period have incomplete semicircular canals.

The pigmentation of the body surface begins from stage 38.

3.2.2. Subsequent Development Leading to the Transition to Active Feeding

The second period of prelarval development is characterized by a number of changes providing the transition to the definitive way of feeding. These changes concern not only the digestive system, but also the visceral apparatus, the changes in which make possible the seizure of prey, and the sense organs, the differentiation of which allows the search for food. In this period the yolk sac and opercular half-gill lose their initial respiratory value. Rhythmic movements of the visceral apparatus are established in correlation with those of the operculum, which is markedly reduced in sturgeons and develops at a slower rate than the gill filaments. Correspondingly, the gill filaments remain uncovered for a long time, especially in the upper part of the branchial arches, thus ensuring constant water change at their surfaces. Rudiments of secondary filaments appear on those of the first two gills and they become pectinate. In all other branchial arches muscles and skeleton develop, rudiments of filaments appear, and the definitive branchial blood circulation is established.

The mandibular arch is not connected directly with the skull, as in teleosteans, but hangs on the hyomandibular and is connected with the jaws by a separate cartilaginous element. A joint develops between the maxilla, mandible, and hyoid arch.

The hyomandibular connects with the ear capsule, and the bottom part of the branchial apparatus chondrifies. Muscle and connective tissue cords are attached to corresponding skeletal parts. At the end of this period the mouth becomes mobile. Such a structure of the mouth allows sturgeons to feed on benthic organisms. Teeth erupt and the prelarva begins to seize food. Respiration is already achieved using the principle of a force pump [5, 60, 65, 66]. Such a method of respiration cannot be achieved with the mouth closed, but, due to the external position of the gills, respiration is not interrupted at the moment of food swallowing, although the rhythmic movements of the visceral apparatus are arrested.

Further differentiation of the digestive system consists of preparation for definitive digestion [51]. Rudiments of tubular glands appear in the stomach, first as buds consisting of a few cells; a lumen arises in them, and alveoli appear which open into the stomach cavity. Their number and size increase rapidly. The pyloric part of the stomach is devoid of tubular glands, which develop in the walls of all the other stomach parts. A rudiment of the pyloric appendage is separated from the anterior part of the intestine. By the end of the second period it already consists of a few lobes. As the yolk is digested, a large amount of fat appears in the intestinal and liver cells; it begins to disappear from the cells only after the transition to active feeding. The differentiation of the intestinal epithelium is complete. By the end of this period the melanin plug is displaced from the spiral gut cavity in the hindgut. The anal opening breaks through and the pigment plugs are discarded. The septum between the pharyngeal cavity and esophagus is then resorbed. All the digestive system parts begin to function prior to full yolk resorption but after the transition to stage 45. The structure of the visceral apparatus and digestive system by the time of transition to active feeding allows the prelarva to swallow rather large prey, and they are immediately capable of feeding on small chironomid larvae.

The following changes are observed in the sense organs. The lobes partitioning off the primary olfactory opening approach one another and then fuse. The suture at the place of fusion is soon resorbed and anterior and posterior nostrils form. A rosette of folds develops on the olfactory sac bottom.

By the beginning of active feeding the whole iris is intensely pigmented; the fundus oculi is fully pigmented as well. Retinal layers and definitive photoreceptor cells are developed all over the fundus oculi; the eye muscles function [4].

Differentiation of the seismosensory system proceeds [17, 18, 21, 25, 27, 28]. Rudiments of the third (dorsal) line appear in the trunk. During this period neuroepithelial follicles develop. Differentiating sensory pits of the opercular group are seen at stage 41, and all other sensory pits of the head, including those on the lower side of the rostrum, at stage 42. At stage 45 most follicles are open to the outside through pores. At this stage the prelarvae are already intensely pigmented, especially in the tail region.

Thus, by the end of the second period the prelarvae are prepared for the transition to active feeding by the level of development of the organs of food seizure and digestion, as well as of the sense organs. A certain amount of yolk is still retained in them and there are large reserves of fat.

Some authors distinguish one more developmental period, that of mixed feeding, lasting from the onset of active feeding to complete yolk resorption. Sturgeon prelarvae are, however, capable of rather long-term starvation [9–12] and can begin active feeding after the resorption of their endogenous reserves [14]. The mixed feeding stage is not, therefore, an obligatory stage of their development.

3.3. DEFECTS OF PRELARVAL DEVELOPMENT

The processes of morphogenesis and differentiation proceed during the prelarval period, and disturbances of these processes lead to developmental defects.
Some of them are similar to those observed at the late embryonic stages and are due
to epiboly disturbances. At the neurula stage such embryos preserve a more or less
large yolk plug. The further fate of such embryos depends on the size of the yolk
plug. The most abnormal embryos perish and less defective embryos hatch; some
of them have typical structure, others have shortened and bent tails. The tail can be
absent almost to the anus or, at sufficient length, be sharply bent, most often dorsally. Such larvae survive usually to complete yolk resorption, and some of them,
under laboratory conditions, begin active feeding.

Such defects can arise in the prelarval period as well if the coordinated growth
of the body parts is affected. It is usually due to a delay of yolk resorption. In
such cases the digestive organs are underdeveloped and the axial ones continue to
grow so that the tail bends ventrally and even forms a spiral. These defects are observed in the prelarvae occurring in an environment containing traces of heavy metals [16, 54; see Fig. 3.4]. A similar teratogenic effect of copper was shown for
teleosteans as well [8].

To determine the causes of these defects, it is enough to follow the development closely. The origin of some other defects was studied experimentally. It was
thus shown that the frequent structural defects of the olfactory organ, in particular
underdevelopment or absence of a partition between the anterior and posterior nostrils, which were assumed to arise under unfavorable influences on the eggs upon
their artificial insemination and on the embryos during the period of incubation [33,
43], can be reproduced experimentally if the prelarvae are subjected to unfavorable
influences during the period of formation of this organ. In prelarvae developing in
the normal environment such defects occur extremely rarely [50].

The presence in water of various harmful chemical agents exerts a teratogenic
effect on many other organs. In particular, the presence of traces of heavy metals
induces, besides the already mentioned defects of the axial organs, a delay of yolk
resorption and, as a result, inhibition of growth and differentiation accompanied by
a decrease in viability of the prelarvae. Hence, body proportions are disturbed,
most organs are underdeveloped, and degenerative changes of muscles are observed. The eyes are usually of smaller size and abnormalities of their morphogenesis are seen on histological sections: no stretching and attenuation of the retina
take place, unlike in normal development; the lens adjoins the retina tightly, and the
latter remains much bulkier than in the normal prelarvae [54].

Phenol induces a number of regular defects. Prelarvae are not pigmented, and
hemorrhages and defects of the visceral system arise sometimes. The absence of fat
can be observed in histological sections of the late prelarvae, unlike in the normal
prelarvae. Degenerative changes of the skeletal muscles and eye retina are often
observed [52–55].

In some cases, the structural defects of prelarvae are, therefore, due to unfavorable influences acting directly on the morphogenetic processes [16].

REFERENCES

1. L. A. Alyavdina, "A contribution to biology and taxonomy of sturgeon fish at the early developmental stages," *Tr. Saratovsk. Otd. Kaspiiskogo Filiala VNIRO* **1**, 33–73 (1951).
2. E. A. Baburina, "Development of eyes and their functions in the Russian and stellate sturgeons," *Dokl. Akad. Nauk SSSR* **106**, 359–361 (1956).
3. E. A. Baburina, "Development of eyes and their function in the Russian and stellate sturgeons," *Tr. Inst. Morfol. Zhivotn. Akad. Nauk SSSR* **20**, 148–185 (1957).
4. E. A. Baburina, *Development of Eyes in Cyclostomata and Pisces with References to Their Ecology* [in Russian], Nauka, Moscow (1972).
5. P. P. Balabai, "Studies of respiratory system in the Acipenseridae," *Tr. Inst. Zool. Biol. Akad. Nauk URSR* **5**, 111–145 (1939).
6. E. K. Balon, "Terminology of intervals in fish development," *J. Fish. Res. Board Can.* **32**, 1663–1670 (1975).
7. D. W. Bath, J. M. O'Connor, J. B. Alber, and L. G. Arvidson, "Development and identification of larval Atlantic sturgeon (*Acipenser oxyrhynchus*) and shortnose sturgeon (*A. brevirostris*) from the Hudson River estuary," *Copeia* **3**, 711–717 (1981).
8. W. J. Birge and J. A. Black, "Effects of copper on embryonic and juvenile stages of aquatic animals," in *Copper in the Environment,* Part 2, J. O. Nriagu, ed., Wiley, New York (1979), pp. 373–399.
9. L. S. Bogdanova, "Comparative characteristics of the transition to active feeding in larvae of different species and ecological forms of sturgeons," *Tr. TsNIORKh* **1**, 196–201 (1967).
10. L. S. Bogdanova, "Capacity of larvae of the Russian and Siberian sturgeons for long-term starvation," in *Materialy Nauchnoi Sessii TsNIORKh, Posvyashchennoi 100-Letiyu Osetrovodstva,* Astrakhan (1969), pp. 24–26.
11. L. S. Bogdanova, "Ecological plasticity of larvae and fry of sturgeons," in *Sturgeons and Problems of Sturgeon Pisciculture* [in Russian], Pishchevaya Promst., Moscow (1972a), pp. 244–250.
12. L. S. Bogdanova, "Comparison of transition to active feeding of larvae of the Russian and Siberian sturgeons at different temperatures," *Tr. TsNIORKh* **4**, 217–223 (1972b).
13. M. Caloianu-Iordachel, "Histophysiological characteristics of digestive system in the Russian, stellate, 'bastard,' and giant sturgeons with reference to their ecology at the early ontogenetic stages," Candidate's Thesis, Leningrad (1959).
14. L. S. Chusovitina, "Postembryonic development of the Siberian (*Acipenser baeri* Brandt) sturgeon," *Tr. Ob'-Tazovskogo Otdeleniya GosNIORKh* **3**, 103–114 (1963).
15. T. A. Dettlaff and A. S. Ginsburg, *Embryonic Development of Sturgeons (Stellate, Russian, and Giant Sturgeons) with Reference to Problems of Their Breeding* [in Russian], Izd. Akad. Nauk SSSR, Moscow (1954).
16. T. A. Dettlaff, A. S. Ginsburg, and O. I. Schmalhausen, *Development of Acipenserid Fish* [in Russian], Nauka, Moscow (1981).

17. N. N. Disler, "Development of cutaneous sense organs of lateral system in the stellate sturgeon *Acipenser stellatus*," *Tr. Inst. Morfol. Zhivotn. Akad. Nauk SSSR* **1**, 333–362 (1949).
18. N. N. Disler, *Organs of Sense of the Lateral Line System and Their Role in Behavior of Fish* [in Russian], Izd. Akad. Nauk SSSR, Moscow (1960).
19. N. I. Dragomirov, "Development of the stellate sturgeon larvae in the period of yolk feeding," *Tr. Inst. Morfol. Zhivotn. Akad. Nauk SSSR* **10**, 244–263 (1953a).
20. N. I. Dragomirov, "Species peculiarities of sturgeon larvae at the stage of hatching," *Dokl. Akad. Nauk SSSR* **93**, 551–554 (1963b).
21. N. I. Dragomirov, "Development of cutaneous receptors on the lower head side in the sturgeon larvae starting the benthic way of life," *Dokl. Akad. Nauk SSSR* **97**, 173–176 (1954).
22. N. I. Dragomirov, "Larval development of the Volga–Caspian sturgeon *Acipenser güldenstädti* Brandt," *Tr. Inst. Morfol. Zhivotn. Akad. Nauk SSSR* **20**, 187–231 (1957).
23. N. I. Dragomirov, "Age correlations in the development of lateral system organs in the Aral 'bastard' sturgeon (*Acipenser nudiventris* Lov.) larvae," *Dokl. Akad. Nauk SSSR* **124**, 489–492 (1959a).
24. N. I. Dragomirov, "Development of the system of neuroepithelial follicles in the larvae of the Aral 'bastard' sturgeon *Acipenser nudiventris* Lov.," *Dokl. Akad. Nauk SSSR* **125**, 1374–1377 (1959b).
25. N. I. Dragomirov, "Ecological–morphological peculiarities of larval development of the giant sturgeon *Huso huso* L.," *Tr. Inst. Morfol. Zhivotn. Akad. Nauk SSSR* **33**, 72–93 (1961).
26. N. I. Dragomirov, "Development of the lateral system of organs in the giant sturgeon larvae," *Zh. Obshch. Biol.* **22**, 273–280 (1961).
27. N. I. Dragomirov, "Gradients of differentiation of temporal follicles in the Russian sturgeon," *Dokl. Akad. Nauk SSSR* **181**, 762–764 (1968).
28. N. I. Dragomirov, "Ways and patterns of integration of organogenesis in the supraoptic part of the lateral system in Russian sturgeon larvae," *Dokl. Akad. Nauk SSSR* **196**, 478–481 (1971).
29. N. I. Dragomirov and O. I. Schmalhausen, "Ecological–morphological peculiarities of the shovelnose sturgeon (*Pseudoscaphyrhynchus*)," *Dokl. Akad. Nauk SSSR* **85**, 1399–1402 (1952).
30. N. L. Gerbilsky, "Histological analysis of transitions between stages of larval development in fish," in *Problems of Modern Embryology* [in Russian], Izd. LGU, Leningrad (1956), pp. 122–129.
31. N. L. Gerbilsky, "Histophysiological analysis of the digestive system of sturgeon and teleostean fish in the early period of development and methods of rearing larvae in pisciculture," in *Proceedings of Conference on Pisciculture* [in Russian], Izd. Akad. Nauk SSSR, Moscow (1957), pp. 89–94.
32. O. L. Gordienko, *Rearing of the Giant Sturgeon* [in Russian], Legpishchepromizdat, Moscow (1953).
33. E. M. Kokhanskaya, *Method of Reproduction of Fish with Sticky Egg Membranes without Preliminary Removal of Mucilage* [in Russian], Nauka, Moscow (1980).
34. P. A. Korzhuev, "Consumption of oxygen by eggs and fry of the Russian (*Acipenser güldenstädti*) and stellate (*Acipenser stellatus*) sturgeons," *Izv. Akad. Nauk SSSR, Biol.* **2**, 291–302 (1941).

35. L. S. Krayushkina, "Histophysiological characteristics of the digestive system organs in the stellate sturgeon (*Acipenser stellatus*) larvae at different developmental stages," *Dokl. Akad. Nauk SSSR* **117**, 542–544 (1957).
36. S. G. Kryzhanovsky, "Development of paired fins in *Acipenser, Amia,* and *Lepidosteus* and problems of the theory of limbs," *Tr. Nauchn.-Issled. Inst. Zool. MGU* **1**, Part 1, 1–63 (1925).
37. S. G. Kryzhanovsky, "Organs of respiration in larvae of fish (Teleostomi)," *Tr. Lab. Evolyuts. Morfol. Akad. Nauk SSSR* **1**(2), 1–104 (1933).
38. B. S. Matveev, "On biological stages in postembryonic development of sturgeons," *Zool. Zh.* **32**, 249–255 (1953).
39. A. A. Ostroumoff, "Zur Entwicklungsgeschichte des Sterlets (*Acipenser ruthenus*). 4. Das Gefässystem des Kopfes," *Zool. Anz.* **32**, 404–407 (1907).
40. T. S. Rass, *Geographical Parallelisms in the Structure and Development of Teleostean Fish of the Northern Sea* [in Russian], Izd. MGU, Moscow (1941).
41. T. S. Rass, "Steps of ontogenesis of teleostean fish," *Zool. Zh.* **25**(2), 137–146 (1946).
42. A. G. Ryndzyunsky, "Development of the form of myotome in fish," *Tr. Inst. Evolyuts. Morfol. Akad. Nauk SSSR* **2**(4), 11–111 (1939).
43. I. A. Sadov and E. M. Kokhanskaya, "Incubation of sturgeon eggs in trays," *Tr. Inst. Morfol. Zhivotn. Akad. Nauk SSSR* **37**, 5–66 (1961).
44. O. I. Schmalhausen, "Development of gill and mouth apparatus in the stellate sturgeon," *Dokl. Akad. Nauk SSSR* **80**, 681–684 (1951).
45. O. I. Schmalhausen, "Development of gills and blood vessels of the visceral apparatus in the stellate sturgeon," *Dokl. Akad. Nauk SSSR* **86**, 193–196 (1952).
46. O. I. Schmalhausen, "Development of gills in the Volga sturgeon larvae," *Dokl. Akad. Nauk SSSR* **100**, 397–400 (1955).
47. O. I. Schmalhausen, "Development of gill vessels in the Volga sturgeon larvae," *Dokl. Akad. Nauk SSSR* **100**, 605–608 (1955).
48. O. I. Schmalhausen, "Ecological–morphological peculiarities in development of the gill apparatus in the Volga sturgeon larvae," *Dokl. Akad. Nauk SSSR* **100**, 837–839 (1955).
49. O. I. Schmalhausen, "A morphological study of olfactory organs in fish," *Tr. Inst. Morfol. Zhivotn. Akad. Nauk SSSR* **40**, 157–187 (1962).
50. O. I. Schmalhausen, "Developmental defects of olfactory organ in sturgeons under certain conditions of rearing," *Tr. Inst. Morfol. Zhivotn. Akad. Nauk SSSR* **40**, 188–218 (1962).
51. O. I. Schmalhausen, "Development of digestive system in sturgeons," in *Morphoecological Studies of Fish Development* [in Russian], Nauka, Moscow (1968), pp. 40–70.
52. O. I. Schmalhausen, "Influence of phenol on prelarval development in sturgeons. 1. External structure and behavior," *Sov. J. Dev. Biol.* **2**, 603–610 (1971).
53. O. I. Schmalhausen, "Influence of phenol on prelarval development in sturgeons. 2. Digestive system," *Sov. J. Dev. Biol.* **3**, 491–497 (1972).
54. O. I. Schmalhausen, "Development of prelarvae of sturgeon fish and its defects under the influence of phenol and ions of heavy metals," Candidate's Thesis, Institute of Developmental Biology, Moscow (1973).

55. O. I. Schmalhausen, "Influence of phenol on prelarval development in sturgeons. 3. Skin pigmentation, eye structure," *Sov. J. Dev. Biol.* **4**, 32–39 (1973).
56. O. I. Schmalhausen, "The Russian sturgeon *Acipenser güldenstädti colchicus*. Development of prelarvae," in *Objects of Developmental Biology* [in Russian], Nauka, Moscow (1975), pp. 264–277.
57. A. N. Severtsov, *Origin and Evolution of Lower Vertebrates* [in Russian], Izd. Akad. Nauk SSSR, Moscow (1948).
58. S. G. Soin, "Materials on development of sturgeons of the Amur River," in *Trans. Amur Ichthyological Expedition, 1945–1949*, Vol. 2 [in Russian], Izd. MOIP, Moscow (1951), pp. 223–232.
59. V. I. Strelkovsky, "Development of the Russian sturgeon *Acipenser güldenstädti* Brandt," Candidate's Thesis, Moscow State University (1940).
60. K. I. Tatarko, "Apparatus of operculum and its ligaments with hyoid and mandibular arches in Acipenseridae," *Tr. Inst. Zool. Biol. Akad. Nauk URSR* **10**, 5–53 (1936).
61. O. B. Timofeev, "Circulatory system in Acipenseridae at the early ontogenetic stages," *Tr. TsNIORKh* **3**, 306–316 (1971).
62. L. K. Titova, "Development of olfactory organ in fish and amphibians," *Dokl. Akad. Nauk SSSR* **107**, 749–751 (1956).
63. M. F. Vernidub, E. N. Kudryashova, G. I. Nishchaeva, and G. I. Ratnikova, "Age changes in the structure and function of the digestive system in the Russian (*Acipenser güldenstädti* Brandt) and stellate (*Acipenser stellatus* Pall.) sturgeons at the early ontogenetic stages," *Tr. TsNIORKh* **3**, 77–113 (1971).
64. Ya. A. Vinnikov and L. K. Titova, *Morphology of Olfactory Organ* [in Russian], Medgiz, Moscow (1957).
65. M. M. Voskoboinikov, "Apparatus of gill respiration in fish (an experience of synthesis in morphology)," in *Trans. Third Conf. of Russ. Zoologists, Anatomists, and Histologists, 1927*, Leningrad (1928), pp. 103–105.
66. M. M. Voskoboinikov (Woskoboinikoff), "Der Apparat der Kiemenatmung bei den Fischen," *Zool. Jahrb., Abt. Allgem. Zool. Physiol. Tiere* **55**, 315–488 (1932).
67. I. V. Yakovleva, "Histogenesis of the thyroid gland and pituitary in the Russian sturgeon with reference to stages of larval development," Candidate's Thesis, Leningrad (1952).
68. I. V. Yakovleva, "Development of the Russian sturgeon teeth with reference to stages of larval development," *Dokl. Akad. Nauk SSSR* **94**, 775–778 (1954).
69. V. V. Zalensky, "Developmental history of the sterlet (*Acipenser ruthenus*). 2. Postembryonic development and development of organs," *Tr. Obshch. Estestv. Ispyt. Kazan. Univ.* **10**(2), 227–245 (1880).
70. E. B. Zaryanova, "Morphobiological characteristics of the Russian sturgeon at the early developmental stages with reference to different methods of egg incubation," *Tr. Saratovsk. Otd. Kaspiiskogo Filiala VNIRO* **1**, 113–131 (1951).
71. E. B. Zaryanova, "Morphobiological characteristics of the Russian (*Acipenser güldenstädti* Brandt) and stellate (*Acipenser stellatus* Pallas) sturgeons at the early developmental stages," *Tr. Saratovsk. Otd. Kaspiiskogo Filiala VNIRO* **3**, 294–355 (1954).

Chapter 4

THE RAINBOW TROUT *Salmo gairdneri*

G. M. Ignatieva

The rainbow trout has been used for a long time for descriptive and experimental embryology, developmental biochemistry, and other biological disciplines. It has become especially popular (partially due to its ever-increasing use in pisciculture) during recent decades: the bibliography [157] on the reproduction of some salmonid species during the period from 1963 to 1979 has 1399 references; they concern only the development of gonads, ovulation, spermatogenesis, spawning, and fertilization. This list contains many references (over 300) to work done on the rainbow trout. The aim of this chapter is not to give an extensive review of the published data on the embryonic and prelarval development of the rainbow trout. A few references only are cited, mainly those of Soviet authors which may be unknown outside the USSR, to demonstrate the diversity of developmental biology problems studied using this species.

As mentioned above, the processes of gametogenesis have been intensively studied (for references see [97, 157]). Great attention is paid to the elaboration of methods for long-term preservation of the capacity of gametes for fertilization and of the conditions of artificial fertilization (for references see [20, 21]) and to studies on *in vivo* and *in vitro* oocyte maturation [58, 66, 89, 90, 91, 142]. Other studies include investigations of blastoderm potencies and competence, inductive interactions, morphogenetic movements during gastrulation, and the distribution of presumptive organ rudiments at pregastrular stages [5–12, 45–52, 62, 108–111, 127–129, 163]. The respiration of embryos and different processes of metabolism have also been studied, as well as the composition and utilization of yolk during development, specific macromolecular syntheses, etc. [3, 18, 23, 24, 26–29, 41, 42, 53–55, 61, 74, 76, 78, 87, 88, 92, 107, 115, 116, 118, 119, 125, 126, 131, 134, 138, 140, 141, 147, 149–151, 159–162, 164–166, 182, 185, 186].

A number of papers describe the influence of environmental factors on embryonic development, especially those of temperature [63, 71, 80, 81, 84, 85, 93, 103, 104, 135, 136, 167], O_2 content [60, 70, 71, 124, 148, 180, 181], and illumination ([19, 102, 103, 113]; for references see also [177]), as well as the influence of some ions [122, 172, 173, 188].

89

Various chemical agents, including mutagens, and physical agents (temperature, UV and x rays, etc.) have been used by many authors to study developmental defects induced by these agents, to obtain polyploid embryos and artificial gynogenesis [16, 17, 34–38, 59, 114, 169–171, 175, 179]. The rainbow trout has also been used in some recent genetic engineering studies [39].

Estimation of the quality of eggs and spermatozoa, of typical development, and the survival of progeny are used also as criteria for the selection and estimation of quality of the spawners [4, 32, 33, 69, 105, 106, 130, 144, 155].

One of the advantages of the rainbow trout (along with other salmonid species) as a subject for studying the early developmental stages in teleostean fish consists in the slow rate of their development at low temperatures, which are optimal for them. While in warmth-loving species the duration of embryogenesis is estimated in days, in the salmonids it lasts several months. At the same time, the temperature limits of development in the rainbow trout are higher than in other salmonids, and this makes it suitable for studying the embryogenesis of salmonid fish.

4.1. TAXONOMY AND DISTRIBUTION

The rainbow trout belongs to the family Salmonidae, order Clupeiformes, superorder Teleostei, class Pisces. There is no universally accepted single opinion on the species name of the rainbow trout; at present American and most European ichthyologists consider the rainbow trout as a freshwater form of the anadromous steelhead trout described in 1836 by Richardson and named, after the American naturalist Gairdner, *Salmo gairdneri*, and call both forms (freshwater and anadromous) *Salmo gairdneri* Rich. [24, 25, 120]. Some scientists (see, e.g., [123, 137]) consider the rainbow trout as an independent species and retain for it the name given by Gibbons in 1855: *S. irideus* Gibb. Such a name is due to a bright iridescent band (with a predominance of red or orange-yellow tints) running along the body of sexually mature individuals.

Various local forms of the rainbow trout were repeatedly described as independent species. But Needham and Gard [120] provided convincing evidence, on the basis of studying many meristic features in these "species," that there are no reliable differences between them, and proposed that these species were different forms (races) of the rainbow trout *S. gairdneri*. The same conclusion was drawn by Borovik [25], who studied the Minsk population of the rainbow trout. At the same time, many European scientists (see [117]) stress the hybrid origin of the trout introduced in Europe, obtained by hybridization of different races of this species.

The genus *Salmo* includes anadromous and freshwater salmon (the freshwater forms are called trout). There are 7 to 10 species in all. Four or five among them occur in the USSR: the Atlantic salmon *Salmo salar* L., the sea trout *S. trutta* L. with different river forms, the brook trout *S. trutta* L., *Morpha fario* L., and others [123]. All representatives of the genus *Salmo* are very similar in their ecology, structures of gametes, and embryonic development and are often used for embryological studies, along with the rainbow trout. In many of the papers cited above the studies were carried out in parallel on several species of salmonid fish. The morphological similarity of the embryos of different *Salmo* species allows for the use of the accompanying description of the normal stages of the rainbow trout (see Table 4.1 and Figs. 4.5–4.10) for the other *Salmo* species as well.

Like other salmonid fish, the rainbow trout is considered to be tetraploid with respect to other salmoniformes [132], $2n = 60$ [31]; see also [94, 132, 174].

TABLE 4.1. Description of Stages of Embryonic and Larval Development of the Rainbow Trout (after Vernier, 1969) (see Figs. 4.5–4.10)

Stage No.	Time from insemination hours, days	τ_n/τ_0	Diagnostic features of stages
	Hours		Embryonic development
[0]	0	0	Mature egg at the moment of insemination (for description see Section 4.3.1)
1	3	1.0	Formation of blastodisc (bipolar differentiation); diameter (D) of blastodisc 1.5–2 mm
1+	6	2.0	Formation of peripheral periblast
2	7.5	2.5	Two blastomeres
3	10.5	3.5	Four blastomeres (furrows of the 2nd cleavage transverse to that of 1st cleavage furrow)
4	13.5	4.5	Eight blastomeres (furrows of the 3rd cleavage parallel to that of 1st cleavage furrow)
5	16.5	5.5	Sixteen blastomeres (furrows of the 4th cleavage transverse to that of 1st cleavage furrow)
6	19.5	6.5	Thirty-two blastomeres; thickness (H) of blastodisc 0.5 mm
	Days		
[7–]	1.5	13	Onset of the period of interphase lengthening; fall of mitotic index
7	2	16	Diameter of blastoderm = 1.2 mm, H = 0.5 mm. The surface of blastoderm is formed by numerous but still quite discernible cells
8	2.5	20	Blastoderm is thickened. Its margins descend steeply; the surface is granular. H = 0.8 mm
9	3	24	Beginning of blastoderm flattening: D = 1.7–1.8 mm, H = 0.6–0.7 mm. Cells are hardly discernible. The blastoderm margins cover the periblast
[9+]	3.25	26	Beginning of morphogenetic nuclear function. This is determined by the method of radiation inactivation of the nuclei
10	3.5	28	[Beginning of gastrulation.] D = 2 mm, H = 0.5 mm. The posterior blastoderm margin is thickened. A subgerminal cavity (blastocoel) appears in some eggs
11	4	32	Appearance of terminal node. D = 2.2 mm, H = 0.5 mm

TABLE 4.1 (continued)

| Stage No. | Time from insemination | | Diagnostic features of stages |
	days	τ_n/τ_0	
12	5	40	[Wide germ ring with a broadened part, the germ disc.] Beginning of formation of the axial structures. D = 3 mm
13	6	48	Convex embryo is distinctly seen. Epiboly by 1/3 [germ ring displaced by 60° from animal pole]. D = 3.5 mm. [Formation of the 1st pair of somites.] Length (L) of the embryo 2.0 mm
14	[6.5]	52	D = 4.5 mm, L = 2.5 mm. Epiboly is by 1/2 [germ ring reaches the egg equator]. Up to 5 pairs of somites. From stage 14 on the main diagnostic features are indicated by arrows on figures.
[14+]	7	56	Ten pairs of somites. Eye vesicles have begun to form.
15	7.5	60	L = 3.1 mm. Eye vesicles are distinct. Epiboly by 2/3. Up to 15 pairs of somites
16	8	64	L = 3.5 mm. Epiboly by 3/4. Up to 20 pairs of somites, 3 cerebral vesicles. Ear placodes have begun to form at the level of the 1st branchial groove
17	9	72	L = 3.8 mm. Blastoderm has fully overgrown the yolk. Up to 29 pairs of somites. 5 neuromeres in rhomben-cephalon. Lens is delineated; ear vesicles are formed
18	10	80	L = 4.2 mm. Undifferentiated tail rudiment. Mesen-cephalon is separated. Intestine rudiment is seen. Up to 40 pairs of somites
19	11	88	L = 4.5 mm. Up to 50 pairs of myotomes (see below) in all. The tail rudiments have less than 10 pairs of myotomes. Olfactory placodes are formed. [Heart is a slightly curved tube. Bendings of tail and body are rare-ly observed.] The contours of neuromeres disappear gradually
20	12	96	L = 5 mm. The 1st gill slit is broken through; the 3rd gill groove is delineated. Up to 58 pairs of myotomes in all, 20 in the tail rudiment. [Blood colorless. Heart pulsates. Active tail movements.]

TABLE 4.1 (continued)

Stage No.	Time from insemination		Diagnostic features of stages
	days	τ_n/τ_0	
21	14	112	L = 6.5 mm. The tail rudiment has over 25 pairs of myotomes, up to 64 pairs in all. The external margin of the eye vascular coat is pigmented. 4 gill slits. The pectoral fin rudiments appear. The posterior intestine region is separated from the yolk. The Wolffian duct is seen. The yolk sac is vascularized by 1/4. [Blood is red.] 2 aortic arches, tail artery and vein, subintestinal vein, afferent yolk vein, 2 efferent yolk veins and anterior cardinal veins. [Soin [153] provides a more detailed description of vascular system development in the embryo of the grayling (a representative of the family Thymallidae, close to the Salmonidae).]
22	16	128	L = 7.5 mm. The eye vascular coat is fully pigmented. Liver is delineated. From the dorsal side the internal eye margin is masked by the mesencephalon lateral wall. Yolk sac vascularized by 2/3. Posterior cardinal veins are formed
23	17	136	L = 7.9 mm [the embryo bends actively]. Blood from intestinal artery enters subintestinal vein. Portal hepatic vein is formed
24	18	144	L = 8.3 mm. A mesenchyme mass is seen in the region of future tail fin (light area on the drawing). Head is separated from yolk. 6 aortic arches. A vascular network in mesencephalon and vessels in segments are formed. A large lumen is seen in the intestine. Yolk sac vascularized by 3/4
25	20	160	L = 8.5 mm. Dorsal notochord flexure at the level of tail fin rudiment. A large mesenchyme mass is seen in the region of future anal fin (light zone on the drawing). Tail artery and vein reach the tail end. Urinary bladder is forming
26	22	176	L = 10 mm. A mesenchyme mass is seen in the region of future tail fin (light zone on the drawing). Hindgut and ureter fully separated. Vascular loop in pectoral fin. Operculum partially covers the 1st gill arch. Bile pigment appears in intestinal lumen
26+ (without drawing)	24	192	First muscle buds appear in the middle of anal fin rudiment

TABLE 4.1 (continued)

| Stage No. | Time from insemination | | Diagnostic features of stages |
	days	τ_n/τ_0	
27	25	200	L = 10.5 mm. First muscle buds appear in the middle of dorsal fin rudiment. Muscle buds of anal fin reach the mesenchyme mass. Vascular network is delineated at the level of tail fin. Single melanophores appear on the dorsal head side and along the dorsal margin of myotomes
28	27	216	L = 11.5 mm. Ventral fin rudiments. Incision on the dorsal margin of fin fold marks a boundary with the growing tail fin. 8–10 muscle buds in dorsal fin. Melanophores are more numerous; some of them reach the level of intestine. Operculum partially covers the 2nd gill arch
29	31	248	L = 13 mm. Rudiments of cartilage rays (pterygiophores) are forming between muscle buds in the middle of anal and dorsal fin rudiments. First lepidotrichia appear in tail fin. Melanophores appear on the ventral embryo side, at the dorsal fin base, and in the upper part of yolk sac. Aorta gill arches are split. Rudiments of first gill leaflets are forming. Operculum covers the first two gill arches and a part of the 3rd

Larval development

30	35	–	Hatching occurs on the 34th–35th day [(272–280 τ_0). In my experiments the beginning of hatching at 10°C was noted on the 31st day after insemination ($244\tau_0$). The same relative duration of embryogenesis was obtained when the data of other authors (in days at different temperatures) were recalculated (2°C, after [167]; 5.5°C, Chaplygin, personal communication; 12°C, after [63]). The embryos hatched were 14–14.5 mm long and were characterized by features indicated by Vernier for stage 30.] Larva is 15 mm long. Pterygiophores formed between muscle buds all along the anal and dorsal fin rudiments. Numerous lepidotrichia in tail fin. Operculum covers all gill arches. Melanophores are oriented along the posterior margin of myotomes and begin to migrate to the yolk sac surface and dorsal and tail fins
31	39	–	Larva is 16 mm long. Appearance of first lepidotrichia in anal and dorsal fins. The upper lobe of the olfactory rudiment approaches the lower one. Melanophores cover the whole body. Lepidotrichia in the pectoral fins and over the dorsal and anal fin pterygiophores. Muscle buds are not seen

TABLE 4.1 (continued)

Stage No.	Time from insemination		Diagnostic features of stages
	days	τ_n/τ_0	
32	42	–	Larva is 18 mm long. Adipose fin is forming behind the dorsal one. Lepidotrichia only are seen in dorsal fin. Melanophores are spread all over the upper half of yolk sac
33	46	–	Larva is 19 mm long. Anal fin in which lepidotrichia only are seen is separated distinctly from fin fold. Lepidotrichia appear in abdominal fins, the upper margin of which reaches that of pre-anal fin fold. The upper lobe of olfactory organ adjoins the lower one. Single melanophores on anal fin. Tail fin is distinctly separated from fin fold
34	52	–	Larva is 20 mm long. Adipose fin is distinctly separated from fin fold. 2 apertures in olfactory organ. Anal fin is wider than fin fold. Single chromatophores on pectoral fins
35	59	–	Larva is 21 mm long. The pectoral fin margin rounds and becomes serrate. Pigment masses form specific pattern (pigment spots). Lepidotrichia reach the posterior end of tail fin. The margins of abdominal and tail fins become serrate
36	70	–	Yolk sac almost fully covered by chromatophores. The margins of dorsal and anal fins become serrate. Ventral fin twice as wide as fin fold
37	85	–	Yolk sac resorbed

Note. Vernier [177] uses a feature such as the number of somites for characterization of developmental stages. One should bear in mind, however, that as the embryo develops the somites are partly differentiated into myotomes; therefore, from stage 18 on, when some somites are already differentiated, the term "somites" is replaced by "myotomes" in all descriptions. In addition, Vernier calls the embryonic cellular material at the stages of mitotic index fall and later "blastodisc," even when the axial organs (embryo proper) develop. We call all cellular material at these stages "blastoderm."

The rainbow trout occurs in the waters of the Pacific coast from Mexico to Alaska [120]. It was introduced in Europe at the end of the last century and, due to its valuable food quality, became the main object of pond-fish farms, having gradually replaced all other species of trout. The rainbow trout was also used in experiments on acclimatization and was introduced into small bodies of water [112, 137, 143, 156]. Developing embryos and fry of rainbow trout for fish farms and for experimental studies in the USSR are provided mainly by the Central Experimental Station of the State Research Institute of Lake and River Fisheries "Ropsha" (Leningrad Region), the trout fish farm "Pylula" (Estonia), and others.

4.2. BIOLOGY OF REPRODUCTION

4.2.1. *In vivo* Oocyte Maturation

The rainbow trout becomes sexually mature at the age of 2 years, but the best results are obtained with 4- to 5-year-old males and 4- to 6-year-old females; one female gives 2-3 thousand eggs per kg of body weight [2, 130]. Rainbow trout eggs are benthic, like those of all salmon, and the development of eggs proceeds on the substratum covered by water in spawning redds, that is, in nests in sandy–stony ground [79]. Unfortunately, data on spawning temperatures of the rainbow trout in its natural habitats are unavailable. Only the months when spawning takes place are usually indicated [79]. For example, Agersborg [1] reports that spawning in California takes place from February until April, in Colorado from May until June, and in Virginia from September until February, but provides no data on water temperature during the spawning season. The problem of spawning temperatures is discussed mainly with respect to the observations of spawning in trout fish farms and to the experimental studies of the effect of different temperatures on development.

In trout fish farms, mature eggs and milt (sperm) from naturally matured spawners are obtained in different regions of the USSR from December until April. In the Chernorechensky trout fish farm the first ripe females appear in January–February at 7–8°C and in the Central Experimental Station "Ropsha" the eggs are obtained in March–April at 4–8°C (Gracheva, personal communication). According to Krupkin and Chaplygin [98], egg incubation and rearing of spawners in thermal waters gives accelerated sexual maturation; spawning begins earlier and at an earlier age, and the fecundity of females is higher.

4.2.2. *In vitro* Oocyte Maturation

A medium has been developed for *in vitro* oocyte maturation which has a composition of cations equivalent to that of the blood plasma serum of the rainbow trout [89]. The composition of this medium (per liter of distilled water) is: 8.6 g NaCl, 0.23 g KCl, 0.07 g $MgCO_4(7H_2O)$, 0.2 g $MgCl_2$, 0.5 g $CaCl_2(2H_2O)$, 0.12 g NaH_2PO_4, 0.57 g $NaHCO_3$, 1 g glucose [91]; the pH is brought to 7.3 by adding sodium bicarbonate. The eggs are incubated at 15°C (or 10°C according to [66]) in the air with 1% CO_2, 50% N_2, and 49% O_2. Extracts of salmon and carp gonadotropic pituitary hormones induce the maturation of follicular oocytes of the rainbow trout in doses of 5 and 25 µg/ml, respectively. At 15°C the germinal vesicle breaks down within 30–35 h, but no ovulation follows. In the solutions of various progesterones (1 µg/ml) the oocytes do not ovulate either, but mature earlier than under the effect of pituitary extracts. The most effective is 17α-hydroxy-20β-dihydroprogesterone, which induces maturation at a dose of 22 ng/ml (85 h of incubation at 10°C [66] or $28\tau_0$ (recalculated from Fig. 4.4). It has recently been shown [22, 43] that synthetic analogs of luliberin and pimozide may accelerate oogenesis.

4.3. STRUCTURE OF GAMETES

4.3.1. Egg

It is well known (see [73]) that in most chordates, and in all teleostean fish in particular, meiosis is blocked at metaphase II. Data are not available for the rainbow trout, but in a closely related species, the Sevan trout *S. ischchan*, the mature egg is blocked at metaphase II [121].

The rainbow trout egg is telolecithal by the distribution of yolk, as in all teleosteans, and polylecithal (oligoplasmatic according to the classification of Kryzhanovsky [99]) by the amount of yolk, as in all representatives of the genus *Salmo*, and this determines the egg size. As shown by Galkina [67, 68], the mean weight of the unfertilized egg varies in different females from 32 to 100 mg and the mean diameter from 3.7 to 5.3 mm; the smallest eggs are observed in young females and the largest eggs in 5- to 6-year-old females, but in the females of the same age the egg size varies markedly as well. The eggs of different females are of different color intensity: from slight yellow to light orange, due to the different content of carotenoids in the egg. The color intensity may depend not only on individual properties of a female but also on food composition; under normal conditions it does not affect the development of embryos (see also [24, 145]).

The mature egg is covered by a translucent envelope, the zona radiata, the size of which varies from 33 to 37 μm in the brook trout; the envelope is pierced by channels opening onto its surface with pores less than 1 μm in diameter, the distance between which is 1.37 μm [15]. There is a fine 10-μm jelly coat over the zona radiata [65].

In the animal region the zona radiata has one micropyle (see Fig. 4.1B), a funnel-shaped depression opening in the egg cytoplasm via the terminal channel, the diameter of which corresponds to the size of a sperm head [72]. The proteins of the zona radiata in salmon (including the trout) are similar to those of the cuticle in invertebrates and are resistant to protein solvents [30, 186]. The unfertilized egg envelope of the lake trout resists a load of 120–160 g [189]. The envelope adjoins the cortical layer of cytoplasm, which contains the cortical alveoli (see Fig. 4.1B); these are 20 μm in diameter [100, 101]. The surface layer of cytoplasm in the animal pole region forms a thickening, the blastodisc rudiment (see [73]). The yolk in the rainbow trout egg consists of dense, viscous fluid called ichtulin [158] with numerous suspended lipid drops (in the animal region of the egg) ranging in diameter from 20 to 260 μm in the brook trout [75]. The yolk coagulates in water and dissolves in NaCl solution at a concentration of not less than 1/8 M [158]. Estimates of the amount of RNA in the unfertilized rainbow trout egg range from 20 [85] to 77 μg [187].

4.3.2. Spermatozoon

An electron microscope study of spermatozoa of the rainbow and brook trout and the American brook trout has revealed no differences between them, and a common description is given for spermatozoa of all three species [64]. The spermatozoa of salmon do not differ from those of other teleostean species with external insemination. They are primitive flagellar spermatozoa with an oval-round or heart-shaped head, $1.7 \times 2 \, \mu m$ in size. A dense middle part is not always distinguished. The tail length is $25–35 \, \mu m$. The tail is inserted by $0.5 \, \mu m$ into the head, where it terminates by the proximal centriole. For the brook trout, however, Ballowitz [13] described two centrioles which are exposed upon the sperm head swelling. Figure 4.1A presents an electromicrograph of the lake trout spermatozoon. In salmonid fish [72], as well as in all teleosteans (with a few, not fully proved exceptions; see [73]), the acrosome is absent. The rainbow trout ejaculate contains 19.9–28.1 billion spermatozoa per cm^3 [40]. Good quality milt has the consistency of dense cream.

4.4. ARTIFICIAL INSEMINATION AND FERTILIZATION

In all teleostean fish, fertilization is monospermic [73]: the egg is fertilized by the spermatozoon which reaches the egg first. In salmon, including the trout, the spermatozoa are activated not only in water but also in body-cavity fluid and Ringer

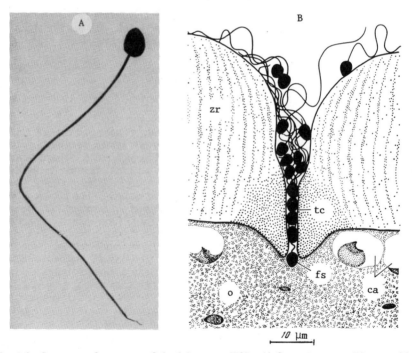

Fig. 4.1. Structure of gametes of the lake trout [72]. A) Spermatozoon; B) egg with the fertilizing spermatozoon inside the micropylar channel (upon fertilization in the body fluid or Ringer solution). ca) Cortical alveoli; fs) fertilizing spermatozoon; o) ooplasm; tc) terminal channel of the micropyle; zr) zona radiata.

solution; they preserve their motility in the body-cavity fluid longer than in water. Upon insemination in body-cavity fluid or Ringer solution, the fertilizing spermatozoon penetrates through the micropylar channel to the cortical egg layer and the sperm head is half submerged in the cytoplasm (see Fig. 4.1B). The egg is not activated, but if it is placed in water (even on the 4th day after insemination) just for a short period it becomes activated and development starts. Thus, in the trout, the processes of gamete fusion and egg activation can be artificially separated in time. The spermatozoa preserve their motility in water only for 2–3 min. The unfertilized eggs are also activated in water [72]. These properties of the gametes of salmonid fish make it necessary to use the "dry" method of insemination upon artificial reproduction of these fish or in laboratory experiments. In so doing one has to take care that no water penetrates the vials with the semen or eggs, since otherwise the spermatozoa very rapidly lose their fertilizing capacity and the eggs are activated and cannot be fertilized.

In trout fish farms (including "Ropsha") artificial insemination is used for trout reproduction. The eggs of spontaneously matured females are drawn off into a pan while running a hand over the abdomen and slightly squeezing it. The milt is drawn off onto the eggs in the same way, and the eggs and milt are mixed carefully with a feather, and sufficient water is added to cover the eggs. Mixing of the eggs is continued for some time. After 5–10 min, the water with milt is poured off, the eggs are washed several times, and left in the pan with water for several hours so that the envelopes swell and then spread in a thin layer on lattice frames. These frames are placed in a trough with a constant water flow.

For laboratory experiments (if the eggs are to be transported some distance) it is better to put unfertilized eggs and milt (from each male separately) in separate dry vials which are firmly closed, and to inseminate eggs artificially at the place of experiment. Trout eggs are slightly sticky and their washing after insemination is enough to remove this stickiness. Eggs kept at 4–8°C and milt kept in test tubes on ice do not lose their capacity for fertilization over a 29-h period [81]. The spermatozoa keep their capacity for activation upon water addition for up to 2 days at 8°C [146].

After the addition of water to the inseminated eggs, the contents of the cortical alveoli are liberated, water enters beneath the egg membrane, the zona radiata is separated from the egg plasma membrane, and the perivitelline space forms (see [73]). The diameter of the egg in the envelope amounts to 5–5.5 mm [7, 177]. The gradual process of egg-membrane hardening begins soon. Within 4–5 days, the egg membrane becomes so strong that a load of 2–2.5 kg is needed to crush one egg [189]. In parallel with the formation of the perivitelline space, the process of bipolar differentiation of the egg takes place: the cytoplasm is concentrated at the animal pole and the blastodisc forms. It is well known that in teleostean fish, including the trout, only the blastodisc cleaves and the rest of the egg, the yolk, covered by a thin cytoplasmic membrane, does not take part in cleavage. In the fertilized rainbow trout egg the blastodisc occupies a very small part of the egg: by the beginning of cleavage its volume is 1/50 that of the yolk volume [87]. A unique cytoplasmic structure, the periblast (yolk syncytium), which plays an important role in the process of epiboly and the assimilation of yolk by the embryo (see [168]), forms under the blastodisc.

Noninseminated eggs are activated in water and undergo the same changes as fertilized eggs, but do not cleave. Therefore, the percentage of fertilization cannot be estimated until the onset of cleavage. Such activated eggs remain alive during almost the whole embryogenesis, and up to the late developmental stages they are hardly distinguishable from fertilized eggs with the naked eye [152].

4.5. EXTERNAL CONDITIONS ESSENTIAL FOR NORMAL DEVELOPMENT

Rainbow trout development proceeds at a gradually increasing temperature, and the hatching takes place in Ropsha (Gracheva, personal communication) within 1.5–2 months at 11–12°C (7–9°C on average). Gracheva states that rainbow trout eggs develop successfully within the temperature range 4–12°C; the best results are obtained at mean temperatures of 6–10°C.

In the laboratory embryogenesis may be completed at a wider range of temperatures, from 1.5 to 18°C [60, 63, 71, 93, 104]. Different authors provide different values of the range of optimal temperatures (by such indices as the survival of embryos, percentage of defects, change in the character of dependence of the developmental duration with temperature). Temperatures between 3° and 12°C may be considered as optimal for rainbow trout embryogenesis (for discussion see [82]). These data are close to those recently obtained by estimating the minimum daily O_2 consumption (Figs. 74 and 75 in [125]; [126]), which also suggest that as the development proceeds the upper limit of the zone of optimal temperatures is shifted toward higher temperatures.

At the early developmental stages (until the end of epiboly) trout embryos are very sensitive to mechanical damage, since the cytoplasmic membrane surrounding the yolk is easily destroyed; the yolk flows out under the envelope and coagulates, and the egg whitens and dies. During this period the eggs must be protected from mechanical damage.

The oxygen regime is very important for rainbow trout embryogenesis. The early embryos are less sensitive to oxygen deficiency and, until the end of epiboly, the rate of development of embryos developing in water under a layer of paraffin oil does not lag behind that of control embryos developing in water alone; but later the development is delayed and affected [180, 181]. Garside [71] showed that at 2.5 mg O_2/liter the development of embryos is delayed (depending on the water temperature at different stages) as compared with that at the normal saturation of water with oxygen. Ostroumova [124] observed the death of embryos at the onset of gastrulation at 3.5 mg O_2/liter; at 5–7 mg/liter the embryos hatch, but the mortality increases and defects arise. At the normal saturation of water with oxygen (12–14 mg/liter at 8–9°C) the development proceeds normally, and increased oxygen content (20–25 mg/liter at the same temperature) stimulates the development at later stages (after circulatory system formation).

There are also data (see above) on the effect of illumination on the development of salmonid fish, but they are difficult to interpret unambiguously (see also [17]).

4.6. INCUBATION OF THE EGGS IN THE LABORATORY

As trout eggs show low sensitivity to oxygen deficiency in early developmental stages, during short-term experiments they can be kept in dechlorinated water (which is carefully changed once or twice a day) in Petri dishes or basins. However, this method is impractical for long-term observations of embryogenesis. The incubator constructed by Yu. N. Gorodilov of the Biological Research Institute, Leningrad State University, is highly recommended for such studies. It allows incubation of the eggs of salmonid fish in running, well-aerated water at a strictly specified temperature which can be changed during the experiment, along with the composition of the water, according to experimental needs required to study the effect of different factors on development, and so on.

Gorodilov's incubator is described below (Fig. 4.2). It is an incubator of cascade type with a water supply in a closed cycle. It consists (Fig. 4.2A) of an upper supply tank 1 with a float device 5, a cascade for the eggs 2, and a lower collecting tank 3 connected with a pump 4. From the supply tank, water flows by gravity through rubber tubes (the movement of water is shown by arrows). The flow rate is regulated by clamps on the rubber tubes.

The cascade is of stepwise construction from acrylic plastic (Fig. 4.2B); it is divided into cells by cross partitions. The water flows from one step to another by gravity through the holes or cuts in the wall limiting the cell in front. The size of the holes or cuts should be less than the diameter of an egg so that the eggs cannot pass through with the water flow. The number of steps in the cascade and of partitions in each step can be varied depending on experimental purposes. For example, in one of the modifications of the incubator in the Biological Research Institute, the cascade consists of 14 steps with 10 cells in each, allowing up to 140 variants of the experiment to be performed, or the utilization, in parallel, of many groups of eggs with different experimental variants. Such a cascade is 2 m long and 0.7 m wide. From the last cascade step the water flows by grooves into the lower tank. A filter can be installed here to purify the water. However, good results may be obtained without a filter, provided that part of the water (up to 10%) is changed daily.

The lower tank connects with the pump (Kama-3 or Agidel) with an intake valve. When some of the water flows out of the upper tank, the float device which is installed in the upper tank operates (Fig. 4.2A, 5). The float device is on the same electric circuit as the pump: the pump is switched on and pumps the water through a tube again into the upper tank.

The incubator is installed in a thermally insulated chamber which is cooled by a refrigerator unit (Fig. 4.2A, 6). The desired temperature is set by a contact thermometer.

There are a few incubators in the Biological Research Institute designed for different purposes, which differ in the size of tanks and cascade, the size and number of steps in them, and systems of float and water flow; the possibilities of adaptation of this device for experimental purposes are practically unlimited. The cascade can be made with two or three steps or even replaced by a tray. The volumes of the upper and lower tanks are the same and vary from 30 to 85 liters. The volume of recirculated water is somewhat greater. The tanks are made of welded or bonded acrylic plastic.

In an incubator with a volume of recirculated water of about 100 liters, Gorodilov incubated 30–40 thousand salmon eggs (as shown in Fig. 4.2C). The larvae can be reared in the cells until the transition to active feeding (Fig. 4.2D).

4.7. NORMAL DEVELOPMENT

The description of the early stages of rainbow trout development was given in 1936 by Pasteels [128], who studied morphogenetic movements. Knight [95],

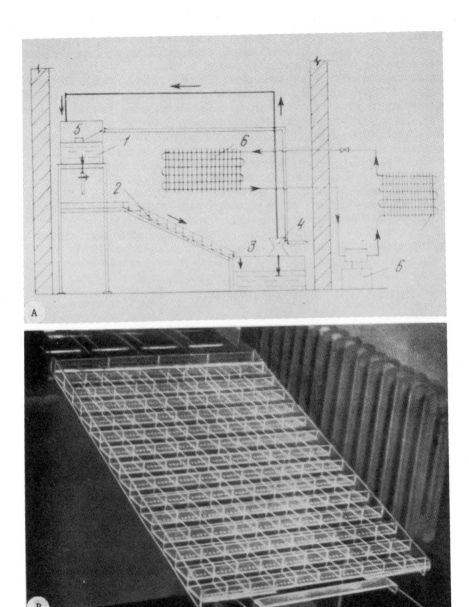

Fig. 4.2. Device designed by Y. N. Gorodilov for the incubation of eggs of salmonid fish with a water supply in a closed cycle. A) Diagram of incubator (explanation in text); B) general view of cascade; C) salmon eggs in the cascade cells; D) salmon larvae in the cascade cell.

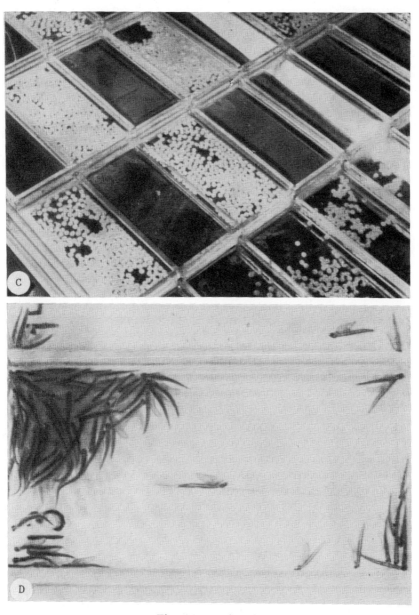

Fig. 4.2, continued.

Vernier [177], and Ballard [7] describe the consecutive stages of the whole embryogenesis, and Vernier also describes the larval development of the rainbow trout.

There are also normal tables for the other *Salmo* species: *S. fario*, early development [77, 96] and the whole embryogenesis, without drawings [70], and *S. salar*, from fertilization to hatching [14, 133]. The embryogenesis of *S. salar* is also described by Vernidub and Yandovskaya [176], who divided the whole period of embryogenesis into nine stages and indicated the duration of these stages in days and degree-days at certain temperatures (varying within the limits of 2–3°C).

Witschi [183, 184] divided the period of embryogenesis of salmonid fish into 25 stages but gave no drawings of these stages. Stefanov and Dencheva [154] used his classification when studying the rainbow trout.

Different authors have distinguished different numbers of stages and have not always indicated the duration and temperature of development or studied the development at varying temperatures. The tables are not always well illustrated, and this creates difficulties for the comparison of results of different authors.

Vernier [177] gave the most detailed and abundantly illustrated description of successive developmental stages of the rainbow trout. He not only gave us permission to use his *Table chronologique du développement embryonnaire de la truite arc-en-ciel, Salmo Gairdneri Rich., 1836* but also kindly sent photocopies of drawings for the table as well as additional, not-yet-published drawings of sections through the embryos at the early stages of embryogenesis. They were published in *Ontogenez (Soviet Journal of Developmental Biology)* in Russian translation [178]. I am very grateful to Dr. Vernier for his permission to publish this table. I would also like to thank Dr. W. W. Ballard for his kind permission to use some illustrations from his papers concerning morphogenetic movements in teleostean fish during gastrulation and the distribution of presumptive organs in salmonid fish (rainbow trout) prior to gastrulation. Figure 4.3 presents the map of organ rudiments [9] at the beginning of gastrulation (stage 10 after Vernier).

Ballard [5–12] has shown that, in teleostean fish, invagination as postulated by Pasteels [128] is absent. The movement of cells during gastrulation to their final position consists of epiboly and convergence, with displacement of deep blastomeres, from the blastodisc periphery to the axial rudiment. Therefore, the germ ring can in no way be homologized with the blastopore in amphibians. It performs only two of the blastopore functions: covers the plasma membrane during epiboly and brings the tail mesoderm to its final position. The periblast plays an important role in epiboly (see [168]). The material of the nervous system, notochord, somites, and endoderm is separated by coordinated movements of the deep blastomeres underlying the cell surface membrane, the periderm, which does not take part in the formation of these structures.

Ballard particularly stressed the three-dimensional arrangement of the presumptive organs prior to the onset of gastrulation in the blastodisc of the rainbow trout (and other teleostean fish). The notochord material underlies the mesoderm material, which is distributed all over the blastodisc (Fig. 4.3A), and the endoderm material is still deeper. The presumptive mesoderm (Fig. 4.3B) underlies the nervous system material, and the cellular envelope, the periderm, is above all these regions (Fig. 4.3C).

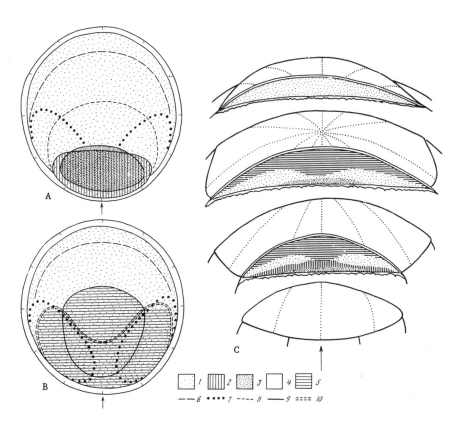

Fig. 4.3. Fate map of the rainbow trout blastoderm at the early gastrula stage (prior to, or just after, the appearance of the margin node) after Ballard. A) View from below; B) view from above; C) fate map as a series of cross sections through the blastoderm at the same developmental stage. Arrow indicates axis of symmetry. 1) Presumptive mesoderm; 2) presumptive notochord; 3) presumptive endoderm; 4) superficial cellular envelope, or periderm (A, B along the blastoderm margin); 5) presumptive nervous system; 6–8) anterior boundaries of the material of the posterior somites (6), anterior somites (7), head mesoderm (8); 9, 10) boundaries of the material of prosencephalon, mesencephalon (9), and metencephalon (10).

4.8. CHRONOLOGICAL TABLES OF DEVELOPMENT

Vernier's tables result from vital observations over the embryonic and larval development; he includes in larval development the whole period from hatching to the end of yolk resorption. In the Soviet literature the embryos, when hatched, are called prelarvae until the transition to active feeding [139]. The normal tables were constructed from observations of the development of ten batches of eggs and embrace the whole period of development. These batches were laid by females over 3 kg in weight, and each of them had about 5000 eggs with an average diameter of 5 mm. These eggs were placed in running water and developed at a mean temperature of 12°C (±1°C). One thousand eggs from each batch were placed in a cold chamber at 10°C in nonrunning water which was artificially oxygenated. A time switch assured a photoperiod corresponding to the natural one. After hatching, the larvae were placed in running water at 12°C. The results of observations concern mainly the external morphology, the development and differentiation of organs easily observable due to their transparency on a white or black back-

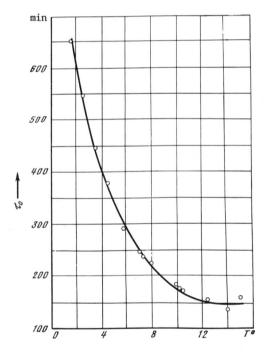

Fig. 4.4. Dependence of the duration of one mitotic cycle during synchronous cleavage divisions (τ_0) on temperature in the rainbow trout. Each point on the curve is the mean of 2–9 determinations.

ground using a binocular microscope. For these observations, mild anesthesia of the embryo or larva is often necessary (MS 222, 1:5000).

Since the eggs used by Vernier were incubated at a constant temperature (10°C) and the time of beginning of the stages (τ_n) indicated in the tables was determined in hours or days after insemination, the present author could establish dimensionless (relative) characteristics of developmental duration for the rainbow-trout embryos [44]. For this purpose the results of Vernier were recalculated by τ_0 (the duration of one mitotic cycle during the synchronous cleavage divisions), the value of which at 10°C is 180 min. The curve expressing the dependence of τ_0 on temperature in the rainbow trout is shown in Fig. 4.4. It has been made more precise especially for this volume, and is given in a different scale from the previous one [80, 81, 84]. The value of τ_0 was determined by observations of the time of appearance of the 2nd–4th cleavage furrows at intervals over not more than $0.03–0.04\tau_0$. The value of τ_0 corresponds to one-half the interval between the appearance of furrows of the 2nd–4th cleavage divisions [80–82, 84, 86].

The envelope of the fertilized rainbow trout egg is only slightly transparent. One can follow the exact time of appearance of the first four cleavage furrows by fixing the eggs with an acetic acid:ethanol (1:3) mixture in which the envelopes become transparent within several minutes, the blastodisc whitens, and the furrows become distinct [81]. The envelope, which is very solid (see above), can be removed for vital observations by microsurgery, using sharp scalpels and forceps in double-strength Holtfreter's solution [80].

The present author had determined the duration of some early developmental periods previously [80, 81, 83, 85]. To compare these data with those of Vernier, eggs from three females were specially incubated in a Gorodilov incubator at 10°C. For the early developmental stages (up to the stage of 10 somites), incorporated in Vernier's table (reproduced here with the permission of the author) are more precise data on the absolute and relative duration of these stages (Table 4.1). Since Vernier made his observations of the embryos at the subsequent developmental stages at intervals of 1–2 days (time of observations not indicated), and 24 h at 10°C is equivalent to $8\tau_0$, the precise determination of the relative duration of any stage can vary within these limits. This explains, in particular, why Vernier observed the end of epiboly at the age of $72\tau_0$ (recalculated by his data), when the embryos have 29 pairs of somites. My observations indicate that the end of epiboly in the advanced embryos is much earlier, at the age of $65\tau_0$, when the embryo has about 21 pairs of somites. I did not determine the relative duration for stages of larval development which in Vernier's observations were carried out at temperatures other than the strictly constant one of 10°C.

I have introduced into the chronological table of rainbow trout development (see Table 4.1) some additional stages (without drawings), as follows: stage 0 (the egg at the moment of insemination), which is necessary as a zero reading for the duration of development; stage 7– (onset of fall of mitotic index); stage 9+ (onset of morphogenetic nuclear function) [87]; and stage 14+ (10 pairs of somites) [81, 83, 85]. All my additions to Vernier's table are shown in square brackets. The numbers of drawings (see Figs. 4.5–4.10) correspond to the numbers of stages. If the embryo at the same stage is drawn at two angles, the drawings are given the same number with the letters *a* and *b*. All changes in the text of the table were submitted for Dr. Vernier's approval. I would like to note also that, on the drawings of the first five cleavage divisions, eggs which were not cleaving very typically appear to be depicted. The blastomeres at these stages are usually more regular both in form and positions (see, for example, [14, 95]).

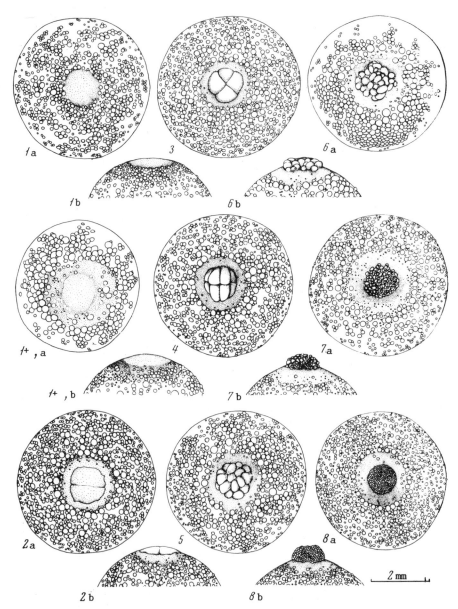

Figs. 4.5–4.10. Stages of normal development of the rainbow trout *Salmo gairdneri* Richardson [177]. Drawing numbers (from 1 to 37) correspond to stage numbers (see description in Table 4.1). a, b) Different views of the same stage.

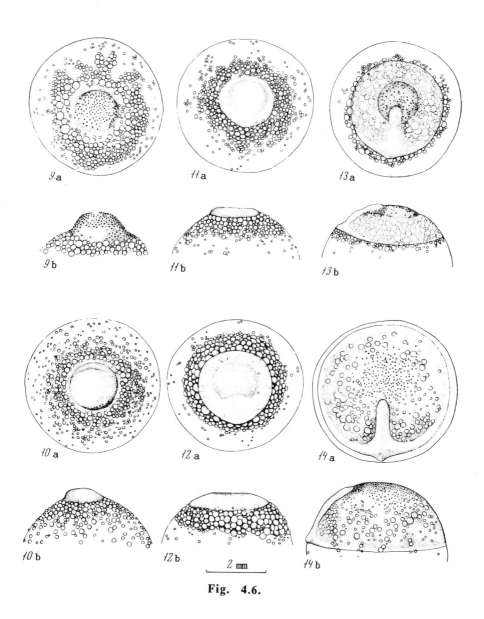

Fig. 4.6.

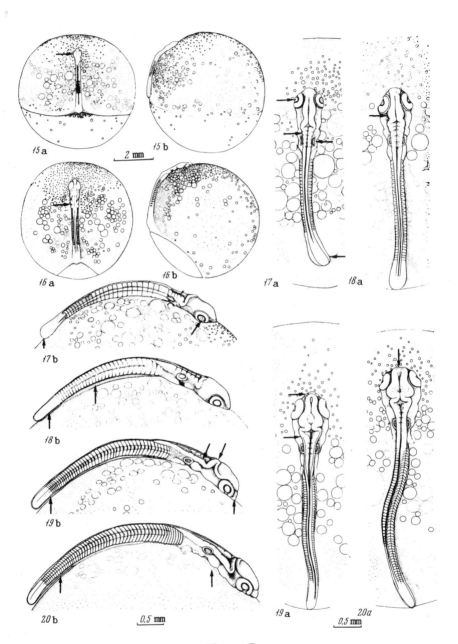

Fig. 4.7.

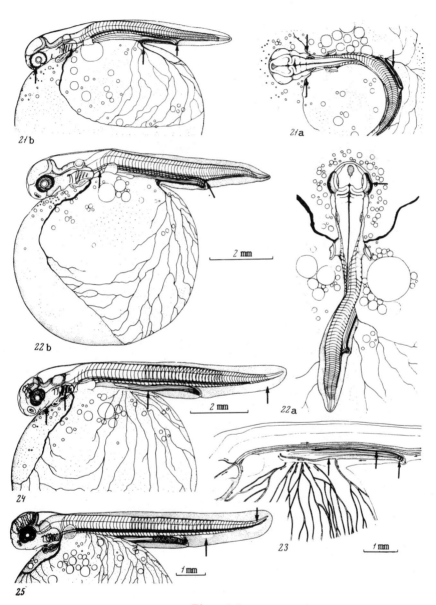

Fig. 4.8.

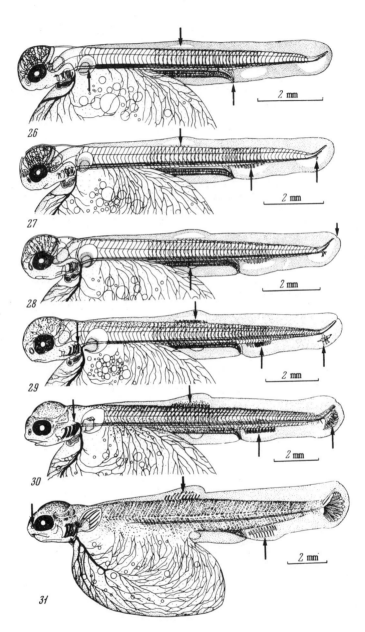

Fig. 4.9.

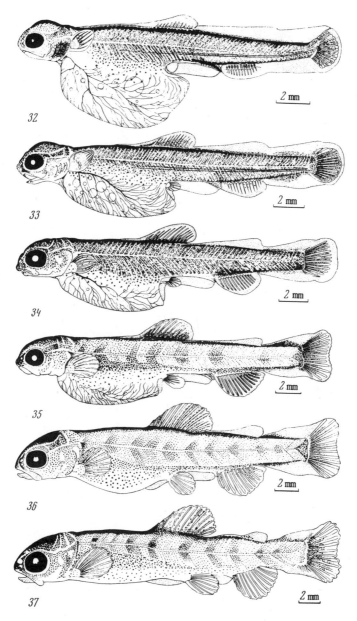

Fig. 4.10.

In conclusion, I would like to most cordially thank Prof. S. G. Soin, Dr. O. F. Sakun, and Dr. G. G. Savostianova, who read the manuscript of this chapter and made some very valuable and useful comments.

REFERENCES

1. H. P. Agersborg, "When do the rainbow trout spawn?" *Trans. Am. Fish. Soc.* **64**, 167–169 (1934).
2. G. H. Allen and G. A. Sanger, "Fecundity of rainbow trout from actual count of eggs," *Copeia* 3, 260–261 (1960).
3. K. Ando, "Ultracentrifugal analysis of yolk proteins in rainbow trout egg and their changes during development," *Can. J. Biochem.* **43**, 373–379 (1965).
4. Y. P. Babushkin, "Correlation between sperm quality in rainbow trout males and condition of the spawners," *Izv. GosNIORKh* **113**, 8–11 (1976).
5. W. W. Ballard, "Morphogenetic movements in teleost embryos," *Am. Zool.* **4**, 12 (1964).
6. W. W. Ballard, "Formative movements in teleost embryos," *Am. Zool.* **5**, 83 (1965).
7. W. W. Ballard, "Normal embryonic stages for salmonid fish, based on *Salmo gairdneri* Richardson and *Salvelinus fontinales* (Mitchill)," *J. Exp. Zool.* **184**, 7–26 (1973).
8. W. W. Ballard, "Morphogenetic movements in *Salmo gairdneri* Richardson," *J. Exp. Zool.* **184**, 27–48 (1973).
9. W. W. Ballard, "A new fate map for *Salmo gairdneri*," *J. Exp. Zool.* **184**, 49–73 (1973).
10. W. W. Ballard, "A re-examination of gastrulation in teleosts," *Rev. Roum. Biol. Ser. Zool.* **18**, 119–136 (1973).
11. W. W. Ballard, "Morphogenetic movements and fate maps of vertebrates," *Am. Zool.* **21**, 391–399 (1981).
12. W. W. Ballard and L. M. Dodes, "The morphogenetic movements at the lower surface of the blastodisc in salmonid embryos," *J. Exp. Zool.* **168**, 67–84 (1968).
13. E. Ballowitz, "Uber die Samenkörper der Forellen," *Arch. Zellforsch.* **14**, 185–192 (1915).
14. H. J. Battle, "The embryology of the Atlantic salmon (*Salmo salar*)," *Can. J. Res. D.* **22**, 105–125 (1944).
15. H. V. Becher, "Beitrag zur feineren Struktur der zona radiata des Knochenfisheries und über ein durch die Struktur der Eihülle bedingtes optisches Phänomen," *Z. Mikrosk.-Anat. Forsch.* **13**, 591–624 (1928).
16. R. G. Belcheva and L. V. Dimcheva-Grozdanova, "Chromosome aberrations and malformations in *Salmo irideus* Gibb. after irradiation of early cleavage embryos with lasers of different wavelengths," *Dokl. Bolg. Akad. Nauk* **26**, 1077–1080 (1973).
17. R. Beltscheva, L. Dimtschewa, and M. Topaschka, "Cytogenetische und embryologische Untersuchungen über die Regenbogenforelle (*Salmo irideus* Gibb.) nach Bestrahlung der Vatergameten mit UV-Licht," *Annu. Univ. Sofia* **65**, 143–155 (1971).
18. D. R. Beritashvili, I. S. Kvavilashvili, N. N. Rott, and G. M. Ignatieva, "The accumulation of potassium in developing blastoderm of the trout," *Sov. J. Dev. Biol.* **1**, 464–465 (1970).

19. K. Bieniarz, "Effect of light and darkness on incubation of eggs, length, weight, and sexual maturity of sea trout (*Salmo trutta* L.), brown trout (*Salmo trutta fario* L.), and rainbow trout (*Salmo irideus* Gibbons)," *Aquaculture* **2**, 299–315 (1973).

20. R. Billard, "L'insémination artificielle de la Truite *Salmo gairdneri* Richardson. V. Effets de la dilution et définition du rapport optimum gametes/dilueur," *Bull. Fr. Piscic.* **257**, 121–135 (1975).

21. R. Billard, "Short-term preservation of sperm under oxygen atmosphere in rainbow trout (*Salmo gairdneri*)," *Aquaculture* **23**, 287–293 (1981).

22. R. Billard, P. Reinaud, M. G. Hollebeeq, and B. Breton, "Advancement and synchronization of spawning in *Salmo gairdneri* and *S. trutta* following administration of LRH-A combined or not with pimozide," *Aquaculture* **43**, 57–66 (1984).

23. E. A. Borovik, "Respiration of the eggs of *Salmo irideus*," *Vestsi Akad. Nauk BSSR, Ser. Biol.* **3**, 111–115 (1963).

24. E. A. Borovik, "Importance of carotenoids in the embryonic development of the trout," *Vestsi Akad. Nauk BSSR, Ser. Biol.* **1**, 132–133 (1965a).

25. E. A. Borovik, "Morphological characteristics of the Byelorussian population of rainbow trout," in *Ecology of Vertebrate Animals* [in Russian], Minsk (1965b), pp. 150–161.

26. H. Boulekbache, "Energy metabolism in fish development," *Am. Zool.* **21**, 377–389 (1981).

27. H. Boulekbache, C. Devillers, A. J. Rosenberg, and C. Joly, "Correlation entre la consommation d'oxygène et l'activité de la L.D.H. et de la glucose-6-phosphate déshydrogénase au cours des premiers stades du développement de l'oeuf de truite *Salmo irideus*," *C. R. Acad. Sci. Paris* **D268**, 2211–2214 (1969).

28. H. Boulekbache, C. Devillers, A. J. Rosenberg, and C. Joly, "Action de l'acide oxamique, inhibiteur spécifique de la lactico-déshydrogénase sur les premiers stades de développement de l'oeuf de la Truite (*Salmo irideus* Gibb.)," *C. R. Acad. Sci. Paris* **D272**, 114–116 (1971).

29. H. Boulekbache, P. Roubaud, C. Devillers, and C. Joly, "Localisation histochimique de la lacticodéshydrogénase dans l'embryon de la Truite (*Salmo irideus* Gibb.) au cours de l'organogenése précoce de la corde et du tube neural," *C. R. Acad. Sci. Paris* **D284**, 381–384 (1977).

30. C. H. Brown, "Egg capsule proteins of Selachians and trout," *Quart. J. Microsc. Sci.* **96**, 483–488 (1955).

31. C. M. Bungenberg de Jong, "Cytological studies on *Salmo irideus*," *Genetica* **27**, 472–483 (1955).

32. V. M. Chaplygin, "The chief types of developmental disturbances in the embryogenesis of progeny of first- and repeatedly spawning trout females," *Izv. GosNIORKh* **107**, 144–151 (1976a).

33. V. M. Chaplygin, "On results of incubation of the eggs obtained from rainbow trout females of different age," *Izv. GosNIORKh* **113**, 42–45 (1976b).

34. D. Chourrout, "Thermal induction of diploid gynogenesis and triploidy in the eggs of the rainbow trout (*Salmo gairdneri* Richardson)," *Reprod. Nutr. Dév.* **20**, 727–733 (1980).

35. D. Chourrout, "Tetraploidy induced by heat shocks in the rainbow trout (*Salmo gairdneri* R.)," *Reprod. Nutr. Dév.* **22**, 569–574 (1982a).

36. D. Chourrout, "Gynogenesis caused by ultraviolet irradiation of salmonid sperm," *J. Exp. Zool.* **223**, 175–181 (1982b).

37. D. Chourrout and E. Qullet, "Induced gynogenesis in the rainbow trout: sex and survival of progeny production of all-triploid populations," *Theor. Appl. Genet.* **63**, 201–205 (1982).

38. D. Chourrout, B. Chevassus, and F. Herioux, "Analysis of a Hertwig effect in the rainbow trout (*Salmo gairdneri* Richardson) after fertilization with γ-irradiated sperm," *Reprod. Nutr. Dév.* **20**, 719–726 (1980).

39. D. Chourrout, R. Guyomard, and L.-M. Houdebine, "High-efficiency gene transfer in rainbow trout (*Salmo gairdneri* Rich.) by microinjection into egg cytoplasm," *Aquaculture* **51**, 143–150 (1986).

40. H. P. Clemens and F. B. Grant, "The seminal thinning response of carp (*Cyprinus carpio*) and rainbow trout (*Salmo gairdneri*) after injections of pituitary extracts," *Copeia* **2**, 174–177 (1965).

41. R. Coupe, P. Roubaud, H. Boulekbache, and C. Devillers, "Étude de la dissociation cellulaire *in vitro* sur les premiers stades du développement embryonnaire de la truite (*Salmo irideus* Gibb.)," *C. R. Acad. Sci. Paris* **D277**, 353–356 (1973).

42. R. Coupe, P. Roubaud, H. Boulekbache, and C. Devillers, "Évolution de la 'dissociativité' des blastomères de l'embryon de Truite (*Salmo irideus* Gibb.) au cours des premiers stades du développement," *C. R. Acad. Sci. Paris* **D279**, 1549–1552 (1974).

43. L. W. Crim, A. M. Sutterlin, D. M. Evans, and C. Weil, "Accelerated ovulation by pelleted LHRH analog treatment of spring-spawning rainbow trout (*Salmo gairdneri*) held at low temperature, "*Aquaculture* **35**, 299–307 (1983).

44. T. A. Dettlaff and A. A. Dettlaff, "On relative dimensionless characteristics of the development duration in embryology," *Arch. Biol. (Liège)* **72**, 1–16 (1961).

45. C. Devillers, "Explanations *in vitro* de blastoderms de poisson (*Salmo, Esox*)," *Experientia* **3**, 71–74 (1947).

46. C. Devillers, "Le cortex de l'oeuf de truite," *Ann. Stat. Centr. Hydrobiol.*, Suppl. **2**, 229 (1948).

47. C. Devillers, "Mecanisme de l'épibolie gastrulenne," *C. R. Acad. Sci. Paris* **230**, 2232–2234 (1950).

48. C. Devillers, "Les mouvements superficiels dans la gastrulation des poissons," *Arch. Anat. Microsc., Morphol. Exp.* **40**, 298–312 (1951).

49. C. Devillers, "Coordination des forces épiboliques dans la gastrulation de *Salmo*," *Bull. Soc. Zool. Fr.* **77**, 304–309 (1953).

50. C. Devillers, "Les aspects caractéristiques de la prémorphogénèse dans l'oeuf des téléostéens. L'origine de l'oeuf télolecithique," *Année Biol.* **32**, 437–456 (1956).

51. C. Devillers, "Mouvements cellulaire dans le développpment de l'embryon des amphibiens et des poissons," *Exp. Cell. Res.*, Suppl., **8**, 201–233 (1961).

52. C. Devillers, "Structural and dynamic aspects of the development of the teleostean egg," *Adv. Morphogen.* **1**, 379–428 (1961).

53. C. Devillers, "Respiration et morphogénèse dans l'oeuf des téléostéens," *Année Biol.* **4**, 157–186 (1965).

54. C. Devillers and L. Rajchman, "Quelques données sur l'utilisation du vitellus au cours de la gastrulation dans l'oeuf de *Salmo irideus,*" *C. R. Acad. Sci. Paris* **D247**, 2033–2035 (1958).

55. C. Devillers, A. Thomopoulos, and J. Colas, "Différenciation bipolare et formation de l'espace périvitellin dans l'oeuf de *Salmo irideus,*" *Bull. Soc. Zool. Fr.* **78**, 462–470 (1953).

56. C. Devillers, J. Colas, and A. M. Cantacuzène, "Gastrulation de l'oeuf de truite (*Salmo irideus*) en l'absence d'oxygène ou en présence d'inhibiteurs du métabolisme," *C. R. Acad. Sci. Paris* **D245**, 1461–1462 (1957).

57. C. Devillers, J. Colas, and L. Richard, "Différenciation *in vitro* de blastodermes de truite (*Salmo irideus*) dépourvus de couche énveloppante," *J. Embryol. Exp. Morphol.* **5**, 264–273 (1957).

58. L. Dimcheva-Grozdanova, "Embryonic development of *Salmo irideus* Gibb.," *Izv. Zool. Inst. Muz. Balg. Akad. Nauk* **26**, 83–101 (1968).

59. L. Dimcheva-Grozdanova and R. Belcheva, "Embryological and cytogenetic investigations of rainbow trout embryos (*Salmo irideus* Gibb.) after exposure in early cleavage to different doses of γ-rays," *Annu. Univ. Sofia, Fac. Biol.* **69**, 41–52 (1976).

60. J. Domurat, "Zaburzenia wymiany wodnej a tempo wzrostu zarostu zárodków pstraga teczowego (*Salmo gairdneri* Rich.)," *Zesz. Nauk. Wyzszej Szkoly Rolnieczej Olsztynie* **21**, 563–567 (1966).

61. G. V. Dontzova, G. M. Ignatieva, N. N. Rott, and I. I. Tolstorukov, "Nucleic acids in early embryogenesis of the trout," *Sov. J. Dev. Biol.* **1**, 340–345 (1970).

62. R. M. Eakin, "Regional determination in the development of the trout," *Wilhelm Roux's Arch. Entwicklungsmech. Org.* **139**, 274–281 (1939).

63. G. S. Embody, "Relation of temperature to the incubation periods of eggs of four species of trout," *Trans. Am. Fish. Soc.* **64**, 287–289 (1934).

64. H. Fischer, O. Hug, and W. Lippert, "Elektronmikroskopische Studien an Forellenspermatozoen und ihren Zellkernen," *Chromosoma* **5**, 69–80 (1952).

65. H. Flüger, "Electron microscopic investigations on the fine structure of the follicular cells and the zona radiata of trout oocytes during and after ovulation," *Naturwissenschaften* **51**, 564–565 (1964).

66. A. Fostier, B. Jalabert, and M. Terqui, "Action prédominante d'un dérivé hydroxylé de la progestérone sur la maturation *in vitro* des ovocytes de la truite arc-en-ciel *Salmo gairdneri,*" *C. R. Acad. Sci. Paris* **D277**, 421–424 (1973).

67. Z. I. Galkina, "The effect of size and intensity of color of eggs on the embryonal development and the growth of young *Salmo irideus* Gibb.," *Izv. GosNIORKh* **68**, 173–186 (1969a).

68. Z. I. Galkina, "Heterogeneity of ripe eggs of *Salmo irideus* Gibb.," *Izv. GosNIORKh* **68**, 187–196 (1969b).

69. G. E. Gall, "Influence of size of eggs and age of female on hatchability and growth in rainbow trout," *Calif. Fish Game* **60**, 26–35 (1974).

70. E. T. Garside, "Some effects of oxygen in relation to temperature on the development of embryos of lake trout embryos," *Can. J. Zool.* **37**, 689–698 (1959).

71. E. T. Garside, "Effects of oxygen in relation to temperature on the development of embryos of brook trout and rainbow trout," *J. Fish. Res. Board Can.* **23**, 1121–1134 (1966).

72. A. S. Ginsburg, "Sperm–egg association and its relationship to the activation of the egg in salmonid fish," *J. Embryol. Exp. Morphol.* **11**, 13–33 (1963).
73. A. S. Ginsburg, *Fertilization in Fish and the Problem of Polyspermy*, Israel Program for Scientific Translations, Ltd., IPST Cat. No. 600418 (1972).
74. J. Gras, R. Reynand, J. Frey, and J. C. Henry, "Modifications biochimique au cours du développement des oeufs et des alevins de la truite arc-en-ciel (*Salmo gairdneri* Rich.)," *C. R. Soc. Biol.* **160**, 1262–1264 (1966).
75. Z. Grodzinski, "Fat drops in the yolk of the sea trout *Salmo trutta* L.," *Bull. Int. Acad. Pol. Sci. Lett., Ser. B, II* **1/3**, 59–78 (1949).
76. H. E. Hagenmeier, "Der Nukleinsäure-bzw. Ribonuklein-proteid-Status Während der Frühentwicklung der Fischkeimen (*Salmo irideus* and *Salmo trutta fario*)," *Wilhelm Roux's Arch. Entwicklungsmech. Org.* **162**, 19–40 (1969).
77. F. Henneguy, "Embryogénie de la truite," *J. Anat. Physiol.* **24**, 413–502 (1888), cited by Vernier (1969).
78. S. Hirao, J. Yamada, and R. Kikuchi, "Relation between chemical constituents of rainbow trout eggs and the hatching rate," *Bull. Jpn. Soc. Sci. Fish.* **21**, 270–273 (1955).
79. D. F. Hobbs, "Natural reproduction of quinnat salmon and brown and rainbow trout in certain New Zealand waters," *N. Z. Mar. Dept. Fish. Bull.* **6**, 7–104 (1937).
80. G. M. Ignatieva, "The relative duration of some processes of early embryogenesis in salmonid fish," *Dokl. Akad. Nauk SSSR* **188**, 1418–1421 (1969).
81. G. M. Ignatieva, "Regularities of early embryogenesis in salmonid fish as revealed by the method of dimensionless characterization of development time," *Sov. J. Dev. Biol.* **1**, 20–32 (1970).
82. G. M. Ignatieva, "Temperature dependence of cleavage rates in carp, pike, and whitefish," *Sov. J. Dev. Biol.* **5**, 24–28 (1974a).
83. G. M. Ignatieva, "Relative durations of corresponding periods of early embryogenesis in teleosts," *Sov. J. Dev. Biol.* **5**, 379–386 (1974b).
84. G. M. Ignatieva, "Regularities of early embryogenesis in teleosts as revealed by studies of the temporal pattern of development. I. The duration of the mitotic cycle and its phases during synchronous cleavage divisions," *Wilhelm Roux's Arch. Dev. Biol.* **179**, 301–312 (1976).
85. G. M. Ignatieva, "Regularities of early embryogenesis in teleosts as revealed by studies of the temporal pattern of development. II. Relative duration of corresponding periods of development in different stages," *Wilhelm Roux's Arch. Dev. Biol.* **179**, 313–325 (1976).
86. G. M. Ignatieva and A. A. Kostomarova, "The duration of the mitotic cycle at the period of synchronous cleavage division (τ_0) and its dependence on temperature in loach embryos," *Dokl. Akad. Nauk SSSR* **168**, 1221–1224 (1966).
87. G. M. Ignatieva and N. N. Rott, "The temporal pattern of interphase prolongation and nuclear activities during early embryogenesis in Teleostei," *Wilhelm Roux's Arch. Dev. Biol.* **165**, 103–109 (1970).
88. G. M. Ignatieva and N. V. Smirnova, "The determination of RNA content in eggs of some fishes and amphibians with respect to the regulation of onset of RNA synthesis in the embryo," *Ontogenez* **14**, 45–51 (1983).

89. B. Jalabert, B. Breton, and C. Bry, "Maturation et ovulation *in vitro* des ovocytes de la truite arc-en-ciel *Salmo gairdneri*," *C. R. Acad. Sci. Paris* **D275**, 1139–1142 (1972).

90. B. Jalabert, "*In vitro* oocyte maturation and ovulation in rainbow trout (*Salmo gairdneri*), northern pike (*Esox lucius*), and goldfish (*Carassius auratus*), *J. Fish. Res. Board Can.* **33**, 974–988 (1976).

91. B. Jalabert, C. Bry, D. Szöllösi, and A. Fostier, "Étude comparée de l'action des hormones hypophysaires et stéroides sur la maturation *in vitro* des ovocytes de la truite et du Carassin (poissons téléostéens)," *Ann. Biol. Anim. Biochim. Biophys.* **13**, hors-Série, 59–72 (1973).

92. D. W. Jared and R. Wallace, "Comparative chromatography of the yolk proteins of teleosts," *Comp. Biochem. Physiol.* **24**, 437–443 (1968).

93. M. Kawajiri, "On the optimum temperature of water for hatching the eggs of rainbow trout (*Salmo irideus* Gibbons)," *J. Imp. Fish. Inst.* **23**, 59–65 (1927).

94. V. S. Kirpichnikov, *Genetic Basis of Fish Selection*, Springer-Verlag, Berlin (1981).

95. A. E. Knight, "The embryonic and larval development of the rainbow trout," *Trans. Am. Fish. Soc.* **92**, 344–355 (1963).

96. F. Kopsch, "Die Entwicklung der äusseren Form des Forellembryos," *Arch. Mikr. Anat.* **51**, 181–213 (1898).

97. A. A. Kostomarova, "Transcription and translation in spermatogenesis," in *Modern Problems of Spermatogenesis* [in Russian], Nauka, Moscow (1982), pp. 160–190.

98. B. Z. Krupkin and V. M. Chaplygin, "Fish culture and biological characteristics of rainbow trout females reared under conditions of a warm-water farm," *Izv. GosNIORKh* **150**, 127–140 (1980).

99. S. G. Kryzhanovsky, "The yolk-sac surface size of teleostean eggs and organogenesis," *Zool. Zh.* **19**, 456–470 (1940).

100. M. Kusa, "Studies on cortical alveoli in some teleostean eggs," *Embryologia* **3**, 105–129 (1956).

101. M. Kusa, "Electron microscopy of cortical cytoplasm in the rainbow trout egg," *Annot. Zool. Jpn.* **39**, 67–70 (1966).

102. W. Kwain, "Embryonic development, early growth, and meristic variation in rainbow trout (*Salmo gairdneri*) exposed to combinations of light intensity and temperature," *J. Fish. Res. Board Can.* **32**, 397–402 (1975).

103. W. Kwain, "Effect of temperature on development and survival of rainbow trout, *Salmo gairdneri*, in acid water," *J. Fish. Res. Board Can.* **32**, 493–497 (1975).

104. O. A. Lebedeva and M. M. Meshkov, "The alteration of term of formation of organs and duration of embryogenesis of *Salmo irideus* Gibb. depending on temperature conditions," *Izv. GosNIORKh* **68**, 136–155 (1969).

105. N. A. Lemanova, "Comparative analysis of vitellogenesis process in different age females of rainbow trout participating in spawning for the first time and repeatedly," *Izv. GosNIORKh* **97**, 150–154 (1974).

106. N. A. Lemanova, "Peculiarities of vitellogenesis in rainbow trout females attaining sexual maturity," *Izv. GosNIORKh* **113**, 36–41 (1976).

107. W. I. Long, "Proliferation, growth, and migration of nuclei in the yolk syncytium of *Salmo* and *Catostomus*," *J. Exp. Zool.* **214**, 333–343 (1980).

108. W. Luther, "Entwicklungsphysiologische Untersuchungen an Forellenkeim: die Rolle des Organisationszentrums bei der Entstehung der Embryonalanlage," *Biol. Zbl.* **55**, 114–137 (1935).

109. W. Luther, "Potenzprüfungen an isolierten Teilstücken der Forellenkeimscheibe," *Wilhelm Roux's Arch. Entwicklungsmech. Org.* **135**, 359–383 (1937a).

110. W. Luther, "Austausch von präsumptiver Epidermis und Medular-platte beim Forellenkeim," *Wilhelm Roux's Arch. Entwicklungsmech. Org.* **135**, 384–388 (1937b).

111. W. Luther, "Transplantations- und Defektversuche am Organisationszentrum der Forellenkeimscheibe," *Wilhelm Roux's Arch. Entwicklungsmech. Org.* **137**, 404–424 (1938).

112. H. R. MacCrimmon, "World distribution of rainbow trout (*Salmo gairdneri*)," *J. Fish. Res. Board Can.* **28**, 663–704 (1971).

113. H. R. MacCrimmon and W. H. Kwain, "Influence of light on early development and meristic characters in the rainbow trout, *Salmo gairdneri* Richardson," *Can. J. Zool.* **47**, 631–637 (1969).

114. J. P. McGregor and H. B. Newcombe, "Major malformations in trout embryos irradiated prior to active organogenesis," *Radiat. Res.* **35**, 282–300 (1968).

115. M. G. Manfredi Romanini, A. Fraschini, and F. Porcelli, "Enzymatic activities during the development and the involution of the yolk sac of the trout," *Ann. Histochim.* **14**, 315–324 (1969).

116. N. L. Melnikova, M. Y. Timofeeva, N. N. Rott, and G. M. Ignatieva, "Synthesis of ribosomal RNA in early embryogenesis in the trout," *Sov. J. Dev. Biol.* **3**, 67–74 (1972).

117. H. Müller, *Die Forellen. Die einheimischen Forellen und ihre wirtschaftliche Bedeutung*, Wittenberg, Luttenstadt (1956).

118. H. Nakagawa and Y. Tsuchiya, "Studies on rainbow trout egg (*Salmo gairdneri irideus*). V. Further studies on the yolk protein during embryogenesis," *J. Fac. Fish Anim. Husb. Hiroshima Univ.* **13**, 15–27 (1974).

119. H. Nakagawa and Y. Tsuchiya, "Studies on rainbow trout egg (*Salmo gairdneri irideus*). VI. Changes of lipid composition in yolk during development," *J. Fac. Fish Anim. Husb. Hiroshima Univ.* **15**, 35–46 (1976).

120. P. Needham and R. Gard, *Rainbow Trout in Mexico and California with Notes of the Cut-throat Series*, University of California Press, Berkeley–Los Angeles (1959).

121. I. T. Negonovskaya, "Oogenesis and scale of maturity of ovaries in the Sevan trout *Salmo ischchan* (Kessler)," *Biol. Zh. Arm.* **19**, 14, 58–71 (1966).

122. W. I. Nelson, "Physiological observations on developing rainbow trout *Salmo gairdneri* (Richardson) exposed to low pH and varied calcium ion concentrations," *J. Fish Biol.* **20**, 359–372 (1982).

123. G. V. Nikolsky, *Descriptive Ichthyology* [in Russian], Vysshaya Shkola, Moscow (1971).

124. I. N. Ostroumova, "The growth and development of *Salmo irideus* Gibb. embryos under various concentrations of oxygen in water," *Izv. GosNIORKh* **68**, 202–216 (1969).

125. N. D. Ozernyuk, *Bioenergetics in Early Fish Development* [in Russian], Nauka, Moscow (1985).

126. N. D. Ozernyuk and V. G. Lelanova, "Factors of respiration intensity in early ontogenesis of rainbow trout," *Dokl. Akad. Nauk SSSR* **292**, 985–988 (1987).

127. J. R. Pasteels, "La gastrulation et la répatition des territoires dans la moitié dorsale du blastodisque de la truite (*Salmo irideus*), *C. R. Soc. Biol.* **113**, 425–428 (1933).

128. J. Pasteels, "Études sur la gastrulation des vertébrés meroblastiques. I. Téléostéens," *Arch. Biol.* **47**, 205–308 (1936).

129. J. Pasteels, "Dévéloppement embryonnaire," in *Traité de Zoologie*, Vol. 13, P. P. Grassé, ed. (1958), pp. 1685–1754.

130. D. V. Pchelovodova, "Correlation between fat content in eggs and some biological indices in rainbow trout females," *Izv. GosNIORKh* **113**, 12–15 (1976).

131. D. V. Pchelovodova and L. T. Shestakova, "Pattern of fat utilization in rainbow trout early embryogenesis," *Izv. GosNIORKh* **113**, 57–58 (1976).

132. R. A. Pedersen, "DNA content, ribosomal gene multiplicity, and cell size in fish," *J. Exp. Zool.* **177**, 65–78 (1971).

133. D. Pelluet, "Criteria for the recognition of developmental stages in salmon (*Salmo salar*)," *J. Morphol.* **74**, 395–407 (1944).

134. A. K. Pillai and C. Terner, "Studies of metabolism in embryonic development. 6. Cortisol-binding proteins of trout embryos," *Gen. Comp. Endocrinol.* **24**, 162–167 (1974).

135. V. Pleva, "Embryonalmi vijvoj ptruha duhoveho (*Trutta gairdneri irideus*)," *Sb. Vysoké Skoly Zemedel. Lesnické Brno* **6**, 77–86 (1958).

136. A. G. Polyanovskaya, "Influence of low temperature on early developmental stages of salmonid fish," *Uch. Zap. LGU* **113**(20), 63–80 (1949).

137. T. I. Privolnev, "Ecological–physiological and fish-cultural characteristics of *Salmo irideus* Gibb.," *Izv. GosNIORKh* **68**, 3–22 (1969).

138. L. Purko and B. A. Haylett, "Appearance and role of nucleoli in early teleost embryogenesis," *J. Cell. Biol.* **47**, 2 (1970).

139. T. S. Rass, "Phases and stages in the ontogenesis of teleostean fish," *Zool. Zh.* **25**, 137–148 (1946).

140. S. D. Rice and R. M. Stokes, "Metabolism of nitrogenous wastes in the eggs and alevins of rainbow trout, *Salmo gairdneri* Richardson," in *The Early Life History of Fish*, Z. H. S. Blaxter, ed., Springer-Verlag, Berlin–New York (1974), pp. 325–338.

141. P. Roubaud, H. Boulekbache, C. Devillers, and C. Joly, "Localisation histochimique de la lacticodéshydrogénase dans les blastomères au cours du développement précoce de l'embryon de la truite arc-en-ciel (*Salmo irideus* Gibb.)," *C. R. Acad. Sci. Paris* **D283**, 1543–1546 (1976).

142. O. F. Sakun, "Experimental study on oogenesis in Salmonidae and possible ways to guide this process," *Izv. GosNIORKh* **97**, 169–173 (1974).

143. G. G. Savostianova, "Origin, culturing, and selection of the rainbow trout in the USSR and abroad," *Izv. GosNIORKh* **117**, 3–13 (1976).

144. G. G. Savostianova and V. Y. Nikandrov, "Dependence of some biometric indices of eggs on age of rainbow trout females," *Izv. GosNIORKh* **113**, 3–7 (1976).

145. G. G. Savostianova and E. S. Slutsky, "On size variation of rainbow trout eggs," *Izv. GosNIORKh* **97**, 159–168 (1974).

146. L. Scheuring, "Weitere biologische und physiologische Untersuchungen an Salmonidensperma," *Zool. Jahrb., Abt. 1* **45**, 651–706 (1928).
147. L. T. Shestakova, "Lipid composition during the embryonal–larval period of rainbow trout development," *Izv. GosNIORKh* **113**, 59–61 (1976).
148. S. J. Silver, C. E. Warren, and P. Doudoroff, "Dissolved oxygen requirements of developing steelhead trout and chinook salmon embryos at different water velocities," *Trans. Am. Fish. Soc.* **92**, 327–343 (1963).
149. J. B. Smith, J. Maclaughlin, and C. Terner, "Studies of metabolism in embryonic development. IV. Protein synthesis in mitochondria and ribosomes of unfertilized and fertilized trout eggs," *Int. J. Biochem.* **1**, 191–197 (1970).
150. S. Smith, "Studies in the development of the rainbow trout (*Salmo irideus*). I. The heat production and nitrogenous excretion," *J. Exp. Biol.* **23**, 357–378 (1947).
151. S. Smith, "Studies in the development of the rainbow trout (*Salmo irideus*). II. Metabolism of carbohydrates and fats," *J. Exp. Biol.* **29**, 650–666 (1952).
152. S. G. Soin, "On development of unfertilized eggs of salmonid fish," *Rybn. Khoz.* **5**, 55–58 (1953).
153. S. G. Soin, "On the reproduction and development of *Thymallus arcticus baicalensis* Dybowsky," *Zool. Zh.* **42**, 1817–1840 (1963).
154. S. K. Stefanov and L. D. Dencheva, "Rainbow trout (*Salmo irideus* Gibb.) embryo development in the state piscicultural farm, Samokov, Bulgaria," *Annu. Univ. Sofia, Fac. Biol.* **59**, 83–96 (1967).
155. G. A. Stepanova, "Characteristic features of trout spawners with regard to some fish-culture physiological indices," *Izv. GosNIORKh* **139**, 130–137 (1979).
156. I. I. Strekalova, "Some problems of reproduction of salmonid fish," in *Progress in Science, Zoology Series* [in Russian], VINITI, Moscow (1965).
157. C. G. Swann and E. M. Donaldson, "Bibliography of salmonid reproduction from 1963–1979 for family Salmonidae; subfamilies Salmonidae, Coregoninae, and Thymallinae," *Can. Tech. Rep. Fish. Aquat. Sci.,* Vol. 970 (1980).
158. B. Szubinska-Kilarska, "The morphology of the yolk in certain Salmonidae," *Acta Biol. Cracov., Ser. Zool.* **2**, 97–111 (1959).
159. C. Terner, "Studies of metabolism in embryonic development. I. The oxidative metabolism of unfertilized and embryonated eggs of the rainbow trout," *Comp. Biochem. Physiol.* **24**, 933–940 (1968).
160. C. Terner, "Studies of metabolism in embryonic development. III. Glycogenolysis and gluconeogenesis in trout embryos," *Comp. Biochem. Physiol.* **25**, 989–1003 (1968).
161. C. Terner, "Metabolism and energy conversion during early development," in *Fish Physiology, Vol. 8, Bioenergetics and Growth*, W. S. Hoar et al., eds., Academic Press, New York (1970), pp. 261–278.
162. C. Terner, L. A. Kumar, and Choe Tae Sik, "Studies of metabolism in embryonic development. II. Biosynthesis of lipids in embryonated trout ova," *Comp. Biochem. Physiol.* **24**, 941–950 (1968).
163. A. Thomopoulos, "Mouvements d'ensemble à l'intérieur de l'oeuf de truite (*Salmo fario* et *S. irideus*)," *Bull. Soc. Zool. Fr.* **79**, 42–46 (1954).
164. L. A. Timoshina, "Amino acids of rainbow trout embryos," *Vopr. Ikhtiol.* **7**, 626–632 (1967).

165. L. A. Timoshina, "The study of amino acids of rainbow trout embryos using a chromatographic method," *Izv. GosNIORKh* **65**, 201–207 (1969a).
166. L. A. Timoshina, "Free and fixed amino acids of *Salmo irideus* Gibb. embryos," *Izv. GosNIORKh* **68**, 156–172 (1969b).
167. L. A. Timoshina, "The embryonic development of rainbow trout *Salmo gairdneri irideus* Gibb. at different temperatures," *Vopr. Ikhtiol.* **12**, 471–478 (1972).
168. J. P. Trinkaus, "The role of the periblast in *Fundulus* epiboly," *Sov. J. Dev. Biol.* **2**, 323–326 (1971).
169. R. M. Tsoi, "Effects of dimethyl sulphate and nitrosomethylurea on the development of eggs of *Salmo irideus* Gibb. and *Coregonus peled* Gmel.," *Tsitologiya* **11**, 1440–1448 (1969).
170. R. M. Tsoi, "Chemical gynogenesis of *Salmo irideus* and *Coregonus peled*," *Genetika* **8**, 185–188 (1972).
171. R. M. Tsoi, "Problems of artificial mutagenesis in fish culture," *Izv. GosNIORKh* **107**, 109–118 (1976).
172. W. Vahs and H. Zenner, "Der Einfluss von Lithium-Ionen auf die Embryonalentwicklung meroblastischer Wirbeltierkeime (*Salmo irideus*)," *Wilhelm Roux's Arch. Entwicklungsmech. Org.* **155**, 632–634 (1964).
173. W. Vahs and H. Zenner, Der Einfluss von Alkali und Erdalkalisalzen auf die Embryonalentwicklung der Regenbogenforelle. (Ein Vergleich mit der spezifischen morphogenetischen Wirkung des Lithiums.)," *Wilhelm Roux's Arch. Entwicklungsmech. Org.* **156**, 96–100 (1965).
174. V. P. Vasiliev, "On polyploidy in fish and some problems of evolution of salmonid Karyotypes," *Zh. Obshch. Biol.* **38**, 380–408 (1977).
175. O. Vassileva-Dryanovska and R. Belcheva, "Radiation gynogenesis in *Salmo irideus* Gibb.," *Dokl. Bolg. Akad. Nauk* **18**, 359–362 (1965).
176. M. F. Vernidub and N. I. Yandovskaya, *Instructions for Keeping Spawners: Collection, Fertilization, and Incubation of Eggs; Keeping and Rearing of Salmon Larvae in the Northwest USSR* [in Russian], Pishchepromizdat, Moscow (1935).
177. I. M. Vernier, "Table chronoligique du développement embryonaire de la truite arc-en-ciel, *Salmo gairdneri* Rich. 1836," *Ann. Embryol. Morphogen.* **2**, 495–520 (1969).
178. J. M. Vernier, "Histological study of the early stages of development of the rainbow trout (*Salmo gairdneri* Rich.)," *Sov. J. Dev. Biol.* **7**, 35–46 (1976).
179. A. D. Welander, "Some effects of x-irradiation on different embryonic stages of the trout (*Salmo gairdneri*)," *Growth* **18**, 227–257 (1954).
180. A. Winnicki, "Embryonic development and growth of *Salmo trutta* L. and *Salmo gairdneri* Rich. in conditions unfavorable to respiration," *Zool. Pol.* **17**, 45–48 (1967).
181. A. Winnicki, "Respiration of the embryos of *Salmo trutta* L. and *Salmo gairdneri* Rich. in media differing in gaseous diffusion rate," *Pol. Arch. Hydrobiol.* **15**, 23–38 (1968).
182. A. Winnicki and C. Cykowska, "New data on the mechanism of water uptake in salmonid eggs," *Acta Ichthyol. Piscatoria* **3**, 3–9 (1973).
183. E. Witschi, "Proposals for an international agreement on normal stages in vertebrate embryology," *14th Int. Zool. Congr. Copenhagen*, Copenhagen (1953).

184. E. Witschi, *Development of Vertebrates*, Saunders, Philadelphia (1956).
185. K. Yamagami and I. Yasumasu, "Phosphorus metabolism in fish eggs. V. Incorporation of inorganic ^{32}P phosphate into nucleic acids in the separated embryos of rainbow trout," *Sci. Pap. Coll. Gen. Educ. Univ. Tokyo* **14**, 245–254 (1964).
186. R. T. Yong and W. R. Inman, "The protein of the casing of salmon eggs," *J. Biol. Chem.* **124**, 189–193 (1938).
187. I. H. Zeitoun, D. E. Ullrey, W. G. Bergen, and W. T. Magee, "DNA, RNA, protein, and free amino acids during ontogenesis of rainbow trout (*Salmo gairdneri*)," *J. Fish. Res. Board Can.* **34**, 83–88 (1977).
188. H. Zenner, "Untersuchungen über die morphogenetischen Wirkungen des Li$^+$ Ions in der Keimesentwicklung der Regenbogenforelle (*Salmo irideus* Gibb.)," *Biol. Zbl.* **84**, 139–179 (1965).
189. A. I. Zotin, "Early stages of hardening of the egg envelopes in salmonid fish," *Dokl. Akad. Nauk SSSR* **89**, 573–576 (1953).

Chapter 5

THE LOACH *Misgurnus fossilis*

A. A. Kostomarova

The loach *Misgurnus fossilis* L. is widely used in the USSR to study a number of problems of developmental biology, including contemporary biochemical and cytological aspects ([1–3, 6, 8, 9, 13, 14, 17, 25, 30–34, 38, 39, 40, 43, 44, 47–52, 55–57, 60–65, 69, 71, 72, 75–79, 81–85]; see also reviews [35, 37, 38, 45, 63, 70, 74] and several monographs [22, 28, 36, 58, 66–68] which deal specifically with cytological, biochemical, and molecular–biological studies using this species). A nuclear transplantation method has been developed for the loach by Hung and Gasaryan [20, 26], and several cytological, embryological, and cytochemical investigations were carried out using this method [10, 11, 16, 20, 21, 26, 46].

The relatively short duration of development of this species, easily obtainable eggs and spermatozoa, and the absence of special difficulties in keeping these fish in the laboratory account for their wide use in experimental studies.

In the loach, as well as in other teleosteans, the embryos are impermeable to most drugs, thus hampering to a considerable extent the use of inhibitors, labeled precursors, and other substances. Since the impermeability of these embryos is due not only to the properties of the egg membranes (which can of course be removed), but also to those of the external surface of the cells covering the embryo, a method for the separation of the blastoderm from the yolk was developed so that the internal surface of the basal cells became exposed; all those drugs which could not penetrate into the intact embryo easily penetrated into the exposed embryo [50]. The procedure of blastoderm separation from the yolk is as follows. A batch of eggs from one female is placed into 1% trypsin (Difco) in phosphate buffer (pH 7.4–7.5), where the egg membranes dissolve within several minutes. Trypsin solution with the remains of membranes is washed from the eggs with double-strength Holtfreter's solution (isotonic for the loach eggs). After that the eggs are placed into test tubes with double-layered sucrose (8 mm 0.5 M/14 mm 0.75 M) solution. In this density gradient the eggs sink down to the interface between the layers. Then they are centrifuged at 6000g for 2–3 min. Such acceleration is sufficient to isolate the blastoderm from the yolk. Isolated blastoderms float to the upper sucrose layer,

125

while the yolk precipitates to the bottom of the test tube. The isolates are pipetted into Petri dishes and washed free of sucrose with Holtfreter's solution [49].

This method allowed the use of loach embryos in a number of studies which necessitated the preparation of yolk-free embryos [51]. As a result, the loach has become a common laboratory animal and is widely used for developmental biology studies.

The method of mass separation of loach blastoderms from the yolk overcame these difficulties and allowed the use of loach embryos in a number of studies which necessitated the preparative isolation of the embryos from the yolk [51].

5.1. TAXONOMY, DISTRIBUTION, AND SPAWNING

The loach *Misgurnus fossilis* L. belongs to the family Cobitidae, order Cypriniformes, superorder Teleostei, class Pisces.

This species occurs in oxbow rivers. In the USSR the range includes the rivers and tributaries of the Baltic Sea, Pskov Lake, and Ilmen Lake, sometimes the Volkhov and Vuoksu Rivers, the rivers of the Black Sea basin, from the Danube to the Don, and the Volga River basin. It occurs sometimes in the lower Kuban flow, but is absent in the Crimea and Caucasus (except for the lower Kuban flow and Transcaucasia), as well as in Turkmenia and Siberia. A closely related species, *M. anguillicaudatus*, occurs in the Amur River basin [12]; it is often considered as a subspecies *M. fossilis anguillicaudatus*.

In the Moscow Region the loach spawns in the first half of May at a water temperature of 13–14°C [53] and near Leningrad from mid-April until early May [24]. The spawning grounds of the loach have not been found, but its spawning supposedly takes place near the shore among aquatic plants [53]. The loach are caught for laboratory purposes in autumn with the help of special traps, but are easier to catch in winter when the water bodies are covered with ice: holes are made in the ice, and the loach gather around them.

5.2. MANAGEMENT OF REPRODUCTION IN THE LABORATORY

5.2.1. Conditions for Keeping Adult Fish

Loach males and females are kept separately in big aquaria with cold (4–7°C) water which is changed once a week or every 2 weeks, and preferably in a thermo-statically controlled cold room. Under these conditions one need not feed the loach. When the loach are caught, males are usually separated from females, and they are kept in separate aquaria, mainly for convenience of handling them in experimental work. Males are easily distinguished by the presence of marked skin thickenings in the form of paired ridges at the dorsal fin's posterior margin. They are often of orange-reddish color. The presence of ridges can be established by running a finger along the posterior part of the body at the dorsal fin base.

Under laboratory conditions eggs can be obtained almost the whole year round (practically from October–November until August–September).

5.2.2. Stimulation of *in vivo* Oocyte Maturation

Mature eggs are obtained in the laboratory following injections of pituitary hormones to females during the prespawning period. By this means ovulation can be artificially stimulated [15, 41, 42]. This widely known method was modified by Neyfakh [61] and consists of the injection of a chemically pure preparation of chorionic gonadotrophic hormone, choriogonin (G. Richter), rather than the extract of acetonized pituitaries of teleostean fish. Females are injected with 100–250 units of the hormone.

The hormone is injected in the following way. In a standard vial with 1500 units of the hormone powder, 1.5 ml of distilled water is introduced using a syringe, and then an appropriate volume of this solution containing the necessary amount of the hormone for a given female is used. The volume is made up to 0.5 ml with distilled water, and this solution is injected into the female intramuscularly. In the procedure the female is covered by a piece of gauze previously moistened with water and bent in a horseshoe shape, the head and tail downward, and the syringe needle is introduced into the muscle at the white line of the anterior body part in the direction of the head. The hormone should be injected slowly and, as it is injected, the fish should be straightened gradually so that it cannot push the solution out of the entry point of the needle. After the injection the fish is released from the gauze and placed in a Dewar flask with water. Ovulation takes place within 36–38 h at 18°C.

Unlike the females, the males are not injected since the testes have mature spermatozoa from the beginning of winter. Sperm can also be obtained using the injection procedure, but usually the fish is killed and the testes taken out, minced with scissors in a small dish, and kept in a refrigerator for further use.

5.2.3. *In vitro* Oocyte Maturation (after [71, 73])

The ovaries are removed from females kept in aquaria with a water temperature of 4–7°C and placed in Ringer solution with 0.2% egg albumin [54] at a pH of 8.5–9.0. Under a binocular microscope the ovaries are dissected into fragments containing 2 to 20 oocytes, the fragments are transferred in Petri dishes (4 cm in diameter), 50–100 oocytes are placed in each dish, and 5 ml of Ringer solution containing choriogonin is added together with 0.1% egg albumin. At room temperature the best results are obtained after 48-h incubation of the oocytes with the hormone. The highest percentage of oocyte maturation is attained at hormone concentrations of 10 and 25 units/ml. Upon maturation the oocytes become transparent; therefore, the mature oocytes can be distinguished from the immature ones with the naked eye. According to Saat [71], a medium containing Hanks balanced salt solution and 20% of bovine serum has proved to be best for *in vitro* maturation and ovulation of loach oocytes. The highest percentage of ovulated oocytes (over 90%) has been observed under the combined effect of progesterone (0.1–10 μg/ml during 6–0.1τ_0) and 25 units/ml of choriogonin. Oocytes matured and ovulated *in vitro* are capable of development when fertilized. But normal fertilization and subsequent development is rarely achieved in this way, and this procedure cannot be considered routine.

5.3. STRUCTURE OF GAMETES

5.3.1. Eggs

According to the pattern of cytoplasm and yolk distribution, loach eggs are telolecithal, and by the ratio of these components belong to the polyplasmatic type. The mature egg is at metaphase of the second maturation (meiotic) division. Its diameter (without envelopes) is 1.17–1.30 mm. The egg envelope consists of a well-developed zona radiata and an adjacent outer membrane of villiferous structure. The villi are small and protrude slightly over the membrane surface. Due to such a distribution of villi, the membrane looks cellular from outside. When in water, the villi swell and become sticky, so that the eggs become attached to the substrate. The membrane of the loach mature egg, like that of other teleosteans, is of purely protein nature, but there is a thin membrane containing many polysaccharides between the zona radiata and the villiferous layer [4]. The proteins constituting the zona radiata are insoluble in water and saline solutions, but are digested by trypsin [52].

As in all teleosteans, there is a micropyle in the egg envelope in the region of the animal pole; here a funnel-shaped recess of the envelope surface turns into a short terminal channel. The channel opens into the cytoplasm on the inner surface of the zona radiata [23], its diameter corresponding to the width of a loach sperm head.

A layer of cortical alveoli is at the surface of the mature egg cytoplasm. The nucleus at metaphase II is located at the animal pole, near the micropyle. The major (central) part of the egg is filled with yolk inclusions. The yolk protein is accumulated in the oocytes as yolk granules. In the loach the yolk granules do not suffer marked changes; they do not fuse and are preserved as separate entities included in the cytoplasmic network [23]. Individual yolk granules in the mature egg are 4.5–4.8 μm in diameter [80]. Lipid drops are absent.

5.3.2. Spermatozoa

In most teleosts studied the spermatozoa have the simplest structure and the acrosome is absent. The head length is 2–3 μm and the total length 40–60 μm. The loach spermatozoa, as shown by scanning electron microscopy, consists of a round head without an acrosome, a conelike intermediate part, and a tail part ten times longer than the head [16]. The life span of the sperm in water is 3 h [7]. In perivisceral fluid loach spermatozoa are not active, becoming active only after water addition [23].

5.4. INSEMINATION AND FERTILIZATION

Fertilization is external and monospermic. The time during which the capacity for fertilization is preserved by the spawned eggs is no more than 1–2 min (Neyfakh, personal communication) and in the closely related *M. fossilis anguillicaudatus* this capacity is preserved for 5 min at a water temperature of 19–23°C [19].

Five minutes after fertilization (at 18–19°C) the egg nucleus is at anaphase II; 10 min after fertilization the male pronucleus begins to sink into the cytoplasm, and after 15 min the oocyte nucleus completes the second meiotic division and the second polar body is formed. At the same time, the dense male pronucleus begins to swell. Thirty minutes after fertilization the female and male pronuclei come into contact with one another [16].

During the first seconds after fertilization, the vitelline membrane separates from the egg surface and the perivitelline space forms. At the same time, the cytoplasm is drawn upward to the animal pole and the blastodisc forms. Not all the cytoplasm is involved in its formation. A thin layer of cytoplasm surrounds the yolk; in addition, its fine strands penetrate the yolk, anastomose between each other, and form a fine network inside the yolk [79]. After the separation of the vitelline membrane and the formation of the perivitelline space, the diameter of the fertilized egg (with the membranes and perivitelline space) is 1.6–1.8 mm [53, 59], and the diameter of the egg itself (without the membrane) is 1.1–1.3 mm. The height of the blastodisc equals 0.35–0.45 mm.

To obtain eggs, a ripe female is covered with a piece of gauze, taken in the left hand, and the eggs squeezed out into a dry Petri dish by pressing the abdomen from the head to the tail with the right-hand thumb. The fish is then again placed into natural water. A few drops of Ringer solution, together with the sperm, are placed

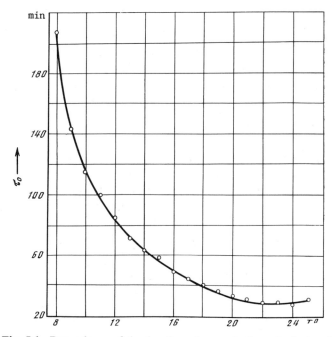

Fig. 5.1. Dependence of the duration of one mitotic cycle during the period of synchronous cleavage divisions (τ_0) on temperature in the loach [29].

in the same Petri dish so that they do not contact the eggs. Some water is added, and the eggs and spermatozoa are brought into contact by gently shaking the dish. The dish is shaken for 3–5 min (depending on the water temperature) so that the eggs, which just after fertilization (activation) become sticky, are not attached to the dish bottom and do not form clumps. Within 3 to 5 min, the water in which the fertilization was carried out is replaced. If the eggs remain sticky, they are separated from the bottom with caution, and gentle shaking is continued. The eggs are then transferred in crystallizing dishes or Petri dishes (100 mm in diameter, with no more than 300–500 eggs in each) with a thin water layer covering the eggs and placed in a thermostat with constant temperature. The water is changed daily, and any dead eggs are removed.

5.5. NORMAL DEVELOPMENT

There are only a few studies of loach development. The successive stages of loach development from fertilization until hatching were studied in more detail by Neyfakh [61] and Mitashov [59].

Kryzhanovsky [53] described some stages of embryonic and larval development of the loach with special reference to the ecological–morphological patterns. With the aid of a film shot by Galustyan and Bystrov, Svetlov et al. [79] studied the early development of the loach from fertilization to morula. The development of the eye [5] and the processes of oocyte maturation [8, 72] are also described. There is also a paper dealing specially with the larval development stages of this species [24].

To describe the successive stages of loach development, we have undertaken an additional study. The eggs obtained, as described above, were fertilized at $18 \pm 0.1°C$ and $21.5 \pm 0.1°C$, and placed in Petri dishes in an ultrathermostat, where they developed. The water in the dishes was changed twice a day. The embryos were fixed every hour at 21.5°C and every 80 min at 18°C, that is, every $2\tau_0$ at each of the temperatures [18]. The embryos were drawn at successive developmental stages under the microscope using an Abbe drawing device, in both the live and the fixed state. The embryos were fixed with Bouin's and Stockard's fluids. The latter does not shrink the object and keeps it transparent.

In embryology the successive stages of animal development are indicated by numbers. To preserve this general principle and, at the same time, in agreement with the already accepted schedule of loach developmental stages in hours of development at 21.5°C, we have assigned to the stages the numbers which, up to and including stage 33, correspond to the time of the onset of a given stage at 21.5°C in hours. For subsequent stages there is no agreement between the stage number and its age in hours at 21.5°C. The time of the onset of stages at different temperatures, within the optimal limits, is indicated in Table 5.1 by the number of τ_0 units [18].

The duration of τ_0 in the loach, and its dependence on temperature, has been determined previously [29]. The graph provided here (Fig. 5.1) allows an easy

---

Figs. 5.2–5.4. Normal developmental stages of the loach *Misgurnus fossilis* L. Drawings made by A. A. Kostomarova and E. A. Smirnova. Drawing numbers correspond to stage numbers.

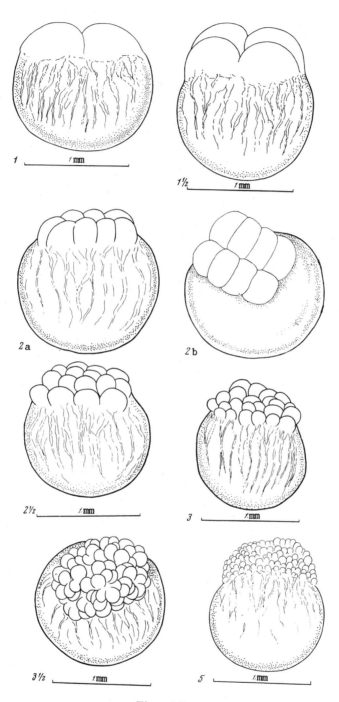

Fig. 5.2.

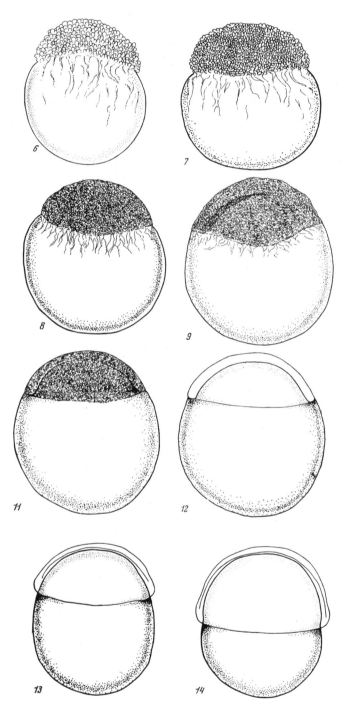

Fig. 5.3.

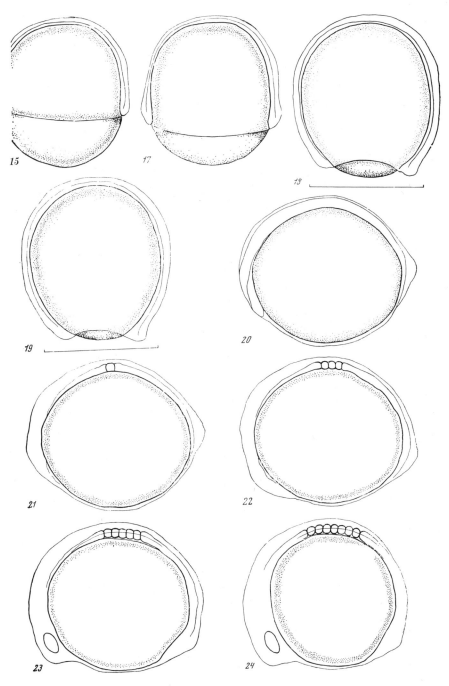

Fig. 5.3, continued.

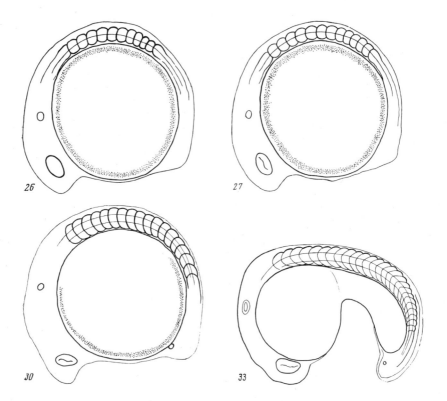

Fig. 5.3, continued.

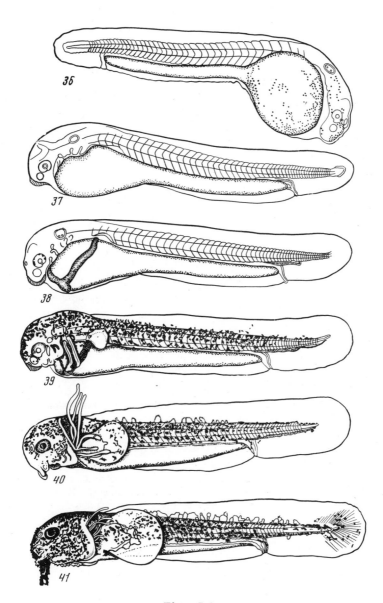

Fig. 5.4.

TABLE 5.1. Developmental Stages of the Loach (*Misgurnus fossilis* L.) at 21.5°C

Stage No.	hours, days	τ_n/τ_0	External and internal diagnostic features of the stages
	Time*		
	Hours		Embryonic development
0	–	0	Mature egg at the moment of insemination
	45 min	$1^1/_2$	Fertilized egg. The vitelline membrane is separated from the egg surface. The perivitelline space has formed and the blastodisc forms on the animal pole
1	1	2	2 blastomeres; 1st cleavage furrow is meridional
$1^1/_2$	$1^1/_2$	3	4 blastomeres; 2nd cleavage furrow is meridional, transverse to 2nd cleavage furrow
2	2	4	8 blastomeres; two 3rd cleavage furrows are meridional, transverse to 2nd cleavage furrow
$2^1/_2$	$2^1/_2$	5	16 blastomeres; 4th cleavage furrows are meridional, transverse to the 3rd cleavage furrows
3	3	6	32 blastomeres; 5th cleavage furrow is parallel to the egg equator
$3^1/_2$	$3^1/_2$	7	64 blastomeres are settled down in 2–3 layers and form a high cap
4	4	8	128 blastomeres
5	5	10	Morula: blastomeres are settled down in several layers and form a tuberous cap. Desynchronization of nuclear divisions
6	6	12	Early blastula: the cap of blastomeres is raised above the yolk or somewhat flattened. The external blastomeres form the epiblast, but its surface is still tuberous. Beginning of asynchronous nuclear divisions, fall of mitotic index. Onset of intensive RNA synthesis and morphogenetic nuclear function
7	7	14	Middle blastula: the blastomeres are smaller than at stage 6; the blastula surface is smooth
8	8	16	Late high blastula: the cap of small blastomeres is still somewhat raised above the yolk. Individual blastomeres cannot be distinguished under low magnification. There is a small recess between the blastodisc basal margin and the yolk
9	9	18	Late epithelial blastula: the blastodisc is flat, somewhat overgrows the yolk, its lower contour is unclear. The recess between the blastodisc and the yolk has disappeared

TABLE 5.1 (continued)

Stage No.	Time* hours, days	τ_n/τ_0	External and internal diagnostic features of the stages
10	10	20	Onset of gastrulation. The lower blastoderm contour is distinct, its diameter increased
11	11	22	Thickening of the blastoderm peripheral parts and formation of germinal ring
12	12	24	Formation of embryonic shield
13	13	26	The blastoderm has overgrown 1/3 of the yolk surface. The embryonic shield has the form of a tongue
14	14	28	The blastoderm has overgown 1/2 of the yolk surface. The embryonic shield has the form of a tongue with a slightly widened anterior part. Separation of notochord from mesoderm
15	15	30	The blastoderm has overgrown 2/3 of the yolk surface. The CNS rudiment forms
16	16	32	The blastoderm has overgrown 3/4 of the yolk surface. The "yolk plug" is large; the embryo has the form of a strongly elongated plate, widened at the head end
17	17	34	The blastoderm has overgrown 4/5 of the yolk surface
18	18	36	The blastoderm has overgrown 7/8 of the yolk surface
19	19	38	The process of epiboly is nearly completed; the "yolk plug" is very small; the embryo is elongated, its anterior end is widened, and the posterior end is somewhat raised above the "yolk plug"
20	20	40	Completion of epiboly. The blastoderm has overgrown the yolk fully; the mesodermal ridges have formed in the head end of the embryo
21	21	42	Appearance of the 1st pair of somites. Each subsequent pair appears at 1 τ_0 interval
22	22	44	Appearance of the 3rd pair of somites. The mesodermal ridges begin to disintegrate into individual cells
23	23	46	Appearance of the 5th and 6th pairs of somites, formation of dense eye rudiments
24	24	48	Appearance of the 7th pair of somites
25	25	50	Appearance of the 9th pair of somites

TABLE 5.1 (continued)

Stage No.	Time*		External and internal diagnostic features of the stages
	hours, days	τ_n/τ_0	
26	26	52	Appearance of the 10th pair of somites. Formation of ear vesicles. The head region increased and raised somewhat over the yolk
27	27	54	Appearance of the 11th and 12th pairs of somites. A slitlike cavity forms in the previously dense eye rudiments. Beginning of lens formation
28	28	56	Appearance of the 13th pair of somites
29	29	58	Appearance of the 15th pair of somites (from the 15th–17th pair onward, the size of somites diminishes somewhat and the rate of their formation increases; they appear at intervals of less than $1\tau_0$)
30	30	60	Appearance of the 17th pair of somites
31	31	62	Appearance of the 19th and 20th pairs of somites. The embryo is more raised above the yolk, the head end increased, the Kupfer vesicle is clearly seen. The foregut and pericardial cavity rudiment have formed. The somites are differentiated in myo-, sclero-, and dermatome
32	32	64	Up to 44 somites in the mesoderm
33	33	66	Appearance of the 25th and 26th pairs of somites. Beginning of movements. The embryo is elongated; onset of separation of the tail region from the body region. The yolk becomes pear-shaped. The Kupfer vesicle is small and does not adjoin the yolk. Beginning of invagination of the front wall of eye cup, formation of lens placodes
34	37	74	Up to 32 pairs of somites in the mesoderm. The tail is separated from the yolk. The tail mesoderm is not segmented. The Kupfer vesicle is located at the level of the 32nd somite. Each ear vesicle has 2 otoliths. Formation of eye cups is completed
35	40	80	34 somites in the mesoderm, 4–5 somites in the tail mesoderm. Appearance of hatching glands, beginning of heartbeat. The yolk is retort-shaped. The blood is colorless
36	46	92	Stage prior to hatching. The organ of attachment has formed; the number of hatching glands has decreased markedly. Over 10 somites in the tail mesoderm; the yolk is elongated; its anterior part is widened. The membranes are loosened. The embryos move vigorously inside the membranes. Beginning of eye pigmentation. The body is not pigmented

TABLE 5.1 (continued)

Stage No.	Time* hours, days	τ_n/τ_0	External and internal diagnostic features of the stages
37	50-52	100-104	Stage of hatching. The head is raised over the yolk; the heart beats weakly; the blood runs weakly; it is colorless. 17 somites in the tail mesoderm; there are rudiments of spiracular and gill slits. The pectoral fin rudiment is large. The hatched prelarvae attach themselves by means of their organs of attachment to the aquatic plants
	Days		Prelarval development
38	1	–	Body length 5.5 mm. Stage of beginning of erythrocytic blood circulation. Completion of segmentation of the mesoderm tail region; 21–22 somites in the tail. Formation of external gills and appearance of pigment. The prelarvae are indifferent to light
39	2	–	Body length 5.8 mm. Stage of appearance of the barbel rudiments. Melanophores are numerous. The pigment is clearly seen in the eyes. The prelarvae do not respond to light
40	5–6	–	Body length 6.9 mm. Stage of the maximal length of external gills. Appearance of intestinal veins, of vessels in the pectoral fin, and of a network of segmental vessels in the unpaired fin fold. The eyes are pigmented, almost black. The yolk is markedly resorbed. The prelarvae do not respond to light
41	~10	–	Body length 7.6 mm. Stage of transition to active feeding (food is searched for by barbels). Appearance of vessels in the anal region of the fin fold. Appearance of mesenchyme clots in the regions of presumptive dorsal and anal fin skeleton

*For embryonic development, time was measured in hours from the moment of insemination. For prelarval development, time was measured in days from time of hatching.

determination of the value of τ_0 at a given temperature and calculation of the time of any developmental stage at different temperatures, within optimal limits.

The mid-range of optimal temperatures for loach development and for obtaining eggs of good quality lies between 13–14 and 20–21°C [19]. The estimation of the relative duration of different periods of the early development in the loach [28] has shown that it does not change reliably for the same developmental periods within the limits of 14–22°C. Table 5.1 presents also the characteristics of diagnostic features for every stage.

The drawings of the embryos and prelarvae are presented in Figs. 5.2–5.4. The numbers of the drawings correspond to the numbers of the stages.

The description of stages and drawings of prelarvae until their transition to active feeding follow those of Kryzhanovsky [53] and Baburina [5].

REFERENCES

1. N. B. Abramova, V. P. Korzh, and A. A. Neyfakh, "Regulation of oxygen consumption and cytochrome oxidase activity after artificial increase in the amount of mitochondria in fish embryos," *Biokhimiya* **45**, 1124–1132 (1980).
2. N. B. Abramova, T. A. Burakova, V. P. Korzh, and A. A. Neyfakh, "Injection of mitochondria into oocytes and fertilized eggs," *Sov. J. Dev. Biol.* **10**, 361–364 (1979).
3. M. A. Aitkhozhin, N. V. Belitsina, and A. I. Spirin, "Nucleic acids at the early developmental stages in fish (the loach taken as an example)," *Biokhimiya* **29**, 169–174 (1964).
4. E. A. Arndt, "Die Aufgaben des Kerns während der Oogenese der Teleosteer," *Z. Zellforsch.* **51**, 356–378 (1960).
5. E. A. Baburina, *Development of the Eye in Cyclostomata and Pisces with Reference to Ecology* [in Russian], Nauka, Moscow (1972), pp. 102–108.
6. N. V. Belitsina, L. P. Gavrilova, A. A. Neyfakh, and A. S. Spirin, "Effect of radiation inactivation of nuclei on synthesis of messenger ribonucleic acid in the loach (*Misgurnus fossilis* L.) embryos," *Dokl. Akad. Nauk SSSR* **153**, 1204–1206 (1963).
7. E. V. Belyaev, "Some aspects of the physiology of the sperm and eggs in fish," *Tr. Mosk. Tekhn. Inst. Rybn. Prom. Khoz.* **8**, 271–277 (1957).
8. V. N. Belyaeva and N. B. Cherfas, "On the processes of maturation and fertilization in loach (*Misgurnus fossilis* L.) eggs," *Vopr. Ikhtiol.* **5**, 82–90 (1965).
9. A. O. Benyumov, A. A. Kostomarova, and K. G. Gasaryan, "A method of microinjections of nucleic acid precursors in loach eggs and embryos as a way to overcome the barrier functions of the egg and blastomere membranes," in *Histohematic Barriers and Neurohumoral Regulation* [in Russian], Nauka, Moscow (1981), pp. 126–132.
10. A. O. Benyumov, A. A. Kostomarova, and K. G. Gasaryan, "Synthesis and transport of nuclear RNA in early loach embryos produced by nuclear transplantation in the eggs," *Zh. Obshch. Biol.* **44**, 694–700 (1984).
11. A. O. Benyumov, A. A. Kostomarova, and K. G. Gasaryan, "RNA migration in the loach (*Misgurnus fossilis* L.) blastula nuclei after their transplantation into the activated egg cells," *Zh. Obshch. Biol.* **45**, 681–686 (1984).
12. L. S. Berg, *Fish of Fresh Waters of the USSR and of Adjacent Countries*, Part II [in Russian], Izd. Akad. Nauk SSSR, Moscow–Leningrad (1949), pp. 900–901.
13. I. B. Bronshtein, G. K. Shakhbazyan, and K. A. Kafiani, "Thermostable DNAse from fish eggs producing specific breaks in supercoiled DNA," *Dokl. Akad. Nauk SSSR* **264**, 1500–1502 (1982).

14. I. B. Bronshtein, G. K. Shakhbazyan, and K. A. Kafiani, "Endodeoxyribonuclease from loach eggs. Purification and properties," in *Macromolecules in Functioning Cells*, F. Salvatore, G. Marconi, and P. Volpe, eds., Plenum Press, New York (1982), pp. 91–102.

15. O. B. Chernyshev, "Obtaining mature sexual products in the loach (*Misgurnus fossilis* L.) in winter," *Dokl. Akad. Nauk SSSR* **33**, 155–158 (1941).

16. N. V. Dabagyan, L. A. Sleptsova, D. R. Baldanova, and K. G. Gasaryan, "Fertilization and early cleavage in the loach (*Misgurnus fossilis* L.)," *Zh. Obshch. Biol.* **42**, 440–447 (1981).

17. A. N. Davitashvili, K. G. Gasaryan, L. A. Strelkov, and K. A. Kafiani, "Nuclear RNA particles in loach embryos," *Dokl. Akad. Nauk SSSR* **250**, 470–473 (1980).

18. T. A. Dettlaff and A. A. Dettlaff, "On relative dimensionless characteristics of development duration in embryology," *Arch. Biol.* **12**, 1–6 (1961).

19. H. Gamo, E. Yamauchi, and R. Suzuki, "The fertilization of dechorionated eggs in two species of fish," *Bull. Aichi Gakugei Univ. Nat. Sci.* **9**, 117–126 (1960).

20. K. G. Gasaryan and N. M. Hung, "Transplantation of somatic nuclei in teleostean eggs (the loach *Misgurnus fossilis* L. taken as an example)," *Dokl. Akad. Nauk SSSR* **240**, 725–728 (1978).

21. K. G. Gasaryan, N. M. Hung, A. A. Neyfakh, and V. V. Ivanenkov, "Nuclear transplantation in teleost *Misgurnus fossilis*," *Nature (London)* **280**, 585–587 (1979).

22. G. G. Gause, *Mitochondrial DNA* [in Russian], Nauka, Moscow (1977).

23. A. S. Ginsburg, *Fertilization in Fish and the Problem of Polyspermy*, Israeli Program for Scientific Translations, Ltd. (1972).

24. A. Grieb, "Die larvale Periode in der Entwicklung des Sclammbeissers (*Misgurnus fossilis*)," *Acta Zool.* **18** (1937).

25. D. B. Gulyamov, E. F. Knyazeva, A. A. Kostomarova, and V. S. Mikhailov, "Localization of DNA-polymerase α in the nucleus and of DNA-polymerase γ in the cytoplasm of maturing oocytes in the loach," *Ontogenez* **13**, 641–644 (1982).

26. N. M. Hung and K. G. Gazaryan, "Transplantation of somatic nuclei into activated eggs of bony fish (using the loach *Misgurnus fossilis* as an example)," *Sov. J. Dev. Biol.* **9**, 67–72 (1978).

27. G. M. Ignatieva, "Relative durations of corresponding periods of early embryogenesis in teleosts," *Sov. J. Dev. Biol.* **5**, 379–386 (1974).

28. G. M. Ignatieva, *Early Embryogenesis of Fish and Amphibians* [in Russian], Nauka, Moscow (1979).

29. G. M. Ignatieva and A. A. Kostomarova, "Duration of the mitotic cycle during synchronous cleavage divisions (τ_0) and its dependence on temperature in loach embryos," *Dokl. Akad. Nauk SSSR* **168**, 1221–1224 (1966).

30. V. V. Ivanenkov, "Esterase-2 in the development of loach (*Misgurnus fossilis* L.). I. Period of expression of esterase-2 genes and duration of maternal enzyme persistence in the embryos," *Isozyme Bull.* **13**, 76 (1980).

31. V. V. Ivanenkov, ""Esterase-2 in the development of loach (*Misgurnus fossilis* L.). II. Differential expression of the allelic esterase-2 genes in the oocytes and eggs," *Isozyme Bull.* **13**, 77 (1980).

32. V. V. Ivanenkov, ""Esterase-2 in the development of loach (*Misgurnus fossilis* L.). III. Absence of feedback in the regulation of esterase-2 gene expression," *Isozyme Bull.* **13**, 78 (1980).
33. V. V. Ivanenkov, "Carboxylesterase-2 in the development of the loach (*Misgurnus fossilis* L.)," *Biochem. Genet.* **18**, 353–364 (1980).
34. V. V. Ivanenkov, "Differential expression of allelic carboxylesterase-2 genes in oocytes of loach (*Misgurnus fossilis* L.) and heterogeneity of loach oocytes and eggs for the expression of allelic carboxylesterase-2 genes," *Biochem. Genet.* **18**, 365–375 (1980).
35. K. A. Kafiani, "Genome transcription in fish development," *Adv. Morphogen.* **8**, 209–284 (1970).
36. K. A. Kafiani and A. A. Kostomarova, *Informational Macromolecules in Early Embryogenesis of Animals* [in Russian], Nauka, Moscow (1978).
37. K. A. Kafiani, L. A. Strelkov, N. B. Chiaureli, and A. N. Davitashvili, "Nucleus-associated polyribosomes in early embryos of loach," in *Macromolecules in Functioning Cells*, F. Salvatore, G. Marconi, and P. Volpe, eds., Plenum Press, New York (1979), pp. 263–282.
38. K. A. Kafiani and M. Ya. Timofeeva, "Molecular and regulatory aspects of early embryogenesis," *Usp. Biol. Khim.* **8**, 138–167 (1967).
39. K. A. Kafiani, M. Ya. Timofeeva, N. L. Melnikova, and A. A. Neyfakh, "Rate of RNA synthesis in haploid and diploid embryos of the loach (*Misgurnus fossilis*)," *Biochim. Biophys. Acta* **169**, 274–277 (1968).
40. K. A. Kafiani, M. Ya. Timofeeva, A. A. Neyfakh, N. L. Melnikova, and Ya. Ya. Rachkus, "RNA synthesis in the early embryogenesis of a fish (*Misgurnus fossilis*)," *J. Embryol. Exp. Morphol.* **21**, 295–308 (1969).
41. B. N. Kazansky, "Peculiarities of the functions of ovary and pituitary in fish with interrupted spawning," *Tr. Lab. Osn. Rybovodstva* **2**, 64–120 (1949).
42. Y. D. Kirshenblat, "Effect of human gonadotrophic hormones on fish females," *Priroda* **4**, 75–76 (1949).
43. O. S. Klyachko, V. P. Korzh, S. I. Gorgolyuk, A. V. Timofeev, and A. A. Neyfakh, "Nonuniform distribution of enzymes in fish eggs," *J. Exp. Zool.* **222**, 137–148 (1982).
44. E. F. Knyazeva, A. O. Benyumov, and A. A. Kostomarova, "Synthesis and funding of nuclear nonhistone proteins in loach oocytes and early embryos," *Ontogenez* **13**, 500–508 (1982).
45. A. A. Kostomarova and E. F. Knyazeva, "Some patterns of metabolism of the sperm nucleus after fertilization," in *Spermatogenesis and Its Regulation* [in Russian], Nauka, Moscow (1983), pp. 194–224.
46. E. F. Knyazeva and A. A. Kostomarova, "Migration of the oocyte nonhistone proteins in the loach blastula and gastrula nuclei transplanted in the egg," *Zh. Obshch. Biol.* **45**, 98–104 (1984).
47. V. P. Korzh, "Microinjection of macromolecules and mitochondria into loach eggs," *Sov. J. Dev. Biol.* **12**, 134–138 (1981).
48. V. P. Korzh, A. V. Timofeev, and A. A. Neyfakh, "The absence of regulation of the activity of glucose-6-phosphate dehydrogenase following its injection into loach eggs," *Isozyme Bull.* **13**, 80 (1980).
49. A. A. Kostomarova, "The differentiation capacity of isolated loach (*Misgurnus fossilis*) blastoderm," *J. Embryol. Exp. Morphol.* **22**, 407–430 (1969).
50. A. A. Kostomarova and A. A. Neyfakh, "A method of blastoderm separation in loach embryos and possibilities of its application," *Zh. Obshch. Biol.* **25**, 386–388 (1964).

51. M. R. Krigsgaber and A. A. Neyfakh, "Protein synthesis in the blastoderm of loach embryos," *Dokl. Akad. Nauk SSSR* **180**, 1259–1261 (1968).
52. M. R. Krigsgaber, A. A. Kostomarova, T. A. Terekhova, and T. A. Burakova, "Synthesis of nuclear and cytoplasmic proteins in the early development of fish and echinoderms," *J. Embryol. Exp. Morphol.* **26**, 611–622 (1971).
53. S. G. Kryzhanovsky, "Ecological–morphological patterns of development of the Cyprinoidei and Siluroidei," *Tr. Inst. Morfol. Zhivotn. Akad. Nauk SSSR* **1**, 186–195 (1949).
54. Y. Masui, "Hormonal and cytoplasmic control of the maturation of frog oocytes," *Sov. J. Dev. Biol.* **3**, 484–495 (1972).
55. V. S. Mikhailov, D. B. Gulyamov, and I. B. Zbarsky, "Differences in the content of DNA polymerases α, β, γ in the liver and mature eggs of the loach *Misgurnus fossilis*," *Dokl. Akad. Nauk SSSR* **254**, 503–506 (1980).
56. V. S. Mikhailov and D. B. Gulyamov, "Two enzymatic systems of DNA synthesis in the nuclei from the loach *Misgurnus fossilis*. The participation of DNA polymerase α in DNA replication in embryonic nuclei and of DNA polymerase β in DNA synthesis in liver nuclei," *Dokl. Akad. Nauk SSSR* **257**, 229–233 (1981).
57. V. S. Mikhailov, A. A. Kostomarova, D. B. Gulyamov, and E. F. Knyazeva, "Compartmentalization of DNA polymerases α and γ in the maturing oocytes of the loach (*Misgurnus fossilis*) oocytes," *Cell Differ.* **13**, 87–91 (1983).
58. L. S. Milman and Yu. G. Yurovitzky, *Regulation of Glycolysis in the Early Development of Fish Embryos*, S. Karger, Basel (1973).
59. V. I. Mitashov, "Embryonic development of the loach (*Misgurnus fossilis*)," Thesis, Moscow State University (1963).
60. A. A. Neyfakh, "Changes in radiosensitivity during fertilization in the loach *Misgurnus fossilis*," *Dokl. Akad. Nauk SSSR* **109**, 943–946 (1956).
61. A. A. Neyfakh, "Application of the method of radioactive inactivation of the nuclei for studying their functions in early development of fish," *Zh. Obshch. Biol.* **20**, 202–213 (1959).
62. A. A. Neyfakh, "Utilization of nucleic acid metabolism inhibitors for studying the periods of morphogenetic nuclear function in development," in *Cell Differentiation and Mechanisms of Induction* [in Russian], Nauka, Moscow (1965), pp. 38–60.
63. A. A. Neyfakh, "Steps of realization of genetic information in early development," in *Current Topics in Developmental Biology*, Vol. 6, A. Monroy and A. A. Moscona, eds., Academic Press, New York (1971), pp. 45–77.
64. A. A. Neyfakh and A. A. Kostomarova, "Migration of newly synthesized RNA during mitosis. I. Embryonic cells of the loach (*Misgurnus fossilis* L.)," *Exp. Cell Res.* **65**, 340–344 (1971).
65. A. A. Neyfakh, A. A. Kostomarova, and T. A. Burakova, "Transfer of RNA from nucleus to cytoplasm in early development of fish," *Exp. Cell Res.* **72**, 223–232 (1972).
66. A. A. Neyfakh and M. Ya. Timofeeva, *Molecular Biology of Development*, Vol. I, *Molecular Events*, Plenum Press, New York (1984).
67. A. A. Neyfakh and M. Ya. Timofeeva, *Molecular Biology of Development*, Vol. II, *Problems of Regulation*, Plenum Press, New York (1984).
68. N. D. Ozernyuk, *Growth and Reproduction of Mitochondria* [in Russian], Nauka, Moscow (1978).

69. Y. A. Rachkus, K. A. Kafiani, and M. Ya. Timofeeva, "Some characteristics of ribonucleic acids synthesized in the loach embryos," *Sov. J. Dev. Biol.* **2**, 222–229 (1971).

70. N. N. Rott, "Cell divisions during the pregastrulation period of development," *Sov. J. Dev. Biol.* **11**, 1–16 (1980).

71. T. V. Saat, "Oocyte maturation and ovulation in the loach in different media and under different hormonal influences," *Ontogenez* **11**, 545–554 (1980).

72. T. V. Saat, "Chronology of oocyte maturation in the loach and the acquisition of developmental capacity," *Sov. J. Dev. Biol.* **13**, 159–166 (1982).

73. M. N. Skoblina, "The maturation of loach oocytes under the *in vitro* influence of chorionic gonadotrophin," *Sov. J. Dev. Biol.* **4**, 284–288 (1973).

74. A. A. Spirin, "On 'masked' forms of messenger RNA in early embryogenesis and in other differentiating systems," in *Current Topics in Developmental Biology*, Vol. 1, A. Monroy and A. A. Moscona, eds., Academic Press, New York (1966), pp. 1–38.

75. A. S. Spirin, N. V. Belitsina, and M. N. Aitkhozhin, "Messenger RNAs in early embryogenesis," in *Cell Differentiation and Mechanisms of Induction* [in Russian], Nauka, Moscow (1965), pp. 18–37.

76. L. A. Strelkov and N. B. Chiaureli, "Distribution of newly formed poly(A)-RNAs in subcellular fractions of early loach embryos," *Ontogenez* **13**, 488–499 (1982).

77. L. A. Strelkov, N. V. Chiaureli, A. N. Davitashvili, and K. A. Kafiani, "A special class of polyribosomes associated with the embryonic nuclei in the loach," *Dokl. Akad. Nauk SSSR* **244**, 479–483 (1979).

78. L. A. Strelkov, N. B. Chiaureli, and K. A. Kafiani, "Characteristics of polyribosome complexes from the cytoplasm of loach embryonic cells," *Biokhimiya* **45**, 1934–1943 (1980).

79. P. G. Svetlov, V. D. Bystrov, and T. F. Korsakova, "On the morphology and physiology of the early developmental stages in teleostean fish (from the film 'Development of the loach *Misgurnus fossilis*' by Sh. D. Galustyan and V. D. Bystrov)," *Arkh. Anat., Gistol., Embriol.* **42**, 22–37 (1962).

80. B. Szubinska, "Further observations on the morphology of the yolk in Teleostei," *Acta Biol. Cracov., Ser. Zool.* **4**, 1–19 (1961).

81. A. V. Timofeev and A. A. Neyfakh, "Expression of glucose-6-phosphate dehydrogenase gene in the development of the loach, *Misgurnus fossilis*," *Isozyme Bull.* **13**, 79 (1980).

82. A. V. Timofeev and A. A. Neyfakh, "The synthesis of glucose-6-phosphate dehydrogenase during the embryogenesis of the loach (*Misgurnus fossilis*)," *Isozyme Bull.* **15**, 57 (1982).

83. M. Ya. Timofeeva and K. A. Kafiani, "Nucleic acids of the loach unfertilized eggs and developing embryos," *Biokhimiya* **29**, 110–115 (1964).

84. M. Ya. Timofeeva, A. A. Neyfakh, and A. A. Strokov, "The escape of RNA in protein-synthesizing complexes of the cytoplasm in haploid loach embryos and haploid hybrid (loach × goldfish) embryos," *Genetika* **7**, 93–102 (1971).

85. A. S. Voronina, S. A. Bogatyrieva, A. A. Kostomarova, and N. N. Rott, "On the existence of free informosomes in the cytoplasm," *FEBS Lett.* **89**, 242–247 (1978).

Chapter 6

THE NEWTS *Triturus vulgaris* AND *Triturus cristatus*

L. D. Liozner and T. A. Dettlaff

Newts of different species are widely employed in developmental biology studies. The normal development of *Triturus vulgaris* has been studied in great detail and tables of normal development have been constructed [12]. The method of vital staining was developed on *T. vulgaris* and *T. alpestris* embryos, whereby the morphogenetic movements of cell layers were studied during gastrulation and neurulation, and fate maps (Fig. 6.1) were constructed [20, 35, 36]. Experiments involving separation and recombination of blastomeres were carried out on *T. vulgaris* embryos, and microsurgical methods were developed for the isolation and grafting of pieces of blastula, gastrula, and neurula. The phenomenon of primary embryonic induction was discovered in an experiment using heteroplastic grafting between the embryos of *T. vulgaris* and *T. cristatus*. Spemann [30] provides a complete review of these results. Later, the sources of inductive influences and stages of determination of the material of the neural plate [29] and of placodes of different sense organs (see [10, 11, 24]) were studied, and experiments with nuclear transplantation and immunochemical studies of differentiation were carried out on newt embryos [25–28]. Newt embryos are widely used to test the effect of heterogeneous inductors (see [23]).

Both larvae and adult newts are used to study regeneration, in particular that of the lens, retina, and other eye parts, and also the limbs, tail, nose, and internal organs (liver, lungs, spleen, gonads) [4, 5, 14, 16–19, 21, 28, 31, 38, 41].

However, during the last few decades *T. vulgaris* ceased to be one of the main subjects of developmental studies and has been replaced by such animals as *Pleurodeles waltlii, Ambystoma mexicanum, A. maculatum,* and *Cynops orientalis*. All these animals may be reared in the laboratory. Nevertheless, the results obtained using newt embryos remain important, and studies on them continue, although to a lesser extent. Due to the fact that in the experiments on *T. vulgaris* embryos, developmental stages were usually determined precisely enough by the tables of Glaesner [12], it is possible to compare them with those in the other species. The relative characteristics of the time of onset of successive developmental stages in *T. vulgaris* given here (Table 6.1) thus allows the use of this species for studying

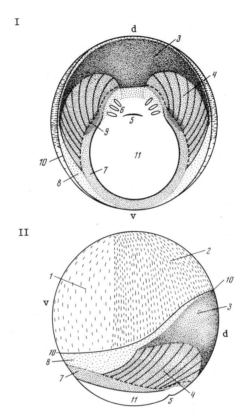

Fig. 6.1. Fate map for the newt embryo at the blastula–early gastrula stage (after Vogt [36]). I) View from dorsal blastopore lip; II) lateral view (right side). 1) Epidermis; 2) neural plate; 3) notochord; 4) somites; 5) dorsal blastopore lip; 6) gill pouches; 7) lateral plates; 8) tail; 9) forelimb; 10) boundary of invaginating material; 11) endoderm. d) Dorsal side; v) ventral side.

temporal patterns of embryogenesis as well (see [6–9]). The main experimental methods involving newt embryos are described in a number of monographs by Vorontsova et al. [37], Wilt and Wessels [40], Hamburger [15], Rugh [22], and Brodsky et al. [3].

6.1. TAXONOMY, DISTRIBUTION, AND REPRODUCTION

Triturus vulgaris L. (= *T. taeniatus*), *T. cristatus* Laur., and *T. alpestris* Laur. belong to the genus *Triturus*, family Salamandridae, order Urodela.

Smooth newts (*T. vulgaris*) are usually 8–11 cm long; the tail occupies half of the whole length. The back is olive-brown and the abdomen yellowish with small yellow spots. There are longitudinal dark bands on the head; the band apparently passing across the eyes is especially discernible.

TABLE 6.1. Stages of Normal Development of the Smooth Newt *Triturus vulgaris* (after Glaesner [12])

Stage No.	Time from appearance of first cleavage furrow hours, days	τ_n/τ_0*	Diagnostic features
1	0.25 h	0.12	2 blastomeres. Diameter (*D*) equals 1.2 mm
2	1.5 h	0.76	4 blastomeres
3	3.5 h	1.78	8 blastomeres
4	6 h	3.05	20 blastomeres, *D* = 1.3 mm
5	10 h	5.08	Morula, ~180 blastomeres. Numerous tangential divisions
6	16 h	8.14	Blastula. Surface smooth
7	20 h	10.17	Dorsal blastopore lip delineated as a crescentlike groove
8	24 h	12.2	Blastopore slit surrounded by accumulation of pigment. Shallow archenteron cavity
9	1 day, 6 h	15.8	Blastopore as a crescentlike slit
10	1 day, 10 h	17.3	Lateral blastopore lip formed. Blastopore occupies about 2/3 of circumference
11	1 day, 12 h	18.3	Embryo somewhat elongated. Ventral blastopore lip formed. Large oval yolk plug. Length (*L*) equals 1.4 mm
12	1 day, 18 h	21.4	Blastopore is smaller. Traces of blastocoele. Embryo becomes egg-shaped
13	2 days, 7 h	28.0	Yolk plug almost not seen. Outlines of neural plate and neural groove delineated
14	2 days, 20 h	34.6	Neural plate becomes fiddle-shaped
15	3 days	36.6	Neural folds elevated and neural plate narrower and intensively pigmented
16	3 days, 4 h	38.6	Neural folds approached closely by 3/5 of their length. Neural plate begins to sink. Embryo cylindrical, flattened ventrally
17	3 days, 6 h	39.6	Neural folds do not touch in anterior 3/5 of their length. Gill area appears
18	3 days, 8 h	40.6	A narrow slit between neural folds, slightly wider anteriorly. Gill area distinctly protruding. Eye vesicles appear

TABLE 6.1 (continued)

Stage No.	Time from appearance of first cleavage furrow		Diagnostic features
	hours, days	τ_n/τ_0*	
19	3 days, 12 h	42.7	Posterior end of embryo separated ventrally, neural folds fused all along. Primary cerebral and eye vesicles first discernible, as well as mandibular and branchial rudiments. Anus a small round opening. $L = 1.5$ mm
20	3 days, 20 h	46.8	Ventral side slightly concave. Three cerebral vesicles and eye vesicles clearly seen. Parietal and occipital curvatures are delineated
21	4 days	48.8	Body more elongated. Ventral side markedly concave. Tail rudiment is outlined. Hyoid arches discernible. Head somewhat separated
22	4 days, 12 h	54.9	Distinct tail bud. Parietal curvature very pronounced, occipital curvature outlined. Ventral side strongly concave. Hyoid arches outlined, ear pits appear. $L = 1.6$ mm
23	5 days	61.0	Body thinner, markedly curved ventrally. Ear vesicles and rudiments of pronephros clearly seen. $L = 1.7$ mm
24	6 days	73.2	First pair of branchial arches outlined. Olfactory pit discernible. $L = 1.8$ mm
25	6 days, 12 h	79.3	Embryo markedly elongated. 10–12 pairs of somites visible. First movements of embryo. Second pair of branchial arches formed. Olfactory pits and lens rudiments clearly seen. $L = 2.2$ mm
26	7 days	85.4	Pronounced body elongation. Tail curved and directed forward. Third pair of branchial arches outlined. $L = 2.4$ mm
27	8 days	97.6	Body elongation complete. Tail curved and directed vertically downward, dorsal fin fold appeared on it. Rudiments of balancers outlined. Pronephros rudiments protrude markedly. $L = 2.9$ mm
28	8 days	103.7	Tail slightly compressed on each side, its end directed backward. Fin fold reaches posterior third of back. Balancers longer than external gills. Heart visible. $L = 3.2$ mm

TABLE 6.1 (continued)

Stage No.	hours, days	τ_n/τ_0*	Diagnostic features
	Time from appearance of first cleavage furrow		
29	9 days	109.8	Tail almost fully straightened, compressed on each side to a greater extent. Fin fold distinctly seen. Pigmentation intensified. Two longitudinal pigment bands formed in dorsolateral area. Gills distinctly seen. $L =$ 3.6 mm
30	9 days, 12 h	116.0	Gills as short processes. Tail completely straightened and strongly compressed on each side. In front of eyes pigment spreads ventrally. Embryo within the envelopes curved in a ring. Heart beat. $L = 3.8$ mm
31	10 days	122.0	Balancers and gills present as short cylindrical processes. Forelimbs as small vesicles. Fin reaches the posterior back half. Second pigment band distinctly seen on flank at the level of pronephros. Egg envelopes stretched. $L = 4.4$ mm
32	11 days	134.2	Buds of forelimbs protrude farther. Gills and balancers as thick short processes. Pulsation of gill vessels. Fin reaches 2/3 along length of back on dorsal side and anus on ventral side. A slit of eye cup is seen. $L = 4.8$ mm
33	12 days	146.4	Body constricted on each side. Rudiments of forelimbs as short processes. Gill processes overlap. Rudiments of secondary gill processes are first outlined. Fin is higher. Pigment is localized mainly in two longitudinal bands, two spots on lower head surface, and two oblique bands in the heart area. Iris pigmented. No eye slit seen. $L = 5.4$ mm
34	13 days	163.2	Gill processes longer, all with one branch, in some places the second one is delineated. Few pigment cells on gills. Fin higher, spreads in front of cloaca. Yolk vessels clearly seen. Lens transparent, pupil dark. $L = 6.0$ mm
35	14 days	170.8	Distinct second gill branches seen, opercular fold and mouth opening outlined ventrally to gills. Head round or flask-shaped, somewhat constricted in gill area. Large subintestinal vein seen. Accumulation of pigment in the heart area. Stellate bright regions in iris. $L = 6.3$ mm

TABLE 6.1 (continued)

Stage No.	Time from appearance of first cleavage furrow hours, days	τ_n/τ_0*	Diagnostic features
36	15 days	183.0	Forelimbs much longer. Yolk markedly reduced. Lateral mouth margin distinctly outlined. Almost all iris of golden shade. Labyrinth otoliths visible. Pigmentation of upper longitudinal band very intensive. $L = 6.6$ mm
37	16 days	195	Gills displaced dorsally. Gill branches longer. Opercular folds outlined more distinctly. Upper pigment bands begin to divide into separate pigment accumulations. Hatched larva. $L = 6.9$ mm

*The values of τ_n/τ_0 in the paper by Dettlaff [8] should be neglected since their calculation was based on the assumption that Glaesner [12] incubated the embryos at 15–16°R, whereas they were actually incubated at 15–16°C, as determined by comparing the data of Glaesner with those of Neyfakh and Gorgolyuk (see [2]).

Crested newts (*T. cristatus*) are 14–18 cm long. The back is black or brown-black, the abdomen orange with black spots. The females are distinguished by the presence of a thin mid-dorsal yellow line.

Alpine newts (*T. alpestris*) are 5–6 cm long and have a low even ridge turning into a tail fold. The back is dark brown or bluish-grey, the abdomen orange without spots. During the period of reproduction the flanks become blue.

In the USSR, there also occur the Carpathian (*T. montandoni* Boul) and banded (*T. vittatus* Jen.) newts. These two species occur, like the Alpine newt, in limited areas: the Carpathian and Alpine newts in the Carpathians and adjacent regions of West Ukraine, and the banded newt in the West Caucasus.

Smooth and crested newts are much more widespread. The smooth newt occurs in the European part of the USSR and West Siberia (up to the Abakan mountain ridge), as well as in the Aral territory and near Lake Balkhash; the crested newt occurs in the central regions of the European part of the USSR, in the Urals, in the Crimea and West Caucasus (see [1, 31]).

Newts occur in broad-leaved and mixed forests, parks, and shrubs, and live for most of the year on land, where they are active and feed mainly on earthworms and insect larvae during the night and when it rains. They hibernate in the burrows of rodents, in cellars and basements, and under fallen leaves, where they usually go in October. They become sexually mature during the 2nd–3rd year of life, and during the period of spawning occur in ponds of different size, small lakes, swamps, and channels overgrown with grass plants. The crested newt usually prefers bodies of water larger than the smooth newt, but they can occur together. They appear in the water early in spring, usually in March–April, the smooth newt even earlier. When in water during the spring they have a nuptial pattern and coloration. The smooth newt males acquire a scalloped dorsal ridge with orange fold and bluish band of pearly shade. Lobed folds appear on the digits of the hind limbs. The color of the females becomes somewhat brighter. The crested newt males acquire a dentate ridge and bluish-white bands on each side of the tail. During the period of spawning the males lay spermatophores, which are accumulations of spermatozoa surrounded by a bell-shaped jelly coat, on substrate. The females pick up the spermatophores with the cloacal lips. They lay 60–700 eggs, attaching them to the leaves of aquatic plants, and fold the leaf in such a way that the egg is placed in the fold angle. But the eggs can be laid on the plants without leaves, or simply in water. After spawning, the newts come out onto land.

To obtain the eggs in a laboratory, the newt males and females displaying nuptial coloration are caught in the beginning of the spawning period. They are captured by dip nets of tulle or gauze on long handles, as they can occur at considerable depths. When caught before spawning and placed into an aquarium with water plants, the newts soon start to spawn. It is essential that they be not subjected to high temperatures during this period. The recent experience of the Working Group on Breeding Rare and Disappearing Amphibian Species in Captivity has shown that *Triturus cristatus* and *T. vittatus*, when kept in captivity for a long time and given

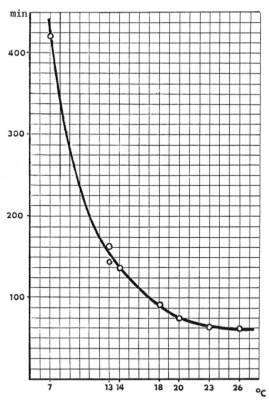

Fig. 6.2. Dependence of the duration of one mitotic cycle during synchronous cleavage divisions (τ_0) on temperature in *Triturus vulgaris* (data of Valouch et al. [34]; see Dettlaff [8]).

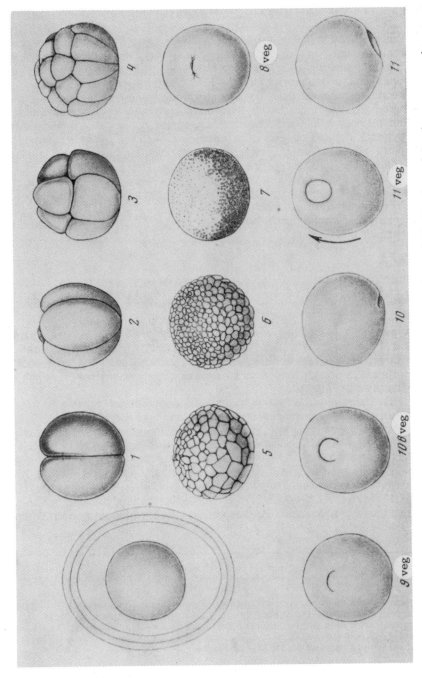

Figs. 6.3–6.7. Normal development stages of the smooth newt, *Triturus vulgaris* [12]. Numbers of drawings correspond to stages. veg) Vegetal view; d) dorsal view; v) ventral view; h) head view; t) tail view. Drawings without letter designations are lateral views.

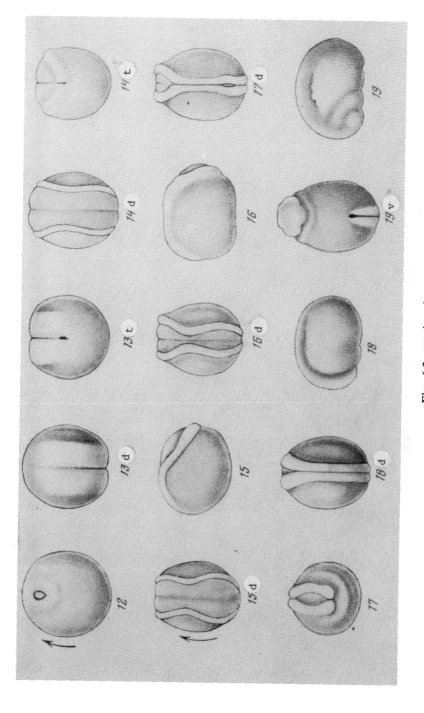

Fig. 6.3, continued.

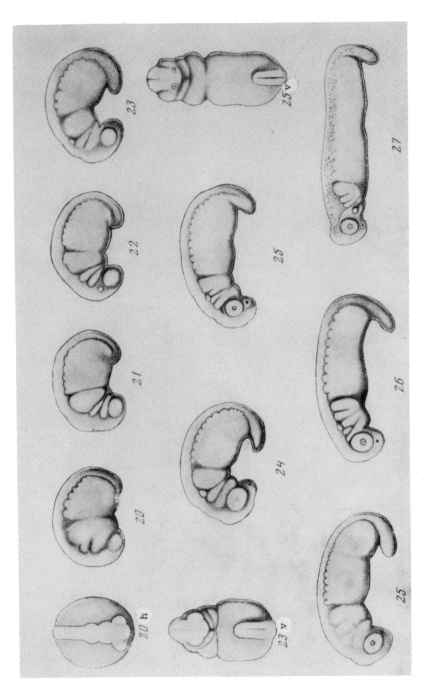

Fig. 6.4.

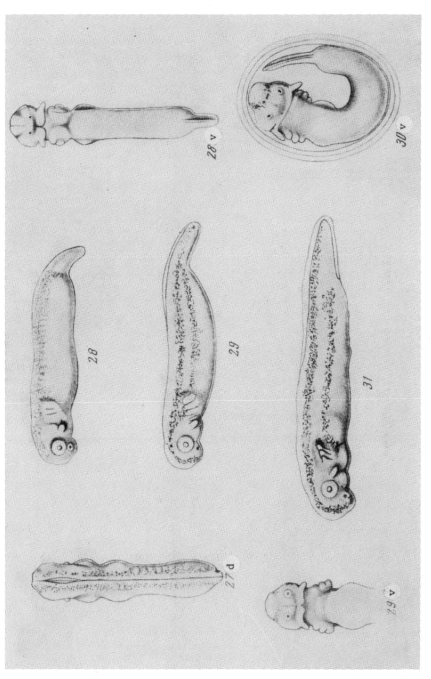

Fig. 6.5.

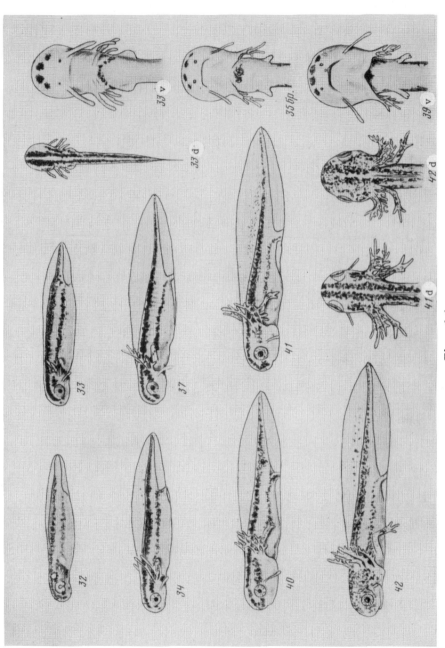

Fig. 6.6.

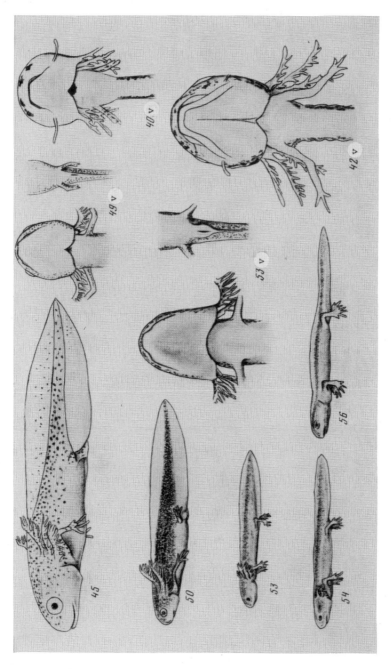

Fig. 6.7.

sufficient food and hibernated in humid moss even for a short time (1–2 months) at 4°C, lay normal fertilized eggs after hormonal stimulation. Human chorionic gonadotropin (100 units/individual with 48-h intervals) or a synthetic analog of luliberin (1–10 μg/individual with 48-h intervals) was used as stimuli. Injections should be given first to the males and then, after they have demonstrated nuptial behavior, to the females. The females lay eggs usually within 4–5 days after the first injection, and spawning occurs practically every day for 20–25 days. The water temperature in the aquarium should be 12–17°C.

6.2. STRUCTURE OF GAMETES AND FERTILIZATION

Newt eggs are spheroid, 1.5 to 3 mm in diameter. They are moderately telolecithal since they contain relatively little yolk. The cytoplasm of the upper, animal region of the egg contains few small yolk granules; their amount and size increase toward the lower, vegetal pole. The nucleus is located at the animal pole. The animal pole is pigmented in the smooth newt and colorless in the crested newt. The yolk imparts a yellowish shade to the vegetal pole. The egg is covered by three envelopes. The internal vitelline membrane adjoins the cytoplasm surface; it is thin and transparent. Two other membranes are jellylike and swell in water. The denser outer coat forms an elastic capsule around the egg.

Fertilization is internal and polyspermic. The spermatozoa penetrate the egg somewhat below the equator, and the place of their penetration remains visible for some time as a small depression. The number of the spermatozoa which have penetrated can be estimated by the number of depressions.

6.3. REARING LARVAE IN A LABORATORY

To rear newts in a laboratory, plants bearing attached eggs or the eggs alone are placed in wide low vessels with water at 10–15°C. The embryos should be protected against direct sunlight and overheating of the water. The water should be changed and dead embryos removed. After hatching, the larvae are fed first with infusorians and then with crustaceans (*Cyclops, Daphnia*). The most difficult period of rearing the newts is just after metamorphosis. During this period smooth newts eat very little and are in poor condition. They get out of the water and can dry out. Crested newts can remain in the water after metamorphosis and survive better. For the period of metamorphosis and the subsequent period the newts are placed in a terrarium with wet sand or moss. They are fed with small worms or larvae. Adult newts are given small earthworms and midge larvae.

Figs. 6.8–6.10. Stages of forelimb development in *Triturus vulgaris* (Figs. 6.8 and 6.9) and *Triturus cristatus* (Fig. 6.10) (after Glücksohn [13]; reproduced from Vorontsova et al. [37]). Drawing numbers correspond to stage numbers 36–61.

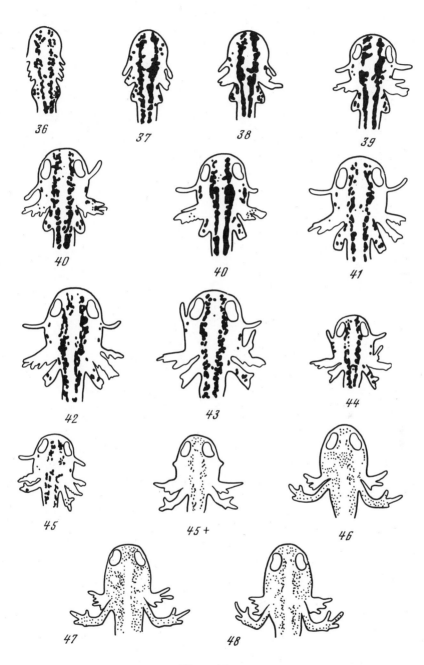

Fig. 6.8.

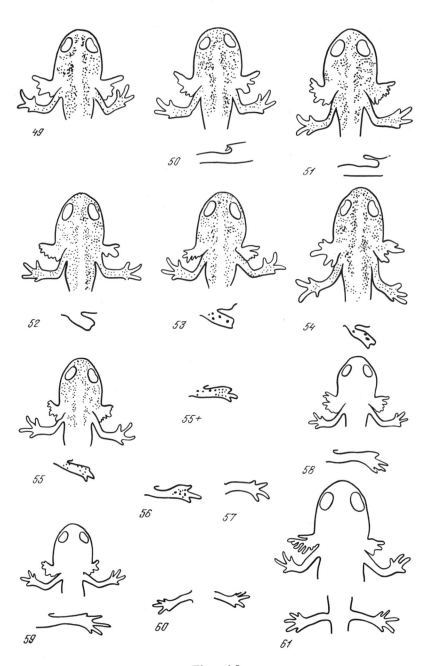

Fig. 6.9.

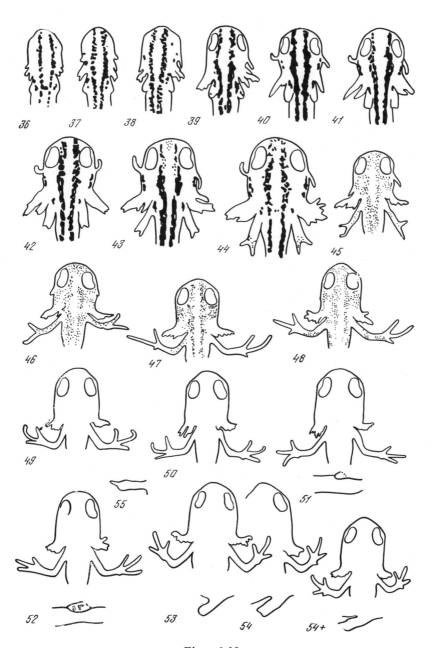

Fig. 6.10.

6.4. THE NORMAL DEVELOPMENT OF *Triturus vulgaris*

The normal development of *T. vulgaris* was studied in detail by Glaesner [12]. He described the external and internal structure of newt embryos at successive developmental stages from fertilization until metamorphosis. These stages were illustrated by corresponding drawings, which are reproduced in Figs. 6.3–6.7.

Table 6.1 lists the stages, from fertilization until hatching, the time of their onset, and gives a brief description of the external features of the embryos at these stages. (For the internal structure of the embryos and larvae at successive developmental stages, see Glaesner [12].)

In addition, the curve shown on Fig. 6.2 expresses the dependence of τ_0 (the duration of one mitotic cycle during the period of synchronous cleavage divisions (see Dettlaff and Dettlaff [9])) on temperature, using, for this volume, the data of Valouch et al. [34] and Neyfakh and Gorgolyuk (see [2]) on the duration of the interval between the appearance of the 1st and 2nd cleavage furrows equal to τ_0. It can be seen that all values of τ_0 fit the curve quite well. With the aid of this curve and the data on the time of successive developmental stages τ_n (Table 6.1) at 15–16°C, one can calculate the tentative time of these stages at the other optimal temperatures. Experimental data on the duration of different developmental periods at various temperatures (8, 12, 16, 30, and 24°C) were also obtained for the *T. vulgaris* embryos [2]. It was also shown that the durations of the early developmental periods (cell cycle during synchronous cleavage divisions τ_0, and periods from the first cleavage stage to the onset of gastrulation and to the end of gastrulation) change in proportion with temperature within the limits from 8 to 20°C and, hence, are measured by an equal number of τ_0 units ($\approx 15\tau_0$ to the onset of gastrulation, stage 8, and $\approx 28\tau_0$ to the end of gastrulation–onset of neurulation, stage 12–13).

To obtain the temporal characteristics of the developmental periods of *T. vulgaris* at different temperatures and in a form comparable to the other animals, Table 6.1 presents the time of onset of different stages (τ_n) not only according to Glaesner, in hours and days at 15–16°C, but also in number of τ_0 units (τ_n/τ_0). To obtain the latter data, the temperature of incubation was assumed to be 15.5°C and τ_0 equal to 118 min (see Fig. 6.2). One has to take into account, however, that in the experiments of Glaesner [12] the temperature varied from 15 to 16°C (and τ_0, respectively, from 120 to 106 min), and that the values of τ_n/τ_0 indicated in Table 6.1 can vary.

In addition to Glaesner's tables, the tables of normal development of the smooth and crested newts after Glücksohn [13] are also included (Figs. 6.8–6.10). The numbers of stages after Glücksohn do not coincide with those of Glaesner, but are widely used in the literature.

A normal table has recently been obtained for *Hynobius keyserlingii* (Dyb.); drawings of embryos at successive developmental stages, and a brief description of external features of the embryos, have been given in [32]. The time of embryonic development at different temperatures was also determined, and the τ_0 versus temperature curve was plotted [2]; this allows a comparison between the rates of development in *H. keyserlingii* and *T. vulgaris*.

REFERENCES

1. A. G. Bannikov, I. S. Darevsky, and A. K. Rustamov, *Amphibians and Reptiles of the USSR* [in Russian], Mysl, Moscow (1971).
2. D. J. Berman, S. I. Gorgolyuk, and A. A. Neyfakh, "Dependence of the rate of embryogenesis on temperature in *Hynobius keyserlingii* and *Triturus vulgaris*," *Ontogenez* **18**, 247–256 (1987).
3. V. Ya. Brodsky, G. G. Gause, and T. A. Dettlaff, eds., *Methods in Developmental Biology* [in Russian], Nauka, Moscow (1974).
4. B. M. Carlson, "The regeneration of skeletal muscle. A review," *Am. J. Anat.* **137**, 119–150 (1973).
5. B. M. Carlson, "Factors controlling the initiation and cessation of early events in the regenerative process," in *Neoplasia and Cell Differentiation*, G. V. Sherbet, ed., Phiebig, New York (1974), pp. 60–105.
6. T. A. Dettlaff, "The species differences in morphogenetic properties of the germ material and the shift of gastrulation processes with respect to cleavage stages (the problem of biological age of the embryo as an interrelation of developmental stages and cell generations)," *Dokl. Akad. Nauk SSSR* **111**, 1149–1152 (1956).
7. T. A. Dettlaff, "Cell divisions, duration of interkinetic states, and differentiation in early stages of embryonic development," *Adv. Morphogen.* **3**, 323–362 (1964).
8. T. A. Dettlaff, "Relative characteristics of development duration in *Triturus vulgaris*," *Ontogenez* **15**, 311–314 (1984).
9. T. A. Dettlaff and A. A. Dettlaff, "On relative dimensionless characteristics of the development duration in embryology," *Arch. Biol.* **72**, 1–16 (1961).
10. D. P. Filatov, *Comparative-Morphological Direction in the Mechanics of Development, Its Object, Aims, and Course* [in Russian], Izd. Akad. Nauk SSSR, Moscow (1939).
11. A. S. Ginsburg, "Specific differences in the determination of the internal ear and other ectodermal organs in certain Urodela," *C. R. (Dokl.) Acad. Sci. URSS* **44**, 557–560 (1946).
12. L. Glaesner, *Normentafeln zur Entwicklungsgeschichte des gemeinen Wassermolches (Molge vulgaris)*, Fischer Verlag, Jena (1925).
13. S. Glücksohn, "Aeussere Entwicklung der Extremitaten und Stadieneinteilung der Larvenperiode von *Triton taeniatus* Laid. und *Triton cristatus* Laur.," *Wilhelm Roux's Arch. Entwicklungsmech. Org.* **125**, 341–405 (1931).
14. E. Guyénot, "Le problème morphogénétique dans régénération des Urodèles, détermination et potentialités des régénérats," *Rev. Suisse Zool.* **34**, 127–154 (1927).
15. V. Hamburger, *A Manual of Experimental Embryology*, University of Chicago Press, Chicago (1960).
16. J. R. Keefe, "An analysis of urodelian retina regeneration," *J. Exp. Zool.* **184**, 185–257 (1973).
17. G. V. Lopashov and O. G. Stroeva, *Development of the Eye. Experimental Studies*, Israel Program for Scientific Translations, Ltd., Jerusalem (1964).

18. V. I. Mitashov, "An autoradiographic assay of melanin synthesis in the pigment epithelium of newts," *Ontogenez* **11**, 246–250 (1980).
19. R. W. Reyer, "The amphibian eye: development and regeneration," in *Handbook of Sensory Physiology*, Springer-Verlag, Berlin (1977), pp. 309–390.
20. K. Röhlich, "Gestaltungsbewegungen der präsumptiven Epidermis während der Neurulation und Kopfbilding bei *Triton taeniatus*," *Wilhelm Roux's Arch. Entwicklungsmech. Org.* **124**, 66–81 (1931)
21. D. Rudnick, ed., *Regeneration,* 20th Symposium, Society for the Study of Development and Growth, New York (1962).
22. R. Rugh, *Experimental Embryology*, Burgess Publishing Company, Minneapolis (1962).
23. L. Saxen and S. Toivonen, *Primary Embryonic Induction*, Logos Press (1962).
24. O. I. Schmalhausen, "A comparative experimental study of the early developmental stages of olfactory rudiments in amphibians," *Dokl. Akad. Nauk SSSR* **74**, 863–865 (1950).
25. F. Sládecek and Z. Mazáková-Stefanová, "Nuclear transplantation in *Triturus vulgaris*," *Folia Biol.* **10**, 152–154 (1964).
26. F. Sládecek and Z. Mazáková-Stefanová, "Intraspecific and interspecific nuclear transplantations in *Triturus*," *Folia Biol.* **11**, 74–77 (1965).
27. F. Sládecek and A. Romanovsky, "Species-specific antigens in nuclear transplantation in newts," in *Genetic Variations in Somatic Cells*, Publishing House of the Czechoslovak Academy of Sciences, Prague (1965), pp. 183–187.
28. F. Sládecek and A. Romanovsky, "Ploidy, pigment patterns, and specific nuclear transplantations in newts," *J. Embryol. Exp. Morphol.* **17**, 319–330 (1967).
29. F. Sládecek, J. Suhana, J. Melichna, and P. Valouch, "The number of cell generations in relation to the interspecific differences in primary embryonic induction in amphibians," *Folia Biol.* **16**, 66–70 (1969).
30. H. Spemann, *Embryonic Development and Induction*, Julius Springer, London (1938).
31. O. G. Stroeva and V. I. Mitashov, "Retinal pigment epithelium: proliferation and differentiation during development and regeneration," *Int. Rev. Cytol.* **83**, 221–293 (1983).
32. L. A. Sytina, I. M. Medvedeva, and L. B. Godina, *Development of Hynobius keyserlingii* [in Russian], Nauka, Moscow (1987).
33. L. V. Terentiev and J. A. Chernov, *A Key to Reptiles and Amphibians* [in Russian], Sov. Nauka, Moscow (1949).
34. P. Valouch, J. Melichna, and F. Sládecek, "The number of cells at the beginning of gastrulation depending on the temperature in different species of amphibians," *Acta Univ. Carol., Biol.* **1970**, 195–205 (1971).
35. W. Vogt, "Gestaltungsanalyse am Amphibienkeim mit örtlicher Vitalfärbung. I. Methodik und Wirkungsweise der örtlicher Vitalfärbung mit Agar als Farbträger," *Wilhelm Roux's Arch. Entwicklungsmech. Org.* **106**, 542–610 (1926).
36. W. Vogt, "Gestaltungsanalyse am Amphibienkeim mit örtlicher Vitalfärbung. II. Gastrulation und Mesodermbildung bei Urodelen und Anuren," *Wilhelm Roux's Arch. Entwicklungsmech. Org.* **120**, 384–706 (1929).
37. M. A. Vorontsova, L. D. Liosner, I. V. Markelova, and E. Ch. Pukhalskaya, *The Newt and the Axolotl* [in Russian], Sov. Nauka, Moscow (1952).

38. M. A. Vorontsova and L. D. Liozner, *Asexual Reproduction and Regeneration* [in Russian], Sov. Nauka, Moscow (1957).
39. H. Wallace, *Vertebrate Limb Regeneration*, Wiley, New York (1981).
40. F. H. Wilt and W. K. Wessels, eds., *Methods in Developmental Biology*, Crowell, New York (1967).
41. T. Yamada, *Control Mechanism in Cell-Type Conversion in Newt Lens Regeneration*, S. Karger, Basel (1977).

Chapter 7

THE SPANISH NEWT *Pleurodeles waltlii*

S. G. Vassetzky

Pleurodeles waltlii is one of the most widespread and convenient subjects for developmental biology studies. Its advantages were widely recognized after L. Gallien published his paper on the conditions for keeping, breeding, and rearing this amphibian in the laboratory [127].

The popularity of *P. waltlii* is accounted for by its evident advantages for experimental studies. Among these are the rapid rate of embryogenesis, the rapid completion of metamorphosis and sexual maturation, the ease with which polyploid individuals are obtained and the possibility of maintaining polyploid strains, the presence of mutations and chromosomal aberrations which are easily identifiable (cytologically or morphologically) and can be used as markers, and, finally, the possibility of experimental phenotypical sex inversion and hence the prospects of studying the mechanisms of sex determination.

Experimental embryological studies of *P. waltlii* form a significant part of the published data. Using vital staining, Vogt [245] elaborated the fate map for the blastula and gastrula stages, which was later modified by Pasteels [205] (see Fig. 6.1, Chapter 6, this volume). Studies of the competence and determination of early embryo regions were carried out with the help of markers (radioactive isotopes, chromosomal markers) and grafting of individual regions of the early embryo [27, 40, 41, 47, 48, 50, 58, 62, 213]. Closely related to these studies are the experiments on parabiosis and the grafting of individual tissues with the aim of studying compatibility [130, 152, 202, 241].

Many studies have been devoted to the development of individual organs and tissues. These include: primary germ cells and morphogenesis of the urogenital system [1, 46, 48, 53, 90–92, 151, 153, 155, 193, 194, 209, 211], development of the heart and circulatory system [10–12, 14, 80, 84, 85, 87, 111, 114–119, 223], skeleton [195], cartilage and bone formation [33, 34, 70–75, 214], head region [49, 50, 226, 227], skin [13], teeth [49, 59], limbs [183–187], nervous tissue [248–250], as well as studies on regeneration [182, 186].

One of the oldest areas of research is directed phenotypical sex inversion under the effect of sex steroids. The first studies in this field were carried out at the beginning of the 1950s [68, 125, 126, 129, 131, 134]. After the treatment of larvae with female steroid hormones, their sex is fully reversed: genotypical males become phenotypical females. Recently, sex inversion has also been obtained using elevated temperatures [93, 156–158].

The problem of heteroploidy and heteroploidization has also been intensively studied [12, 19, 21, 23, 25, 51, 52, 82, 87, 88, 94, 99, 109, 120, 128–132, 140, 162]. Closely related is research on artificial andro- and gynogenesis ([24, 106, 133, 165]; see also [215]).

Studies on nuclear transplantation in *P. waltlii* embryos were initiated by Signoret and Picheral [229] and followed immediately by a large series of experiments on transplantation of embryonic and larval nuclei into unfertilized enucleated eggs ([2–4, 35, 139, 207]; see also [168]). A method of mass extraction of the oocyte nuclei (germinal vesicles) has been developed [198]. Of great interest are studies on interspecific nucleocytoplasmic hybrids [4, 122–124, 149, 190].

P. waltlii has been experimentally crossed with *P. poireti* and representatives of other genera of urodels, and in some cases viable interspecific and intergeneric hybrids were obtained [30, 107–109, 124, 169, 235, 236].

The effect of various agents on embryogenesis of *P. waltlii* has also been studied. These include low temperature [19, 22], ionizing radiation [7, 60, 138, 172], chemical substances [28, 45, 63, 77, 187, 225, 231], and other factors [5, 20]. These investigations were carried out at the cytological, morphological, submicroscopical, and karyological levels.

A number of studies have been devoted to the development of the innervation of grafted nervous tissue and limbs and the "model of the nervous system" [237–240].

Cell differentiation during normal development and under the effect of various environmental influences has also been intensively studied [95–102, 144, 181, 191, 192, 194].

Significant studies have been made on the metabolism of hormones and the activities of the glands of internal secretion, namely: sex hormones and gonads [8, 54, 140, 153, 154, 199, 200, 203, 204], thyroid gland [66, 67, 200, 201], thymus [57, 86, 155, 164], and interrenal gland [103].

Biochemical investigations have been devoted to nucleic acid synthesis during oogenesis [39, 43, 112, 113, 219–222] and embryogenesis [6, 17, 18, 78, 170, 175]. Nucleic acid content has been studied in the mitotic chromosomes of embryonic cells [15, 16] as well as the synthesis of proteins [76, 79, 149], including a special group of serum proteins [55, 56, 141, 150, 212].

Other studies include the dynamics of nuclear processes in fertilized and activated eggs [171, 173], and of the mitotic cycle in the embryos [171, 173] and larvae [61, 64, 65].

7.1. TAXONOMY, KARYOTYPE, AND MUTATIONS

The Spanish newt *Pleurodeles waltlii* Michahelles (= *Molge waltlii*) belongs to the subfamily Salamandrinae, family Salamandridae, suborder Salamandroidea, order Caudata (= Urodela), class Amphibia, and was first described in 1830 [196]. It occurs in Spain and Portugal, as well as in Morocco, where it was first described in 1874 [38]. *P. waltlii* is widely used in the USSR as a laboratory animal.

There exist two races of *P. waltlii*, the Iberian, or European, and the Moroccan. The two races will interbreed and, to judge by all their features, apparently belong to the same species. The genus *Pleurodeles* includes one more species: *P. poireti* Gervais.

The diploid set comprises 24 chromosomes [121]. The karyotype was first studied in detail on spermatocytes [247] and on cells of larval epidermis [31] and was described systematically on the basis of studying the cells of the tail epidermis [173] (Fig. 7.1A). The chromosomes of *P. waltlii* can be divided into three groups which differ both in relative size and centromere index (ratio of long arm to total length). Chromosomes 3 and 11 carry small satellites on both their short and their long arms. A method was developed which allows mitoses to be obtained in a culture of leucocytes using 2–3 drops of blood of the metamorphosed animal, and thus allows the karyotype of a live animal to be studied [159, 189]. A more detailed study of the mitotic chromosomes by histochemical methods revealed more intensely stained (Giemsa) regions of chromosomes, the positions of which are constant and characteristic for different elements of the karyotype [176].

Lampbrush chromosomes have been studied in detail in *P. waltlii* oocytes [178]. They are characterized by many topographical peculiarities, mainly loops. Five types of loops and two types of other structures have been described. All the lampbrush chromosomes were fully described and their map was composed (Fig. 7.1B). These data can be used for the genetic analysis of a number of chromosomal aberrations and for the search for spontaneous mutations at the chromosomal level. There are some data which suggest that a more refined genetic analysis is possible on these chromosomes, as compared with mitotic chromosomes [179]. Recently, mechanisms of transcription were studied in lampbrush chromosomes [39, 167, 219–222, 234].

It is established that in *P. waltlii* the female sex is heterogametic (ZW) and the male sex homogametic (ZZ) ([131]; see also [216]).

Numerous investigations have revealed several mutations and chromosomal aberrations in *P. waltlii* [135, 174]. They are divided into two groups: spontaneous, and those induced by environmental influences such as irradiation, heating, or nuclear transplantation. The first group includes a recessive autosomal mutation *lm* (lethal–mitotique), which is lethal in the homozygous state [136]. It is characterized by a large number of abnormal mitoses at the tailbud stage, resulting in the appearance of pyknotic nuclei and mass death of the embryos at hatching. Beetschen and Jaylet [32] described a recessive autosomal mutation *ac* (ascite caudale) which, in the homozygous state, is characterized by the formation of edema in the tail region. As a result, general edema develops in most parts of the mutant embryo, and defects of blood circulation arise resulting in its death. However, about a quarter of the embryos retain viability. This mutation has been quite extensively studied [26, 29, 37, 104, 105]. A recessive autosomal mutation *u* (ulcer), lethal in the homozygous state, was also described in *P. waltlii* [230]. In the homozygous mutants, the intestine and the abdominal surface become ulcered and, as a result, the larvae die soon after hatching. All these mutations manifest themselves during embryogenesis. However, there is a mutation resulting in a change of pigment pattern of the larvae [180]. This recessive mutation is characterized by the absence of iridiophores and an increased number of equally distributed melanophores. Finally, at the chromosomal level six types of trisomies, two types of double trisomies [145–148, 177], and a triple translocation [69] have been found.

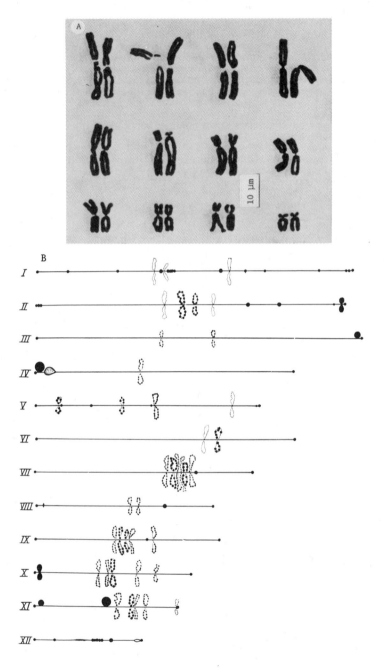

Fig. 7.1. Karyotype of *Pleurodeles waltlii*. A) Mitotic chromosomes in the larval tail epidermis; B) lampbrush chromosomes in oocytes [173, 178].

The second group, induced chromosomal aberrations, includes three translocations and one deletion found in the somatic cells of larvae developed from γ-irradiated eggs [173, 174], six chromosomal aberrations (translocations and deletions) in
the progeny from intact females and males with *in situ* irradiated sperm [163], a
heterozygous reciprocal translocation induced by γ rays [175], and a defect of
melanogenesis induced by x rays and determined by a recessive factor [166].

In addition, there are two strains of animals homozygous by a reciprocal
translocation [160] and a pericentric inversion [161]. These strains were obtained
as a result of studying chromosomal aberrations in the progeny from intact females
and irradiated males. The rearrangements are easily classified under the microscope
and can be used as markers in various experimental studies.

7.2. SEXUAL DIMORPHISM

P. waltlii is the largest of the European newts, attaining a length of 30 cm [42];
Gallien provides the following data [123]:

Age	Length, mm	Weight, g
Stage of metamorphosis	69.6 (25)*	–
9 months	122.0 (31)	–
1 year	158.9 (22)	–
2 years		
males	170.0 (25)	24.3
females	187.0 (25)	39.1

*In parentheses – the number of individuals measured.

Sexual differentiation begins before metamorphosis. After metamorphosis the
sex of animals can be distinguished in sections of gonads [206]. The gonads can
be distinguished macroscopically between 6 and 7 months when the animals attain
85–90 mm in length. The animals become sexually mature at the age of 11–12
months when, in exceptional cases, mating can be observed. However, the first
egg laying is usually obtained at the age of 16 months.

The adult newts have been described in detail by Vorontsova et al. [246]. They
can live in water throughout the whole year, although they do not avoid the land.
In particular, *P. waltlii* is described in Morocco as an aquatic animal [127]. The
Spanish newt reproduces twice a year: in February–March and in July–August.
The male is characterized by a more spotted pattern and more powerful forelimbs on
which a callous body can be seen during the period of sexual activity. During this
period the cloaca is moderately but distinctly swollen, whereas it is almost invisible
during the period of sexual rest.

The female is more bulky. Its limbs are shorter and more spotted, at least in
young animals. At sexual maturity, the trunk in the arm region becomes swollen,
bulky, and no longer reveals the constriction characteristic of immature females.
These features allow sufficiently reliable selection of females capable of egg laying.
The cloaca is usually slightly protruded; it is hypertrophied, but always moderately
as compared with the other newts, during the period of mating and egg laying, and
then returns rapidly to the initial state.

7.3. MANAGEMENT OF REPRODUCTION IN THE LABORATORY

7.3.1. Conditions for Keeping Adults

Adult animals are kept in intensely aerated aquaria (50 × 50 × 25 cm) with a slow water flow; 10 animals may be kept in an aquarium of this size. The bottom is covered with stones large enough for washing. Smooth cement basins (1 × 1 m) can also be used. It is important to have the maximum amount of space. If reproduction is to be controlled, the males and the females are kept separately.

Minced fresh meat is given as food twice a week. The meat is left in the aquaria for about 8 h and thereafter the meat residues, molting residues, and excrements are siphoned out. The early larvae are fed during the first days of life on live *Cyclops*. Then they are fed on chopped and later intact water midge larvae and, finally, pieces of meat [218]. All aquaria are periodically emptied and cleaned. The temperature is maintained at 16–20°C. The aquarium room should be light, but there is no necessity for special illumination or temperature control.

7.3.2. Reproductive Behavior and Collection of Eggs

Under the conditions indicated above, the males are always ready for reproduction from September until May. The callous body is distinct and the cloaca is hypertrophied. In May the males terminate their sexual activity and in May–June spermatozoa do not develop in the testes [206].

To obtain egg laying, a pair of newts is placed in an aquarium (50 × 50 × 25 cm) partitioned off by glass walls. Fertilization is internal and the newts reproduce by means of amplexus. If the male and female are placed together they manifest nuptial behavior very rapidly. Sometimes mating is observed within less than 1 h. The male approaches the female, touches her snout, and then embraces and holds the forelimbs of the female with his own forelimbs. The female remains passive. The pair moves from time to time, but can remain immobile for several minutes. During mating, the male often twists and even embraces the body of the female with his tail or sometimes his hind limbs.

Spermatophores can often be seen in the cloacal lips of the female, but the release of spermatophores by the male is rarely observed. It is observed more often in September, when males are placed together with females after a long period of rest. At that time, rectangular transparent jelly masses 10 mm wide are often found attached to the stones; there are whitish opaque packets on their angles. Sometimes only one of the angles carries such a packet. The male lays up to 6–7 spermatophores. The female picks them up with her cloacal lips. After September, although males continue to mate and fertilize females, the release of spermatophores is observed extremely rarely. The laying of spermatophores is sporadic and takes place at the beginning of the period of sexual activity and for only a short time.

Egg laying usually begins in sexually mature animals within 24–48 h after mating, and lasts for about 48 h. The female, liberated from the male, looks for a stone, plant branch, or some other substrate in the aquarium such as a drainage pipe, on which to attach the eggs laid (10–20 eggs in each cluster). One can see the animal touching the stones with the semi-opened lips of the swollen cloaca. At the moment of egg laying the female bends her back and stretches her hind limbs backward, gliding them along the tail and thus promoting egg laying. After the eggs are laid, the female regains the usual posture and moves around until the next egg laying, which occurs within 10–30 min.

Eggs are easily collected if plant branches are placed in the aquarium. When egg laying terminates, the males are transferred back to the other males and the females placed together in groups – for example, all females which laid eggs during one month. After abundant feeding they again become capable of egg laying within 1 to 2 months. Sometimes a female which laid eggs after mating lays further fertilized eggs after a delay of up to 6 to 8 weeks without further mating. Such eggs can be fertilized by spermatozoa from the original mating, which maintain their fertilizing capacity in the pelvic gland cuts for a long time [188].

The total number of eggs laid varies. During the first egg laying, a young female lays about 150 eggs, but their number increases with the age of the female. Usually, 400–800 eggs can be collected, and old females (at the age of 4 years) lay 700–800 eggs. Several hundred eggs can be obtained from one female in the course of 2 days.

Some females do not lay eggs at all, despite mating. If females capable of egg laying are needed, they are induced to lay eggs by mating 2–3 times a year even when eggs are not needed. The liberation of mature eggs from the ovaries maintains their good functional condition.

7.3.3. Obtaining Gametes and Artificial Insemination

Unfertilized eggs are obtained from the females stimulated by intramuscular injections of anterior pituitary at a dose of 20 units [228]. If females are placed with males after the injection, fertilized eggs can be obtained as well.

A suspension of sperm is obtained by homogenization of vasa deferentia in dechlorinated tap water [30]. The unfertilized eggs are immersed in the suspension. When sperm penetration can be detected on the surface of the eggs they are transferred into clean water and washed. The development of such fertilized eggs is quite normal, although fertilization does not reach 100% [30].

The application of electrical discharge from a capacitor [228] and pricking of the oocyte [175, 6] are convenient methods of activation. Studies on *in vitro* oocyte maturation were also carried out. It was believed that *in vitro* oocyte maturation in a progesterone solution was possible only after hormonal pretreatment of the females [44]. Oocytes were taken from the females preinjected with pregnyl and placed into a Ringer's solution with progesterone (1 ng/ml) for 1 h. Under these conditions, 35 to 100% of oocytes began to mature (0 to 17% in the control). Later it was found that the oocytes (both follicle-enclosed and naked) mature without hormonal pretreatment of the females if the pH of the saline was brought to 8.5. The treatment of the oocytes with progesterone (10^{-6}) in different media (Steinberg's, Merriam's, and Ringer's solutions) resulted in 94, 91, and 81% of maturation, respectively [242]. The mechanisms underlying oocyte maturation were also studied [36, 197, 232, 243, 244, 233].

TABLE 7.1. Stages of Normal Development of the Spanish Newt *Pleurodeles waltlii* Michah. (after Gallien and Durocher [137])

Stage No.	Duration at 18°C		Length, mm	Diagnostic features
	τ_n	τ_n/τ_0		
	Hours ± minutes			
0	0	0	1.7 ± 0.1	Fertilized noncleaved egg. Rotation
1	6 ± 15	4.0	"	2 blastomeres, 1st cleavage plane meridional
2	7.5 ± 15	5.0	"	4 blastomeres, 2nd cleavage plane meridional
3	9 ± 15	6.0	"	8 blastomeres, 3rd cleavage plane subequatorial: 4 micromeres at the animal pole and 4 macromeres at the vegetal pole
4a	10.5 ± 30	7.0	"	8 micromeres on the animal pole. Divisions begin at the animal pole and proceed at a slower rate at the vegetal pole
	Hours			
4b	12 ± 1	8.0	"	16 micromeres at the animal pole. At the end of stage 4 the egg comprises 24 to 32 blastomeres
5	14 ± 1	9.3	"	Early blastula
6	22 ± 2	14.7	"	Middle blastula
7	27 ± 2	18.0	"	Late blastula
8a	31 ± 3	20.7	1.8 ± 0.1	Appearance of dorsal blastopore lip
8b	35 ± 3	23.3	"	Dorsal blastopore lip crescent-shaped
9	38.5 ± 3	25.7	1.8 ± 0.1	Blastopore resembles a basket handle
10	41 ± 3	27.3	"	Blastopore resembles a horseshoe
11	46 ± 3	30.7	"	Circular blastopore. Large yolk plug, its diameter equal to 1/3 of yolk diameter
12	51 ± 4	34.0	"	Small yolk plug. Pigmented strips run from yolk plug, especially in dorsal region
13	56 ± 4	37.3	"	Slitlike blastopore. Neural folds are outlined by dorsal pigmented strips
14	63 ± 5	42.0	"	Formation of neural plate. Neural folds are distinct. Neural fold is rather elongated. Embryo is elongating
15	69 ± 5	46.0	2.0 ± 0.2	Rapprochement of neural folds. Embryo is egg-shaped
16	69.5 ± 5	46.3	"	Contact of neural folds in the trunk region (no drawing)

TABLE 7.1 (continued)

Stage No.	Duration at 18°C		Length, mm	Diagnostic features
	τ_n	τ_n/τ_0		
17	70 ± 5	46.6	2.0 ± 0.2	Closure of neural folds in the trunk region. Neural plate widened in its anterior part (brain) (no drawing)
18	71 ± 5	47.3	"	Closure of neural folds in the head region
19	74 ± 5	49.3	2.3 ± 0.2	Closure of neural folds in the head region. Movement of cilia inside the envelopes. 3 somites are visible
20	77 ± 5	51.3	2.5 ± 0.3	Complete fusion of neural folds. 5 somites are seen. Head is forming
21	80 ± 10	53.3	2.6 ± 0.3	Fully formed neural tube. Head assumes characteristic form. Visual field appears
22	87 ± 10	58.0	3.1 ± 0.3	Ear placode is visible. Rudiments of pronephros appear. Rudiments of gills are seen. Head is bent, angle between ventral head surface and body axis is 70°
23	91 ± 10	60.7	3.4 ± 0.4	Head enlarges. Angle between ventral head surface and body axis is 55° (no drawing)
24	95 ± 10	63.4	3.8 ± 0.4	Head still enlarges. Angle between ventral head surface and body axis is 40°. Stomodeal area outlined by a light strip
25	102 ± 10	68.0	4.4 ± 0.4	Tail elongation. Convexity of dorsal head surface less distinct. Angle between ventral head surface and body axis is 20°. Tail length amounts to 1/2 total body length (no drawing)
26	110 ± 10	73.4	4.4 ± 0.4	Olfactory placode visible. Reaction of muscles to external stimulation. Gill plates clearly seen. Tail distinctly outlined: its length amounts to 1/10 of total length. Angle between ventral body surface and body axis is 10°
27	120 ± 12	80.0	4.9 ± 0.4	Appearance of buds, rudiments of balancers. Embryo straightens, head axis coincides with that of body. Dorsal body surface becomes slightly concave. Gill plates formed by 3 light folds, rudiments of 3 pairs of gills (no drawing)

TABLE 7.1 (continued)

Stage No.	Duration at 18°C		Length, mm	Diagnostic features
	τ_n	τ_n/τ_0		
28	130 ± 2	86.6	5.4 ± 0.5	Balancers as conical buds. Onset of heartbeat. Differentiation of unpaired dorsocaudal fin. Length of tail amounts to 1/9 of total length. A few melanophores appear along somites
29a	137 ± 12	91.3	5.6 ± 0.5	1st pair of gill rudiments appears (no drawing)
29b	147 ± 12	98.0	5.9 ± 0.5	2 pairs of gill rudiments are clearly seen (no drawing)
30	155 ± 14	103.0	6.2 ± 0.5	3 pairs of gill rudiments are clearly seen. Melanophores appear in head region, proctodeum is delineated by dark dots. Tail length amounts to 1/6 of total length. Spontaneous muscle movements
31a	170 ± 14	113.3	6.7 ± 0.5	Blood circulation in 1st pair of gills and anterior tail part. Balancers of cylindrical form, opaque. Stomodeum delineated by a dark line (no drawing)
31b	175 ± 14	116.6	7.1 ± 0.5	Blood circulation in 2nd pair of gills (no drawing)
32a	183 ± 14	122.0	7.6 ± 0.5	Blood circulation in 3rd pair of gills. Anus is a narrow slit. Blood circulation in tail extends (no drawing)
32b	190 ± 14	126.7	8.1 ± 0.5	Common circulation in 3 pairs of gills. Gills of cylindrical forms, not branched. Cornea translucent. Trunk melanophores form longitudinal and transverse band along somites
33a	213 ± 14	142.0	9.0 ± 0.5	Appearance of first gill branching. Forelimb bud appears. Embryo devoid of envelopes can swim. Balancers translucent, slightly club-shaped, blood circulation in tail by 2/3 of its length anteriorly
33b	236 ± 14	157.3	10.4 ± 0.5	2nd and 3rd gill branching. Forelimb bud clearly seen. Tail length amounts to 1/3 of total length. Numerous melanophores on head and somites. Cornea transparent. Iris golden. Pupil black
34	264 ± 24	176.0	11.1 ± 0.6	Hatching. Forelimb bud of conical form. 4–6 gill branchings. Elongation of balancers

TABLE 7.1 (continued)

Stage No.	Duration at 18°C		Length, mm	Diagnostic features
	τ_n	τ_n/τ_0		
	Days			
35	12 ± 1	–	11.3 ± 0.6	Hatching. Forelimb bud assumes cylindrical form (no drawing)
36	13 ± 1	–	11.5 ± 0.6	Hatching. Forelimb bud of cylindrical form. Heart, stomach, visceral arches, ear capsules visible
37	15 ± 1	–	11.8 ± 0.6	Mouth perforation. Forelimb as a space. Yolk resorption. Blood circulation in tail. Balancer club-shaped, very transparent. A few melanophores on tail fin near cloaca
38	17.5 ± 2	–	12.6 ± 0.8	Beginning of active feeding. Forelimb as a spade with incision in the middle. Yolk almost fully resorbed. Tail covered by numerous melanophores
39	20 ± 2	–	12.6 ± 0.8	Two forelimb toes distinctly formed. Yolk fully resorbed. Balancer distal end becomes thinner
40	22 ± 2	–	13.1 ± 1	Movement in knee joint. Elongation of gills, increase in number of their branchings
41	25 ± 2	–	14 ± 1	Appearance of 3rd forelimb toe rudiment. Numerous melanophores on dorsal fin
42	28 ± 2	–	16 ± 1	3 forelimb toes formed. Appearance of hindlimb bud. Onset of balancer reduction
43	33 ± 2	–	17 ± 1	Appearance of 4th forelimb toe rudiment. Gills. Balancers reduced by 1/3 (no drawing)
44	36 ± 4	–	17.5 ± 1	Elongation of two 1st forelimb toes. Hindlimb bud of conical form. Balancer reduced by 1/2
45	40 ± 4	–	18 ± 1.5	4 forelimb toes well formed. Hindlimb becomes cylindrical. Balancer reduced to a short bud (no drawing)
46	43 ± 4	–	18.5 ± 1.5	Hindlimb elongated, cylindrical. Balancer disappeared (no drawing)
47	46 ± 4	–	19 ± 1.5	Hindlimb as a spade (no drawing)
48	50 ± 5	–	19.5 ± 1.5	Hindlimb as a spade with incision in the middle (no drawing)
49	53 ± 5	–	21 ± 1.5	2 hindlimb toes well formed (no drawing)

TABLE 7.1 (continued)

Stage No.	Duration at 18°C		Length, mm	Diagnostic features
	τ_n	τ_n/τ_0		
50	57 ± 5	–	23 ± 2	3rd hindlimb toe rudiment appears (no drawing)
51	61 ± 5	–	24 ± 2	3 hindlimb toes well formed (no drawing)
52a	64 ± 6	–	25 ± 3	Appearance of 4th hindlimb toe rudiment (no drawing)
52b	64 ± 6	–	27 ± 3	4 hindlimb toes well formed (no drawing)
53	72 ± 7	–	30 ± 1	Appearance of 5th hindlimb toe rudiment. Elongation of toes
54	79 ± 7	–	38 ± 5	5 hindlimb toes well formed (no drawing)
55a	90 ± 10	–	44 ± 5	External gills developed maximally (no drawing)
55b	100 ± 10	–	60 ± 8	External gills reduced by 1/2. Fin reduction. Toes thickening. Skin changes in the trunk region
55c	105 ± 10	–	65 ± 8	First moult. Gills markedly reduced. Head thickened. Tail width decreased twice at the expense of fin reduction. Lung respiration
56	110 ± 10	–	72 ± 10	Metamorphosis completed. Fin and gills disappeared fully

Note. The drawings of the stages described are given in Figs. 7.2–7.4. The numbers of the drawings correspond to those of the stages. If drawings are not provided, a corresponding note is given.

7.4. GAMETES AND FERTILIZATION

7.4.1. Structure of the Egg

The mature egg is blocked at the metaphase of the 2nd maturation (meiotic) division [171]. It is telolecithal. The animal hemisphere is pigmented. The egg is surrounded by two envelopes, the vitelline membrane (sometimes termed the chorion) and the jelly coat. In addition, a cluster of simultaneously laid eggs (10–20, usually) is coated by dense jelly. The freshly laid egg is 1.4–1.8 mm in diam-

Figs. 7.2–7.4. Normal developmental stages of the Spanish newt *Pleurodeles waltlii* Michah (stages 0–56). After Gallien and Durocher [137]. Drawing numbers correspond to stage numbers. an) Animal view; veg) vegetal view; d) dorsal view; v) ventral view. Drawings without letter designations are lateral views. Magnification: A) for stages 0–30; B) 32–40; C) 41–48; D) 50–53; E) 54–56.

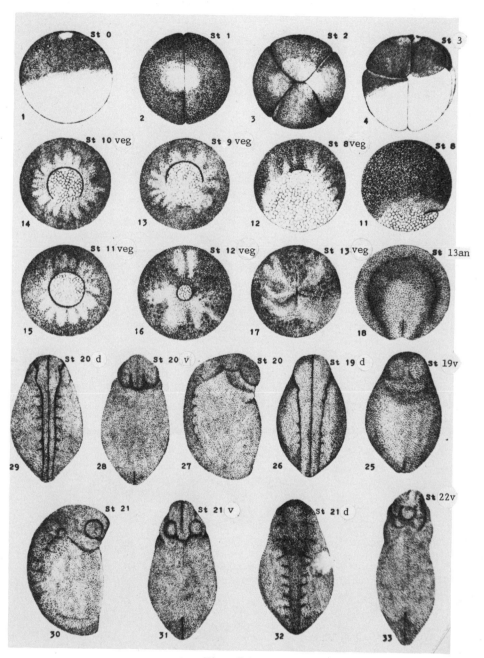

Fig. 7.2.

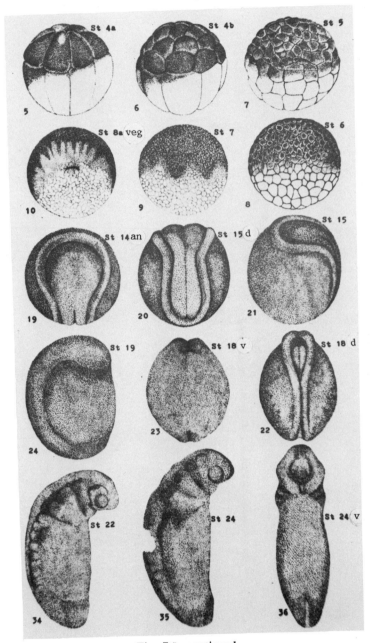

Fig. 7.2. continued.

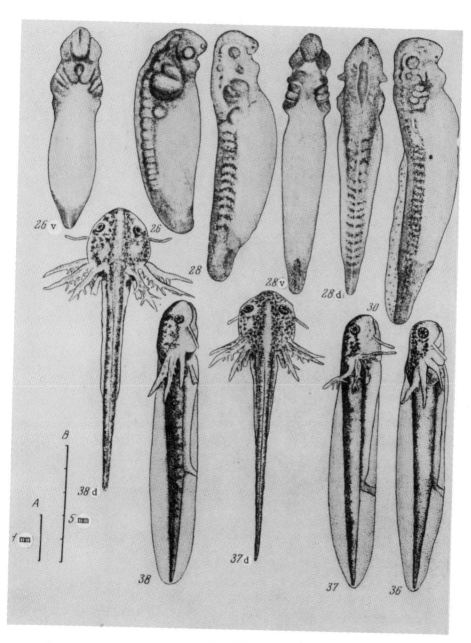

Fig.   7.3.

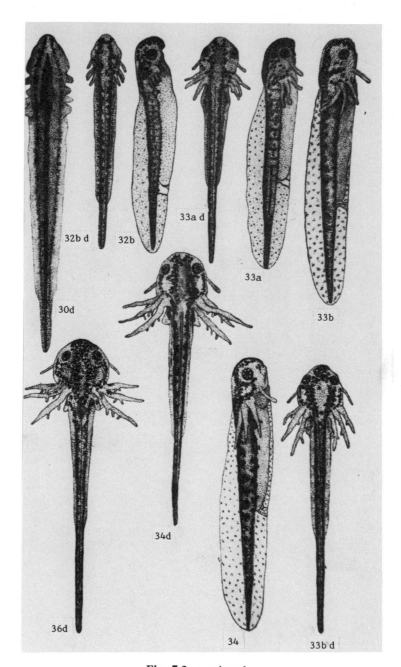

Fig. 7.3, continued.

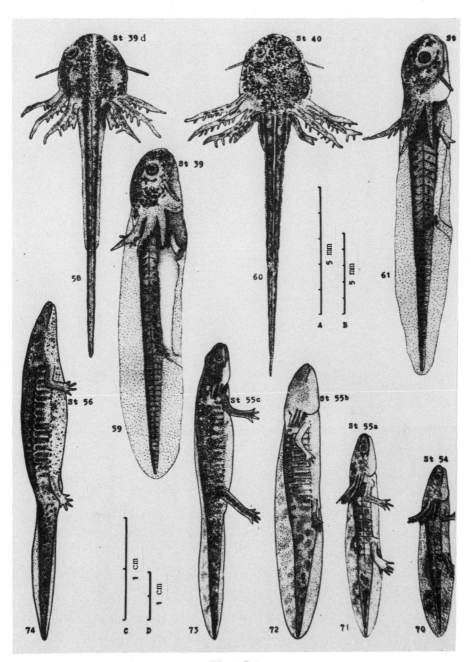

Fig. 7.4.

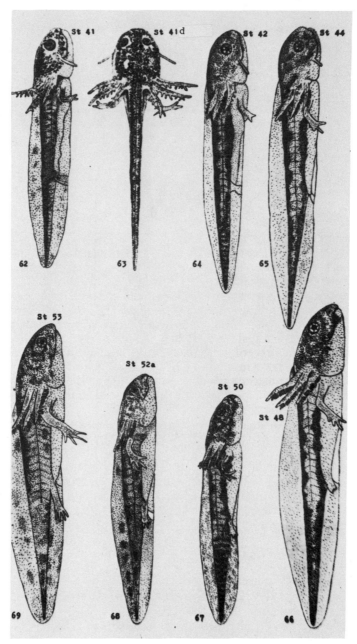

Fig. 7.4, continued.

eter [123, 137, 171]. Both envelopes can be removed by forceps and the eggs then become available for microsurgery or experimentation.

7.4.2. Structure of the Spermatozoon

The fine structure of the spermatozoon was studied by Picheral [208]. The acrosome (approximately 10 μm long) consists of three parts: cap, end process, and hook. Under the acrosomal cap the intranuclear axial stem is located. In the elongated nucleus (70 μm) a chromatin zone can be seen which is limited by a region not stainable with Feulgen. The posterior part of the nucleus is adjoined by a large neck (8–10 μm), at the base of which a pair of centrioles is located. One of them connects with the flagellum by its distal end. The tail (175 μm) comprises, besides the flagellum, the supporting and marginal filaments and the undulating membrane. The total length of the spermatozoon is about 250 μm.

7.4.3. Fertilization

The Spanish newt, like the other species of urodeles, is characterized by physiological polyspermy (for review see [142]); several spermatozoa penetrate the egg (usually 6, but up to 14). Although fertilization is internal, the encounter of the spermatozoa and the eggs takes place at the moment of egg laying. The spermatozoa can be seen in the jelly coat within about 10 min after egg laying and in the egg cortex within 15–20 min. Fifteen to twenty minutes later, the fertilization rotation begins and lasts about 15 min. Thereafter a round light zone, the maturation spot, appears on the darker animal hemisphere. In its center, a small depression (maturation funnel), the first polar body and the egg nucleus are located. Fertilized eggs can be selected by these features.

The number of spermatozoa which penetrated an egg is estimated by the sites of their penetration, which appear as small spots of condensed pigment, as well as by the presence of supernumerary sperm asters in the egg. The spermatozoa penetrate the egg in both the animal and vegetal hemispheres. One of the spermatozoa which penetrates the egg in its equatorial region appears to fertilize the egg [172, 173].

After egg activation induced by the penetration of spermatozoa the 2nd maturation division is completed: within about 40 min after egg laying anaphase II begins and lasts about 5 min; it is followed by telophase II which lasts about 15 min. Then the 2nd polar body is extruded [172, 173].

7.5. NORMAL DEVELOPMENT

The chronology of nuclear division during the first cleavage divisions has been established for *P. waltlii* [172, 224]. The duration of one cleavage division at 18°C is 90–95 min.

The table of normal development for *P. waltlii* (Table 7.1; Figs. 7.2–7.4) is reproduced after Gallien and Durocher [137] who not only gave their permission to use their tables but kindly sent photographs of their original drawings. Experience in rearing Spanish newts in the laboratory has shown that 18°C is the optimal temperature for their development. Therefore, all observations of the development of the eggs from 19 batches on which these tables are based were carried out at this temperature. The mean time of the onset of successive developmental stages is indicated, together with the maximal deviation from these mean values. The duration of development τ_n is given in hours and minutes from stage 0 (freshly laid egg) to stage 34 (hatching) and is recalculated in τ_0 units (τ_n/τ_0; see [89]). We assume the duration of one mitotic cycle during the synchronous cleavage divisions (τ_0) to be 90 min [172]. For *P. waltlii* the values of τ_0 at different temperatures were not determined, and this does not allow prognostication of the time of onset of developmental stages at different temperatures by their relative duration. However, the relative duration of the developmental stages determined in these tables makes possible a comparison of the development of *P. waltlii* with that of other animals (see Chapter 1).

The results of observations relate mainly to external features and some physiological properties. Each stage is determined by one or several distinct criteria easily identifiable under a binocular microscope.

The tables comprise the developmental stages from 0 to 56 (complete metamorphosis). There are two main tables cited by most authors in studies of amphibian embryos: that for *Triturus taeniatus* by Glaesner [143] and that for *Ambystoma maculatum* by Harrison (see [217]). The stages of development of *P. waltlii* were identified according to Glaesner's tables, since the development of *T. taeniatus* and *P. waltlii* is very similar. In addition, Glaesner described development up to the completion of metamorphosis, whereas the tables of Harrison end at a larval stage identified by the appearance of the forelimb's third toe. In some cases, two or three intermediate stages are distinguished within one stage, and they are designated by the letters *a, b,* and *c.*

The development of *P. waltlii*, as well as of other vertebrate animals, is divided into four main periods: cleavage (stages 0–7), gastrulation (stages 8–13), neurulation (stages 14–21), and tailbud period (stages 22–34). Larval development is subdivided into two periods: the initial period (stages 35–38) and the period of active larval life (stages 39–56).

REFERENCES

1. J. H. Abelian, P. Jego, and G. Volotaire, "Effects of 17-β estradiol on the DNA, RNA protein contents and of the DNA, RNA polymerases in the Müllerian duct of the immature female newt (*Pleurodeles waltlii* Michah)," *Gen. Comp. Endocrinol.* **40**, 402–408 (1980).

2. C. Aimar, "Analyse par la greffe nucléaire des propriétés morphogénétiques des noyaux embryonnaires chez *Pleurodeles waltlii* (Amphibien, Urodèle). Application à l'étude de la gémellarité expérimentale," *Ann. Embryol. Morphogen.* **5**, 5–42 (1972).

3. C. Aimar and M. Delarue, "Changes in somatic nuclei exposed to meiotic stimulation in amphibian oocytes," *Biol. Cell.* **38**, 37–42 (1980).

4. C. Aimar, M. Delarue, and C. Vilain, "Cytoplasmic regulation of the duration of cleavage in amphibian eggs," *J. Embryol. Exp. Morphol.* **64**, 259–274 (1981).

5. C. Aimar and J.-M. Olivereau, "Influences de l'ionisation atmosphérique sur le développement et la métamorphose d'un amphibien (*Pleurodeles waltlii* Michah, Urodèle)," *C. R. Acad. Sci. Paris* **278**, 1621–1624 (1974).

6. C. Aimar and J.-P. Labrousse, "DNA synthesis and evolution in presence of a somatic nucleus of the female pronucleus after experimental activation of the egg of *Pleurodeles waltlii*," *Dev., Growth Differ.* **17**, 197–207 (1975).

7. H. Alexandre, "Étude autoradiographique de l'effet des rayons X sur les synthèses de DNA et de RNA au cours de la segmentation et de la gastrulation de *Pleurodeles waltlii* Michah.," *Arch. Biol.* **81**, 139–162 (1970).

8. H. Alexandre and Y. Gerin, "Étude au microscope électronique de l'effet des rayons X sur l'évolution des structures nucléaires et cytoplasmiques au cours de la segmentation des oeufs de *Pleurodeles waltlii*," *Exp. Cell. Res.* **65**, 145–155 (1974).

9. B. Andrieux and A. Collenot, "Hormones gonadotropes et développement testiculaire chez le triton *Pleurodeles waltlii* Michah. hypophysectomisé," *Ann. Endocrinol.* **31**, 531–537 (1970).

10. C. F. Ardavin, A. Zapata, E. Garrido, and A. Villena, "Ultrastructure of gut-associated lymphoid tissues (GALT) in the amphibian urodele *Pleurodeles waltlii*," *Cell Tissue Res.* **224**, 663–671 (1982).

11. C. F. Ardavin, A. Zapata, A. Villena, and M. T. Solas, "Gut-associated lymphoid tissue (GALT) in the amphibian urodele *Pleurodeles waltlii*," *J. Morphol.* **173**, 35–41 (1982).

12. I. Audit, P. Deparis, M. Flavin, and R. Rosa, "Erythrocyte activities in diploid and triploid salamanders (*Pleurodeles waltlii*) of both sexes," *Biochem. Genet.* **14**, 759–769 (1976).

13. J. T. Bagnara and M. Obika, "Comparative aspects of integumental pteridine distribution among amphibians," *Comp. Biochem. Physiol.* **15**, 33–49 (1965).

14. S. Bailly, "Organogénèse du coeur et des arcs aortiques chez le triton *Pleurodeles waltlii* Michah. après fissuration de l'ébauche cardiaque embryonnaire," *Bull. Biol. Fr. Belg.* **97**, 627–642 (1963).

15. S. Bailly, "Étude cytophotométrique de la teneur en acides nucléiques des chromosomes métaphasiques de l'Amphibien Urodèle *Pleurodeles waltlii* Michah.," *Exp. Cell Res.* **48**, 549–566 (1967).

16. S. Bailly, "Analyse cytophotométrique des chromosomes mitotiques chez *Pleurodeles waltlii* (Amphibien, Urodèle)," *Ann. Embryol. Morphogen.* **5**, 75–96 (1972).

17. E. Ballus and J. Brachet, "Le dosage de l'acide désoxyribonucléique dans les oeufs de Batraciens," *Biochim. Biophys. Acta* **61**, 157–163 (1962).

18. W. Beçak, M. L. Beçak, G. Schreiber, D. Lavalle, and F. O. Amorum, "Interspecific variability of DNA content in Amphibia," *Experientia* **26**, 204–206 (1970).

19. J.-C. Beetschen, "Hypomorphoses et altérations du développement embryonnaire, consécutives aux chocs thermiques appliqués à l'oeuf fécondé du triton *Pleurodeles waltlii* Michah.," *C. R. Acad. Sci. Paris* **244**, 1959–1962 (1957).

20. J.-C. Beetschen, "Anomalies morphogénétiques et caryologiques consécutives à l'hypermaturité des oeufs chez le triton *Pleurodeles waltlii* Michah.," *C. R. Acad. Sci. Paris* **245**, 2541–2543 (1957).

21. J.-C. Beetschen, "Quelques aspects cytologiques et morphogénétiques de l'hétéroploidie expérimentale chez le triton *Pleurodeles waltlii* Michah.," *Bull. Soc. Zool. Fr.* **81**, 189 (1957).

22. J.-C. Beetschen, "Modifications des structures nucléaires et cytoplasmiques de l'oeuf fécondé insegmenté soumis à une refrigération prolongée, chez le triton *Pleurodeles waltlii* Michah.," *C. R. Acad. Sci. Paris* **249**, 173–175 (1957).

23. J.-C. Beetschen, "Recherches sur l'hétéroploidie expérimentale chez un Amphibien Urodèle, *Pleurodeles waltlii* Michah.," *Bull. Biol. Fr. Belg.* **94**, 12–127 (1968).

24. J.-C. Beetschen, "Origine androgénétique des germes haploïdes obtenues par réfrigération des oeufs fécondés du triton *Pleurodeles waltlii* Michah.," *C. R. Acad. Sci. Paris* **157**, 1675–1677 (1963).

25. J.-C. Beetschen, "Pentaploïdie expérimentale chez l'Amphibien Urodèle *Pleurodeles waltlii* Michah.," *C. R. Acad. Sci. Paris* **258**, 1641–1643 (1964).

26. J.-C. Beetschen, "Observations préliminaires sur les perturbations de la gastrulation consécutives à l'effet maternel lié à la mutation *ac* chez l'amphibien *Pleurodeles waltlii*," *Bull. Soc. Zool. Fr.* **101**, 57–61 (1976).

27. J.-C. Beetschen and J.-J. Buisan, "Le problème de la détermination mésodermique de la plaque neurale postérieure chez les urodèles. Relations avec l'activité mitotique," *Mém. Soc. Zool. Fr.* **41**, 139–151 (1977).

28. J.-C. Beetschen and A. Dubost, "Action inhibitrice de l'eau lourde sur le clivage et la morphogénèse du germe de *Pleurodeles waltlii* (Amphibien, Urodèle)," *C. R. Acad. Sci. Paris* **254**, 3740–3742 (1962).

29. J.-C. Beetschen and M. Fernandez, "Studies on the maternal effect of the semi-lethal factor *ac* in the salamander *Pleurodeles waltlii*," in *Maternal Effects in Development*, D. R. Newth and M. Balls, eds., Cambridge University Press, Cambridge (1979), pp. 269–286.

30. J.-C. Beetschen and V. Ferrier, "Nouveaux exemples d'hybridation intergénérique letale chez les Salamandridae (Amphibiens Urodèles)," *C. R. Acad. Sci. Paris* **259**, 217–219 (1964).

31. J.-C. Beetschen and A. Jaylet, "Le caryotype somatique de l'Amphibien Urodèle *Pleurodeles waltlii* Michah.," *C. R. Acad. Sci. Paris* **253**, 3055–3057 (1961).

32. J.-C. Beetschen and A. Jaylet, "Sur un facteur récessif semi-létal déterminant l'apparition d'ascite caudal (*AC*) chez le triton *Pleurodeles waltlii*," *C. R. Acad. Sci. Paris* **261**, 5675–5678 (1965).

33. S. Biaggianti and J. Corsin, "Étude en microscopie électronique du rôle des glycoprotéines de la matrice au cours de la différenciation des chondroblastes," *Arch. Anat. Microsc. Morphol. Exp.* **68**, 224 (1979).

34. S. Biaggianti and J. Corsin, "Action du hyaluronate de sodium aux la chondrogénèse embryonnaire chez le Pleurodèle," *Arch. Anat. Microsc. Morphol. Exp.* **70**, 91–107 (1981).

35. M. Bideau, "Manifestations cytologiques et comportement des noyaux au cours de la greffe nucléaire chez l'Urodèle *Pleurodeles waltlii* Michah.," *C. R. Acad. Sci. Paris* **259**, 213–216 (1964).

36. M. P. Bizet, "Aspects ultrastructuraux et métaboliques de la région corticale et du follicule de l'ovocyte de Pleurodèle (Amphibien, Urodèle), *Biol. Cell.* **35**, 36A (1979).

37. J. G. Bluemink and J.-C. Beetschen, "An ultrastructural study of the maternal-effect embryos of the *ac/ac* mutant of *Pleurodeles waltlii* showing a gastrulation defect," *J. Embryol. Exp. Morphol.* **63**, 67–74 (1981).

38. O. Boettger, "Reptilien von Marocco und von Canarischen Inseln," *Abh. Senckenb. Naturforsch. Ges.* **9**, 121–170 (1974).
39. M. Bona, U. Scheer, and E. K. F. Bautz, "Antibodies to RNA polymerase II(B) inhibit transcription in lampbrush chromosomes after microinjection into living amphibian oocytes," *J. Mol. Biol.* **151**, 81–90 (1981).
40. J. C. Boucaut, "Distribution au cours du développement de blastomères isolés marqués par la thymidine tritiée implantés dans le blastocoele chez l'Urodèle *Pleurodeles waltlii* Michah.," *C. R. Acad. Sci. Paris* **268**, 554–557 (1969).
41. J. C. Boucaut, "Capacité migratrice et degré de reconnaissance spécifique de cellules de gastrula greffées au stade blastula chez *Pleurodeles waltlii* (Amphibien Urodèle)," *C. R. Acad. Sci. Paris* **D, 272**, 469–472 (1971).
42. G. A. Boulenger, *Les Batraciens. Encyclopedie scientifique*, Doin, Paris (1910).
43. J. Brachet, "Le contrôle de la synthèse des protéines en l'absence du noyau cellulaire. Faits et hypothèses," *Bull. Cl. Sci., Acad. R. Belg. 5 sèr.* **51**, 257–276 (1965).
44. J. Brachet, "Observations cytologiques et cytochimiques sur la maturation de l'oocyte chez les Urodèles," *Ann. Biol.* **13**, 271–284 (1974).
45. J. Brachet, E. Hubert, and A. Lievens, "The effects of amanitin and rifampycin on amphibian egg development," *Rev. Suisse Zool.* **79** (Suppl.), 47–63 (1972).
46. A. Capuron, "Colonisation de gonades surnuméraires par les cellules germinales primordiales, dans l'embryon induit après greffe de la lèvre dorsale du blastopore chez le triton *Pleurodeles waltlii* Michah.," *C. R. Acad. Sci. Paris* **256**, 4736–4739 (1963).
47. A. Capuron, "Marquage autoradiographique et conditions de l'organogénèse générale d'embryons induits par la greffe de la lèvre dorsale du blastopore chez l'Amphibien Urodèle *Pleurodeles waltlii* Michah.," *Ann. Embryol. Morphogen.* **1**, 271–293 (1968).
48. A. Capuron and J. P. Maufroid, "Rôle de l'endoderme dans la détermination du mésoderme ventro-latéral et des cellules germinales primordiales chez *Pleurodeles waltlii*," *Arch. Anat. Microsc. Morphol. Exp.* **70**, 219–226 (1981).
49. C. Cassin and A. Capuron, "Evolution de la capacité morphogénétique de la région stomodéale chez l'embryon de *Pleurodeles waltlii* Michah. (Amphibien Urodèle). Étude par transplantation intrablastocélienne et par culture *in vitro*," *Wilhelm Roux's Arch. Dev. Biol.* **181**, 107–112 (1977).
50. C. Cassin and A. Capuron, "Buccal organogenesis in *Pleurodeles waltlii* Michah. (Urodele Amphibian). Study by transplantation and *in vitro* culture," *J. Biol. Buccale* **7**, 61–76 (1979).
51. C. Cayrol and P. Deparis, "A comparative study of the levels of glucose-6-phosphate dehydrogenase in the erythrocytes of diploid and polyploid salamanders (*Pleurodeles waltlii*)," *Comp. Biochem. Physiol.* **62B**, 533–537 (1979).
52. C. Cayrol, F. Braconnier, and P. Deparis, "Étude comparée des concentrations de quelques constituants plasmatiques chez des Pleurodèles diploides et polyploides (Amphibiens, Urodèles)," *C. R. Soc. Biol.* **173**, 134–136 (1979).

53. P. Certain, "Organogénèse des formations interrénales chez le Batracien Urodèle *Pleurodeles waltlii* Michah.," *Bull. Biol. Fr. Belg.* **95**, 134–148 (1961).

54. P. Certain, G. Collenot, and R. Ozon, "Mise en évidence biochimique d'une Δ⁵-3β-hydroxystéroide déshydrogénase dans le testicule du triton *Pleurodeles waltlii* Michah.," *C. R. Soc. Biol.* **158**, 1040 (1964).

55. M. T. Chalumeau-Le Foulgoc, "Etude des protéines chez les amphibiens," *Ann. Biol.* **7**, 683–701 (1968).

56. M. T. Chalumeau-Le Foulgoc, "Recherches sur les protéines sériques au cours du développement et chez l'adulte dans le genre Pleurodeles (Amphibien, Urodèle)," *Ann. Embryol. Morphogen.* **2**, 387–417 (1969).

57. J. Charlemagne and C. Houillon, "Effets de la thymectomie larvaire chez l'Amphibien Urodèle *Pleurodeles waltlii* Michah. Production à l'état adulte d'une tolérance aux homogreffes cutanées," *C. R. Acad. Sci. Paris* D, **267**, 253–256 (1968).

58. P. Chibon, "Analyse expérimentale de la régionalisation et des capacités morphogénétiques de la crête neurale chez l'Amphibien Urodèle *Pleurodeles waltlii* Michah.," *Mém. Soc. Zool. Fr.* **36**, 1–107 (1966).

59. P. Chibon, "Étude expérimentale par allations, greffes et autoradiographie de l'origine des dents chez l'Amphibien Urodèle *Pleurodeles waltlii* Michah.," *Arch. Oral Biol.* **12**, 745–753 (1967).

60. P. Chibon, "Effet des doses fortes de thymidine tritiée sur le développement et sur les chromosomes du triton *Pleurodeles waltlii* Michah.," *C. R. Acad. Sci. Paris* D, **266**, 798–801 (1968).

61. P. Chibon, "Étude au moyen de la thymidine tritiée de la durée des cycles mitotiques dans la jeune larve de *Pleurodeles waltlii* Michah. Amphibien Urodèle," *C. R. Acad. Sci. Paris* D, **267**, 203–205 (1968).

62. P. Chibon, "Étude de la régionalization et de la détermination de l'endoderme chez l'embryon de *Pleurodeles waltlii* Michah. (Amphibien, Urodèle) aux stades du bourgeon caudale," *Ann. Embryol. Morphogen.* **2**, 307–315 (1969).

63. P. Chibon and M. Belanger-Barbeau, "Effets du LSD sur le développement embryonnaire et les chromosomes de l'Amphibien Urodèle *Pleurodeles waltlii* Michah.," *C. R. Acad. Sci. Paris* D, **274**, 280–283 (1972).

64. P. Chibon and G. Brugal, "Étude autoradiographique de l'action de la température et de la thyroxine sur la durée des cycles mitotiques dans l'embryon agé et la jeune larve de *Pleurodeles waltlii* Michah. (Amphibien Urodèle)," *C. R. Acad. Sci. Paris* D, **269**, 70–73 (1969).

65. P. Chibon and G. Brugal, "Durée de disponibilité de la thymidine exogène chez la larve et le jeune du triton *Pleurodeles waltlii* Michah.," *Ann. Embryol. Morphogen.* **6**, 81–92 (1973).

66. P. Chibon and A. Verain, "Premiers résultats de l'étude du métabolisme de l'iode chez les Amphibiens Urodèles goitreux ou sains," *C. R. Soc. Biol.* **164**, 2251–2255 (1970).

67. P. Chibon, M. Lussiana, and A. Verain, "Premiers résultats de l'étude de l'hormogenèse thyroidienne chez le Pleurodèle sain ou goitreaux," *Sciences* **3**, 125–133 (1970).

68. A. Collenot, "Recherches comparatives sur l'inversion sexuelle par les hormones stéroides chez les Amphibiens," *Mém. Soc. Zool. Fr.* **33**, 1–14 (1965).

69. A. Conter and A. Jaylet, "Une triple translocation chez le Triton *Pleurodeles waltlii*," *Chromosoma* **46**, 37–58 (1974).

70. J. Corsin, "The development of the osteocranium of *Pleurodeles waltlii* Michahelles," *J. Morphol.* **119**, 209–216 (1966).

71. J. Corsin, "Rôle de la compétition osseuse dans la forme de os du toit cranien des Urodèles," *J. Embryol. Exp. Morphol.* **19**, 103–108 (1968).

72. J. Corsin, "Rôle des ébauches sensorielles au cours de la morphogénèse du chondrocrâne chez *Pleurodeles waltlii* Michah.," *Arch. Anat. Microsc. Morphol. Exp.* **61**, 47–60 (1972).

73. J. Corsin, "Différenciation *in vitro* de cartilage à partir des crêtes neurales céphaliques ches *Pleurodeles waltlii* Michah., *J. Embryol. Exp. Morphol.* **33**, 335–342 (1975).

74. J. Corsin, "Le matériel extracellulaire au cours du développement du chondrocrâne des amphibients: Mise en place et constitution," *J. Embryol. Exp. Morphol.* **38**, 139–149 (1977).

75. J. Corsin and C. Joly, "Effets du dibutyryl-AMP sur la chondrogenèse chez l'embryon de Pleurodèle," *Arch. Anat. Microsc. Morphol. Exp.* **69**, 332 (1980).

76. H. Denis, "La différenciation protéique au cours du développement chez les Amphibiens," *Bull. Soc. Zool. Fr.* **86**, 534–540 (1961).

77. H. Denis, "Effets de l'actinomycine sur le développement embryonnaire. I. Étude morphologique: suppression par l'actinomycine de la compétence de l'ectoderme et du pouvoir inducteur de la lèvre blastoporale," *Dev. Biol.* **9**, 435–457 (1964).

78. H. Denis, "Effets de l'actinomycine sur le développement embryonnaire. II. Etude autoradiographique: influence de l'actinomycine sur la synthèse des acides nucléiques," *Dev. Biol.* **9**, 458–472 (1964).

79. H. Denis, "Effets de l'actinomycine sur le développement embryonnaire. III. Etude biochimique: influence de l'actinomycine sur la synthèse des protéines," *Dev. Biol.* **9**, 473–483 (1964).

80. P. Deparis, "Hématopoièse embryonnaire et larvaire chez l'Amphibien Urodèle *Pleurodeles waltlii* Michah.." *Ann. Embryol. Morphogen.* **1**, 107–118 (1968).

81. P. Deparis, "Modifications hématologiques provoquées par la manipulation et la saignée chez *Pleurodeles waltlii* Michah. (Amphibien, Urodèle)," *J. Physiol. (France)* **64**, 19–30 (1972).

82. P. Deparis and J.-C. Beetschen, "Résultats comparatifs de numérations globulaires faites sur le sang d'individus diploïdes et polyploïdes du triton *Pleurodeles waltlii* Michah.," *C. R. Soc. Biol.* **159**, 1224–1229 (1965).

83. P. Deparis and J.-C. Beetschen, "Influence de la polyploïdie et de l'hybridation intraspécifique sur la formule leucocytaire du triton *Pleurodeles waltlii* Michah.," *C. R. Acad. Sci. Paris* D, **260**, 4269–4272 (1965).

84. P. Deparis and M. Flavin, "Les effets de la splenectomie précoce chez l'Amphibien Urodèle *Pleurodeles waltlii* Michah.," *J. Physiol. (France)* **66**, 19–29 (1973).

85. P. Deparis and A. Jaylet, "Recherches sur l'origine des différentes lignées de cellules sanguines chez l'Amphibien Urodèle *Pleurodeles waltlii*," *J. Embryol. Exp. Morphol.* **33**, 665–683 (1975).

86. P. Deparis and A. Jaylet, "Thymic lymphocyte origin in the newt *Pleurodeles waltlii* studied by embryonic grafts between diploid and tetraploid embryos," *Ann. Immunol.* **127C**, 827–831 (1976).

87. P. Deparis, H. Nouvel, and J.-C. Beetschen, "Taux d'hémoglobine et valeur de l'hématocrite chez des individus diploides et triploides du triton *Pleurodeles waltlii*," *C. R. Soc. Biol.* **160**, 416–419 (1966).

88. P. Deparis, J.-C. Beetschen, and A. Jaylet, "Red blood cells and hemoglobin concentration in normal diploid and several types of polyploid salamanders," *Comp. Biochem. Physiol.* **50A**, 263–266 (1975).

89. T. A. Dettlaff and A. A. Dettlaff, "On relative dimensionless characteristics of the development duration in embryology," *Arch. Biol.* **72**, 1–16 (1961).

90. C. Dournon, "Régulation du nombre de cellules germinales primordiales dans les larves chimères à corps double chez *Pleurodeles waltlii* Michah. (Amphibien Urodèle)," *Mém. Soc. Zool. Fr.* **41**, 61–70 (1977).

91. C. Dournon, "Prolifération des cellules germinales chez les larves de Pleurodèle dans les conditions normales ou expérimentales d'élevage," *Arch. Anat. Microsc. Morphol. Exp.* **69**, 322 (1980).

92. C. Dournon and J. C. Boucaut, "Marquage autoradiographique des cellules germinales primordiales du Pleurodèle," *Ann. Soc. Fr. Biol. Dév. Paris*, p. 28 (1970).

93. C. Dournon and P. Zaborski, "H–Y antigen as a measure for the genotype detection of normal or temperature sex-reversed phenotype in *Pleurodeles*," *Arch. Anat. Microsc. Morphol. Exp.* **68**, 221–222 (1979).

94. A.-M. Duprat, "Étude comparative de l'utilisation des reserves vitellines dans les cellules embryonnaires diploides et triploides d'un Amphibien Urodèle, en culture *in vitro*," *C. R. Acad. Sci. Paris* **D, 258**, 4358–4361 (1964).

95. A.-M. Duprat, "Comportement de cellules embryonnaires vivantes de *Pleurodeles waltlii* (Amphibien Urodèle) cultivées *in vitro* en présence d'actinomycine D," *C. R. Acad. Sci. Paris* **D, 261**, 5637–5640 (1965).

96. A.-M. Duprat, "Action de la puromycine et d'une substance analogue sur la différenciation de cellules embryonnaires d'Amphibiens cultivées *in vitro*," *J. Embryol. Exp. Morphol.* **24**, 119–128 (1970).

97. A.-M. Duprat, "Recherches sur la différenciation de cellules embryonnaires d'Urodèles en culture *in vitro*," *Ann. Embryol. Morphogen.* **3**, 411–423 (1970).

98. A.-M. Duprat and P. Kan, "Stimulating effect of the divalent cation ionophore A23187 on *in vitro* neuroblast differentiation; comparative studies with myoblasts," *Experientia* **37**, 154–157 (1981).

99. A.-M. Duprat, A. Jaylet, and J.-C. Beetschen, "Variations sponanées et expérimentales du nombre de nucléoles des noyaux somatiques chez l'Amphibien Urodèle *Pleurodeles waltlii* Michah.," *C. R. Acad. Sci. Paris* **258**, 1059–1062 (1964).

100. A.-M. Duprat, J.-C. Beetschen, C. Mathieu, J.-P. Zalta, and C. Daguzan, "Effets inhibiteurs de protéines chromatiniennes non histoniques homospécifiques, sur la morphogenèse d'embryons d'amphibiens Urodèles," *Wilhelm Roux's Arch. Dev. Biol.* **179**, 111–124 (1976).

101. A.-M. Duprat, C. Mathieu, and J.-J. Buisan, "Effects of nonhistone chromosomal proteins on primary embryonic induction and cytodifferentiation in salamanders," *Differentiation* **9**, 161–167 (1967).

102. A.-M. Duprat, L. Gualandris, and P. Rouge, "Neural induction and the structure of the target cell surface," *J. Embryol. Exp. Morphol.* **70**, 171–187 (1982).

103. P. Dupuis-Certain, "Glucose 6-phosphate déhydrogénase dans le tissu interrénal de l'Urodèle *Pleurodeles waltlii* Michah.," *Gen. Comp. Endocrinol.* **40**, 308 (1980).

104. M. Fernandez, "Correction of the maternal effect linked to the *ac* mutation, by injury of the egg, in the salamander *Pleurodeles waltlii*," *J. Embryol. Exp. Morphol.* **53**, 305–314 (1979).

105. M. Fernandez and J.-C. Beetschen, "Recherches sur le rôle de la température dans la réalisation du phénotype chez des embryons de l'Amphibien *Pleurodeles waltlii* homozygotes pour la mutation thermosensible *ac* (ascite caudale)," *J. Embryol. Exp. Morphol.* **34**, 221–252 (1975).

106. V. Ferrier, "Gynogenèse diploïde et polyploïde realisée expérimentalement chez le triton *Pleurodeles waltlii*," *C. R. Soc. Biol.* **160**, 1526–1531 (1966).

107. V. Ferrier, "Étude cytologique des premiers stades du développement de quelques hybrides létaux d'Amphibiens Urodèles," *J. Embryol. Exp. Morphol* **18**, 227–257 (1967).

108. V. Ferrier and J.-C. Beetschen, "Étude des chromosomes de *Tylototriton verrucosus* Anderson et de l'hybride viable *Pleurodeles waltlii* × *Tylototriton verrucosus* (Amphibiens, Urodeles, Salamandridae)," *Chromosoma* **42**, 57–69 (1973).

109. V. Ferrier and A. Jaylet, "Induction of triploidy in the newt *Pleurodeles waltlii* by heat shock or hydrostatic pressure," *Chromosoma* **69**, 47–63 (1978).

110. V. Ferrier, J.-C. Beetschen, and A. Jaylet, "Réalisation d'un hybride intergénérique viable entre deux Amphibiens Urodèles européen et asiatique (*Pleurodeles waltlii* × *Tylototriton verrucosus*, Salamandridae)," *C. R. Acad. Sci. Paris* **D, 272**, 3079–3082 (1971).

111. V. Ferrier, A. Jaylet, C. Vayrol, F. Gasser, and J.-J. Buisan, "Étude électrophorétique des protéinases érythrocytaires chez *Pleurodeles waltlii* (Amphibien Urodèle): mise en evidence d'une liaison avec le sexe," *C. R. Acad. Sci. Paris* **290D**, 571–574 (1980).

112. A. Ficq, "Métabolisme de l'oogenèse chez les Amphibiens," in *Sympos. Germ Cell Development,* Inst. Intern. d'Embryologie, Fondazione A. Baselli (1961), pp. 112–140.

113. A. Ficq, "Sites de méthylation des acides ribonucléiques dans les oocytes d'Urodèles," *Arch. Biol.* **77**, 47–58 (1966).

114. M. Flavin, Y. Blouquit, and J. Rosa, "Structural study of the chain of one haemoglobin from the adult salamander, *Pleurodeles waltlii*," *FEBS Lett.* **67**, 52–57 (1976).

115. M. Flavin, Y. Blouquit, and J. Rosa, "Biochemical studies of the hemoglobin switch during metamorphosis in the salamander *Pleurodeles waltlii*. I. Partial characterization of the adult hemoglobin," *Comp. Biochem. Physiol.* **61B**, 533–537 (1978).

116. M. Flavin, Y. Blouquit, A.-M. Duprat, and J. Rosa, "Biochemical studies of the hemoglobin switch during metamorphosis in the salamander *Pleurodeles waltlii*. II. Comparative studies of larval and adult hemoglobins," *Comp. Biochem. Physiol.* **61B**, 539–544 (1978).

117. M. Flavin, A.-M. Duprat, and J. Rosa, "Ontogenetic changes in the hemoglobins of the salamander, *Pleurodeles waltlii*," *Cell Differ.* **8**, 405–410 (1979).

118. M. Flavin, A.-M. Duprat, and J. Rosa, "Effect of thyroid hormones on the switch from larval to adult hemoglobin synthesis in the salamander *Pleurodeles waltlii*," *Cell Differ.* **11**, 27–33 (1982).

119. M. Flaving, H. Ton That, P. Deparis, and A.-M. Duprat, "Hemoglobin switching in the salamander *Pleurodeles waltlii*. Immunofluorescence detection of larval and adult hemoglobins in single erythrocytes," *Wilhelm Roux's Arch. Dev. Biol.* **191**, 185–190 (1982).

120. G. Gaillard and A. Jaylet, "Mécanisme cytologique de la tétraploidie expérimentale chez le Triton *Pleurodeles waltlii*," *Chromosoma* **51**, 125–133 (1975).

121. M. Galgano, "Evoluzione degli spermatociti di I origine e chromosomi pseudosesuali in alcune specie de anfibi," *Arch. Ital. Anat. Embriol.* **32**, 171–200 (1933).

122. C. L. Gallien, "Hybridation interspecifique par greffe nucleocytoplasmique chez deux espèces de tritons du genre *Pleurodeles* (Amphibien, Urodèle). Premiers résultats," *C. R. Acad. Sci. Paris* **265**, 1640–1643 (1967).

123. C. L. Gallien, "Recherches sur la greffe nucléaire interspécifique dans le genre *Pleurodeles* (Amphibien, Urodèles)," *Ann. Embryol. Morphogen.* **3**, 145–192 (1970).

124. C. L. Gallien, "Hybridization by means of interspecific nuclear transplantations and cross breeding among representatives of the genus *Pleurodeles* (Amphibia, Urodela)," *Sov. J. Dev. Biol.* **2**, 41–49 (1971).

125. L. Gallien, "Inversion du sexe (féminisation) chez l'Urodèle traité le benzoate d'oestradiol," *C. R. Acad. Sci. Paris* **231**, 919–920 (1950).

126. L. Gallien, "Inversion du sexe et effect paradoxal (féminisation) chez l'urodèle *Pleurodeles waltlii* Michah. traité par le propionate de testostérone," *C. R. Acad. Sci. Paris* **231**, 1092–1094 (1950).

127. L. Gallien, "Élevage et comportement du Pleurodèle en laboratoire," *Bull. Soc. Zool. Fr.* **77**, 456–461 (1952).

128. L. Gallien, "L'hétéroploidie expérimentale chez les Amphibiens," *Ann. Biol., 3 ser.* **29**, 5–22 (1953).

129. L. Gallien, "Inversion expérimentale du sexe, sous l'action des hormones sexuelles, chez le triton *Pleurodeles waltlii* Michah. Analyse des conséquences génétiques," *Bull. Biol. Fr. Belg.* **88**, 1–51 (1954).

130. L. Gallien, "Parabiose de larves tératologiques chez le triton *Pleurodeles waltlii* Michah.," *C. R. Soc. Biol.* **151**, 1085–1087 (1957).

131. L. Gallien, "Analyse des effets des hormones stéroides dans la différenciation sexuelle des Amphibiens," *Arch. Anat. Microsc. Morphol. Exp.* **48**, 83–100 (1959).

132. L. Gallien, "Recherches sur quelques aspects de l'hétéroploidie expérimentale chez le triton *Pleurodeles waltlii* Michah.," *J. Embryol. Exp. Morphol.* **7**, 380–393 (1959).

133. L. Gallien, "Haploïdie par exérèse du pronucleus femelle de l'oeuf fécondé chez le triton *Pleurodeles waltlii* Michah. et élevage en parabiose des larves obtenues," *C. R. Acad. Sci. Paris* **250**, 4038–4040 (1960).

134. L. Gallien, "Comparative activity of sexual steroids and genetic constitution in sexual differentiation of amphibian embryos," *Gen. Comp. Endocrinol., Suppl.* **1**, 346–355 (1962).

135. L. Gallien, "Spontaneous and experimental mutations in the newt *Pleurodeles waltlii* Michah.," in *Biology of Amphibian Tumors*, M. Mizell, ed., Springer-Verlag, Berlin (1969), pp. 35–42.

136. L. Gallien and A. Collenot, "Sur un mutant récessif léthal dont le syndrôme est associé à des perturbations mitotiques chez le triton *Pleurodeles waltlii* Michah.," *C. R. Acad. Sci. Paris* **259**, 4847–4849 (1964).

137. L. Gallien and M. Durocher, "Table chronologique du développement chez *Pleurodeles waltlii* Michah.," *Bull. Biol. Fr. Belg.* **91**, 97–114 (1957).

138. L. Gallien and M. Labrousse, "Radiosensibilité aux neutrons de l'oeuf fécondé chez l'Urodèle *Pleurodeles waltlii* Michah.," *C. R. Acad. Sci. Paris* **255**, 371–373 (1962).

139. L. Gallien, B. Picheral, and J. Lacroix, "Transplantation des noyaux triploides dans l'oeuf du triton *Pleurodeles waltlii* Michah. Développement de larves viables," *C. R. Acad. Sci. Paris* **256**, 2232–2233 (1963).

140. D. H. Garnier and C. Cayrol, "Oestrogènes plasmatiques chez *Pleurodeles waltlii*. Comparaison entre les femelles diploïdes et les femelles triploïdes," *C. R. Acad. Sci. Paris* **288**, 1407–1410 (1979).

141. F. Gasser, "Fractionnement des protéines sériques par électrophorèse sur acetate de cellulose chez quelques espèces de Salamandridae," *C. R. Acad. Sci. Paris* **258**, 457–460 (1964).

142. A. S. Ginsburg, *Fertilization in Fish and the Problem of Polyspermy*, Israel Program for Scientific Translations, Ltd., IPST Cat. No. 600418 (1971).

143. L. Glaesner, *Normentafeln zur Entwicklungsgeschichte des gemeinen Wassermolches (Molge vulgaris)*, Fischer Verlag, Jena (1925).

144. L. Gualandris and A.-M. Duprat, "A rapid experimental method to study primary embryonic induction," *Differentiation* **20**, 270–273 (1981).

145. C. Guillemin, "Comparaison de la transmission de trois trisomies par les mâles et par les femelles chez l'Amphibien Urodèle *Pleurodeles waltlii* Michahelles," *Ann. Génét.* **22**, 77–84 (1979).

146. C. Guillemin, "Effets phénotypiques de six trisomies et de deux double trisomies chez *Pleurodeles waltlii* Michahelles (Amphibien, Urodèle)," *Ann. Génét.* **23**, 5–11 (1980).

147. C. Guillemin, "Meiosis in four trisomic and one double trisomic males of the newt *Pleurodeles waltlii*," *Chromosoma* **77**, 145–155 (1980).

148. C. Guillemin and J. Générmont, "Analyse quantitative des morphologies à l'état adulte des eudiploides et de trois types de trisomiques chez *Pleurodeles waltlii* (Amphibien Urodèle)," *Ann. Génét.* **23**, 144–149 (1980).

149. F. Guillet and M. T. Chalumeau, "Apport de l'analyse électrophorétique et immunoélectrophorétique à l'étude de quelques protéines d'Amphibiens Urodèles au cours du développement normal et chez les hybrides nucléocytoplasmiques," *Mém. Soc. Zool. Fr.* **41**, 179–190 (1977).

150. F. Guillet and M. T. Chalumeau, "Cartographie et nomenclature des protéines sériques de l'Amphibien Urodèle, *Ambystoma dumerilii* Dugès," *C. R. Acad. Sci. Paris* **286D**, 1077–1080 (1978).

151. C. Houillon, "Analyse expérimentale des relations entre le canal de Müller et le canal de Wolff chez le triton *Pleurodeles waltlii* Michah.," *Bull. Biol. Fr. Belg.* **93**, 299–314 (1959).

152. C. Houillon, "Tolérance dans les chimères hétéroplastiques et xénoplastiques chez les Urodèles," *Bull. Soc. Zool. Fr.* **89**, 254–258 (1964).

153. C. Houillon, "Différenciation sexuelle des chimères chez les Amphibiens Urodèles," *Bull. Soc. Zool. Fr.* **101** (Suppl. 4), 103–107 (1976).
154. C. Houillon, "Greffes de gonades embryonnaires entre espèces différentes chez les Amphibiens Urodèles," *Bull. Soc. Zool. Fr.* **101** (Suppl. 4), 108–114 (1976).
155. C. Houillon, "Étude de l'immunité de greffe chez le Pleurodèle: restauration des animaux thymectomisés, induction de la réaction du greffon contre l'hôte," *Mém. Soc. Zool. Fr.* **41**, 71–82 (1977).
156. C. Houillon and C. Dournon, "Inversion du phénotype sexuel femelle sous l'action d'une température élevée chez l'Amphibien Urodèle, *Pleurodeles waltlii* Michah.," *C. R. Acad. Sci. Paris* **286D**, 1475–1478 (1978).
157. C. Houillon and C. Dournon, "Intersexualité chez le Pleurodèle consécutive à l'élevage à une température élevée," *Ann. Soc. Fr. Biol. Dév., Paris*, p. 27 (1979).
158. C. Houillon and C. Dournon, "Preuves génétiques et immunologiques de l'inversion du sexe dans les chimères hétérosexuées chez les Urodèles," *Arch. Anat. Microsc. Morphol. Exp.* **69**, 320–321 (1980).
159. A. Jaylet, "Technique de culture de leucocytes pour l'étude chromosomique d'Amphibiens Urodèles diploïdes et hétéroploïdes," *C. R. Acad. Sci. Paris* **260**, 3160–3163 (1965).
160. A. Jaylet, "Modification du caryotype par une inversion péricentrique à l'état homozygote chez l'Amphibien Urodèle *Pleurodeles waltlii* Michahelles," *Chromosoma* **35**, 288–299 (1971).
161. A. Jaylet, "Création d'une lignée homozygote pour une translocation réciproque chez l'Amphibien *Pleurodeles waltlii*," *Chromosoma* **34**, 383–423 (1971).
162. A. Jaylet, "Tetraploïdie expérimentale chez le triton *Pleurodeles waltlii* Michah.," *Chromosoma* **38**, 173–184 (1972).
163. A. Jaylet and C. Bacquier, "Accidents chromosomiques obtenus à l'état hétérozygote dans la descendance viable de mâles irradiés, chez le triton *Pleurodeles waltlii* Michah.," *Cytogenetics* **6**, 390–401 (1967).
164. A. Jaylet and P. Deparis, "An investigation into the origin of the thymocytes of the newt *Pleurodeles waltlii* using grafts between normal animals and animals with a marker chromosome," in *Developmental and Comparative Immunology*, Vol. 3, J. B. Solomon, ed., Pergamon Press, New York (1979), pp. 175–180.
165. A. Jaylet and V. Ferrier, "Experimental gynogenesis in the newt species *Pleurodeles waltlii* and *P. poireti*," *Chromosoma* **69**, 65–80 (1978).
166. A. Jaylet, V. Ferrier, and B. Andrieux, "Sur un facteur récessif apparu après mutagénèse aux rayons X affectant l'adaptation pigmentaire chez le Triton *Pleurodeles waltlii*," *C. R. Acad. Sci. Paris* **291**, 673–675 (1980).
167. I. F. Julien, M. F. Loones, B. Picard, and A. Mazabraud, "Mise en évidence de gènes en activité sur les chromosomes en écouvillon de l'amphibien urodèle *Pleurodeles*," *Biol. Cell.* **35**, 56a (1979).
168. T. King, "Nuclear transplantation in Amphibia," in *Methods in Cell Physiology*, Vol. 2, Academic Press, New York–London (1966), pp. 1–36.
169. C. Koch, "Von meinen altesten Urodelen," *Aquarien Terrarien Z.* **5**, 15–19 (1952).
170. J. P. Labrousse, "Synthèse de l'ADN dans l'oeuf de l'Amphibien Urodèle *Pleurodeles waltlii* Michah.," *Ann. Embryol. Morphol.* **4**, 347–358 (1971).

171. M. Labrousse, "Phénomènes cytologiques de la fécondation chez *Pleurodeles waltlii* Michah.," *Bull. Soc. Zool. Fr.* **84**, 493–498 (1954).

172. M. Labrousse, "Développement de tritons (*Pleurodeles waltlii*) présentant des aberrations chromosomiques provoquées par irradiation de l'oeuf," *Ann. Génét.* **8** (1965).

173. M. Labrousse, "Analyse des effets des rayonnements appliqués à l'oeuf sur les structures caryologiques et sur le développement embryonnaire de l'amphibien Urodèle *Pleurodeles waltlii* Michah.," *Bull. Soc. Zool. Fr.* **91**, 491–588 (1966).

174. M. Labrousse, "Aberrations chromosomiques induites et différenciation embryonnaire chez les Amphibiens," *Ann. Embryol. Morphogen., Suppl.* **1**, 199–210 (1969).

175. M. Labrousse, "Sur la localisation et la transmission d'une mutation cromosomique viable chez l'Amphibient Urodèle *Pleurodeles waltlii* Michah.," *Chromosoma* **33**, 409–420 (1971).

176. M. Labrousse, C. Guillemin, and C. Gallien, "Mise en évidence sur les chromosomes de l'Amphibien Urodèle *Pleurodeles waltlii* Michah. de secteurs d'affinité différente pour le colorant de Giemsa à pH 9," *C. R. Acad. Sci. Paris* **D274**, 1063–1065 (1972).

177. J. C. Lacroix, "Obtention de femelles trisomiques fertiles chez l'Amphibien Urodèle *Pleurodeles waltlii* Michah.," *C. R. Acad. Sci. Paris* **264**, 85–88 (1967).

178. J. C. Lacroix, "Étude déscriptive des chromosomes en écouvillon dans le genre *Pleurodeles* (Amphibien, Urodèle)," *Ann. Embryol. Morphogen.* **1**, 179–202 (1968).

179. J. C. Lacroix, "Variations expérimentales ou spontanées de la morphologie et de l'organisation des chromosomes en écouvillon dans le genre *Pleurodeles* (Amphibien, Urodèle)," *Ann. Embryol. Morphogen.* **1**, 205–248 (1968).

180. J. C. Lacroix and A. Capuron, "Sur un facteur récessif, modifiant le phénotype pigmentaire de la larve chez l'Amphibien Urodèle *Pleurodeles waltlii* Michahelles," *C. R. Acad. Sci. Paris* **D270**, 2122–2123 (1970).

181. J.-P. Lamon and A.-M. Duprat, "Effets de la concanavalline A sur la morphologie et le comportement de cellules embryonnaires d'Urodèles en différenciation *in vitro*," *Experientia* **32**, 1568–1573 (1976).

182. B. Lassale, "Surface potentials and the control of amphibian limb regeneration," *J. Embryol. Exp. Morphol.* **53**, 213–223 (1979).

183. M. Lauthier, "Étude descriptive d'anomalies spontanées des membres postérieurs chez *Pleurodeles waltlii* Michah.," *Ann. Embryol. Morphogen.* **4**, 65–78 (1971).

184. M. Lauthier, "Données histoenzymologiques sur les premiers stades du développement des membres de *Pleurodeles waltlii* Michah. (Amphibien, Urodèle)," *Wilhelm Roux's Arch. Dev. Biol.* **175**, 185–197 (1974).

185. M. Lauthier, "Étude ultrastructurale des stades précoces du développement du membre postérieur de *Pleurodeles waltlii* Michah. (Amphibien, Urodèle)," *J. Embryol. Exp. Morphol.* **38**, 1–18 (1977).

186. M. Lauthier, "Étude du développement des membres postérieurs de *Pleurodeles waltlii* Michah. (Amphibien, Urodèle) après la résection de l'épiderme du bourgeon," *Experientia* **34**, 790–791 (1978).

187. M. Lauthier, "Cell death and abnormalities in limb morphogenesis of *Pleurodeles waltlii* Michah. (Urodela, Amphibia) after nitrogen mustard treatment," *Wilhelm Roux's Arch. Dev. Biol.* **189**, 35–45 (1980).

188. F. Lemaitre-Lutz, "Anatomie des glandes pelviennes de la demelle de *Pleurodeles waltlii* Michah. Leur rôle de réceptacle séminal," *Ann. Embryol. Morphogen.* **1**, 409–416 (1968).
189. M. Lenfant, "Culture à court terme de cellules sanguines et analyse du caryotype chez les Amphibiens," *Ann. Embryol. Morphogen.* **6**, 55–62 (1973).
190. M. Leroux and C. Aimar, "Greffe nucléocytoplasmique intergénérique chez deux Amphibiens Urodèles: *Ambystoma mexicanum* et *Pleurodeles waltlii*," *C. R. Acad. Sci. Paris* **D266**, 1042–1044 (1968).
191. C. Mathieu, A.-M. Duprat, J.-P. Zalta, and J.-C. Beetschen, "Action du facteur de croissance nerveuse (nerve growth factor) sur la différenciation de cellules embryonnaires d'Amphibiens," *Exp. Cell Res.* **68**, 25–32 (1971).
192. C. Mathieu, A.-M. Duprat, J.-P. Zalta, and J.-C. Beetschen, "Effets des protéines nucléaires non histoniques sur des cellules embryonnaires d'Amphibiens cultivées *in vitro*," *C. R. Acad. Sci. Paris* **D273**, 292–294 (1972).
193. J.-P. Maufroid and A. Capuron, "Recherches récentes sur les cellules germinales primordiales de *Pleurodeles waltlii* (Amphibien, Urodèle)," *Mém. Soc. Zool. Fr.* **41**, 43–60 (1977).
194. J.-P. Maufroid and A. Capuron, "Différenciation du mésoderme et cellules germinales primordiales à partir de l'ectoderme de *Pleurodeles waltlii* associé temporairement à l'endoderme ventrale de gastrula," *C. R. Acad. Sci. Paris* **293**, 319–321 (1981).
195. A. Mauger, "Organogénèse de la colonne vertebrale et des côtés chez l'Urodèle *Pleurodeles waltlii* Michahelles," *Bull. Soc. Zool. Fr.* **87**, 163–187 (1964).
196. C. Michahelles, "Neue südeuropäische Amphibien," *Isis* **23**, 189–190 (1930).
197. M. Moreau, J.-P. Vilain, and P. Guerrier, "Free calcium changes associated with hormone action in amphibian oocytes," *Dev. Biol.* **78**, 201–214 (1980).
198. N. Moreau and D. Boucher, "Une méthode rapide d'extraction en masse des noyaux d'ovocytes de Pleurodèle (Amphibien, Urodèle)," *Biol. Cell* **42**, 185–188 (1981).
199. M. Olivereau, C. Bugnon, D. Fellmann, J.-M. Olivereau, and C. Aimar, "Etude cytoimmunologique de l'hypophyse du Pleurodèle," *C. R. Soc. Biol.* **170**, 1147–1151 (1976).
200. M. Olivereau, J.-M. Olivereau, and C. Aimar, "Modification hypohysaires et thyroidiennes chez le Pleurodèle en milieu salin," *Gen. Comp. Endocrinol.* **32**, 195–204 (1977).
201. M. Olivereau, J.-M. Olivereau, and C. Aimar, "Métamorphose et réponses cutanées et thyroïdiennes chez le Pleurodèle soumis à une aéroionisation expérimentale," *Gen. Comp. Endocrinol.* **40**, 149–160 (1980).
202. C. Orfila and P. Deparis, "Evolution des homogreffes cutanées chez la larve de *Pleurodeles waltlii* Michah.," *C. R. Soc. Biol.* **164**, 1124 (1970).
203. R. Ozon, "Analyse, *in vivo*, du métabolisme des oestrogènes au cours de la différenciation sexuelle chez le triton *Pleurodeles waltlii* Michah.," *C. R. Acad. Sci. Paris* **257**, 2332–2335 (1963).
204. R. Ozon and P. Dupuis-Certain, "Biosynthèse des hormones stéroïdes dans le tissue interrénal de l'Amphibien Urodèle *Pleurodeles waltlii* Michah.," *Gen. Comp. Endocrinol.* **9**, Abstr. 130 (1967).

205. J. Pasteels, "New observations concerning the maps of presumptive areas of the young amphibian gastrula (*Amblystoma* and *Discoglossus*)," *J. Exp. Zool.* **89**, 255–281 (1942).

206. C. Pastisson, "Evolution de la gonade juvenile et cycles sexuels chez le triton *Pleurodeles waltlii* Michah.," *Bull. Soc. Zool. Fr.* **88**, 364 (1963).

207. B. Pichéral, "Capacité des noyaux de cellules endodermiques embryonnaires à organiser un germe viable chez l'Urodèle *Pleurodeles waltlii* Michah.," *C. R. Acad. Sci. Paris* **225**, 2509–2511 (1962).

208. B. Pichéral, "Structure et organisation du spermatozoide de *Pleurodeles waltlii* Michah. (Amphibien Urodèle)," *Arch. Biol.* **78**, 193–221 (1967).

209. B. Pichéral, "Les élements cytoplasmiques au cours de la spermiogénèse du triton *Pleurodeles waltlii* Michah. I. La génèse de l'acrosome," *Z. Zellforsch.* **131**, 347–370 (1972).

210. B. Pichéral, "Les élements cytoplasmiques au cours de la spermiogénèse du triton *Pleurodeles waltlii* Michah. II. La formation du cou et l'évolution des organités cytoplasmiques non integrées dans le spermatozoide," *Z. Zellforsch.* **131**, 371–398 (1972).

211. B. Pichéral, "Les élements cytoplasmiques au cours de la spermiogénèse du triton *Pleurodeles waltlii* Michah. III. L'évolution des formations caudales," *Z. Zellforsch.* **131**, 399–416 (1972).

212. J. Rakotoarivony, F. Gasser, J.-C. Beetschen, and M. Flavin, "Mise en évidence d'un polymorphisme au niveau des protéines solubles de l'oeuf de *Pleurodeles waltlii* Michah. (Amphibien, Urodèle)," *C. R. Acad. Sci. Paris* **D273**, 1983–1986 (1971).

213. M. Reys-Brion, "L'effet des rayons X sur les potentialités respectives de l'ectoderme compétent et de son inducteur naturel chez la jeune gastrula d'Amphibien," *Arch. Anat., Microsc. Morphol. Exp.* **53**, 397–465 (1965).

214. S. Risbourg, J. Corsin, and C. Joly, "Rôle des glycoprotéines de la matrice au cours de la différenciation des chondroblastes," *Biol. Cell.* **33**, 7a (1978).

215. N. N. Rott, "Androgenesis in amphibians," *Usp. Sovrem. Biol.* **54**, 355–365 (1962).

216. N. N. Rott, "Sex determination in amphibians," *Byull. MOIP, Otd. Biol.* **68**(4), 118–134 (1963).

217. R. Rugh, *Experimental Embryology. Techniques and Procedures*, Burgess, Minneapolis (1948).

218. D. A. Sakharov, "The Spanish newt: Its keeping, breeding, and utilization as a laboratory animal," *Byull. MOIP, Ser. Biol.* **64**, 157 (1959).

219. U. Scheer, "Structural organization of spacer chromatin between transcribed ribosomal RNA genes in amphibian oocytes," *Eur. J. Cell. Biol.* **23**, 189–196 (1980).

220. U. Scheer, "Identification of a novel class of tandemly repeated genes transcribed on lampbrush chromosomes of *Pleurodeles waltlii*," *J. Cell Biol.* **88**, 599–603 (1981).

221. U. Scheer, "A novel type of chromatin organization in lampbrush chromosomes of *Pleurodeles waltlii*: Visualization of clusters of tandemly repeated very short transcriptional units," *Biol. Cell* **44**, 213–220 (1982).

222. U. Scheer, J. Sommerville, and M. Bustin, "Injected antibodies interfere with transcription of lampbrush chromosome loops in oocytes of *Pleurodeles*," *J. Cell. Sci.*, **40**, 1–20 (1979).

223. K. E. Sean, "Ultrastructure of erythrocytes of triton, *Pleurodeles waltlii*," *Can. J. Zool.* **58**, 1193–1199 (1980).

224. P. Sentein, "Le méchanisme normal de la mitose pendant la segmentation de l'oeuf d'Urodèle," *C. R. Acad. Sci. Paris* **253**, 547–549 (1961).

225. P. Sentein, "Action de la quinoline sur les mitoses de segmentation des oeufs d'Urodèles: le blocage de la centrosphere," *Chromosoma* **32**, 97–134 (1970).

226. J. Signoret, "Déterminisme de l'organisation des formations olfactives chez le triton *Pleurodeles waltlii* Michah.," *C. R. Acad. Sci. Paris* **249**, 1937–1939 (1959).

227. J. Signoret, "Anatomie de la région céphalique de *Pleurodeles waltlii* Michah.," *Bull. Soc. Zool. Fr.* **84**, 33–51 (1959).

228. J. Signoret and J. Fagnier, "Activation expérimentale de l'oeuf de *Pleurodeles*," *C. R. Acad. Sci. Paris* **254**, 4079–4081 (1962).

229. J. Signoret and B. Pichéral, "Transplantation des noyaux chez *Pleurodeles waltlii* Michah.," *C. R. Acad. Sci. Paris* **254**, 1150–1151 (1962).

230. J. Signoret, A. Collenot, and L. Gallien, "Description d'un nouveau mutant récessif léthal (*u*) et de son syndrôme chez le triton *Pleurodeles waltlii*," *C. R. Acad. Sci. Paris* **262**, 966–701 (1966).

231. N. Six and J. Brachet, "Effets biochimiques du bromure d'éthidium sur les oeufs de *Pleurodeles* en voie de développement," *Arch. Biol.* **82**, 193–210 (1971).

232. M. N. Skoblina and O. T. Kondratieva, "*In vitro* oocyte maturation in *Pleurodeles waltlii*. Role of follicle cells and of oocyte nucleus," in *6th All-Union Meeting of Embryologists. Abstracts* [in Russian], Nauka, Moscow (1981), p. 169.

233. M. N. Skoblina, K. K. Pivnitsky, and O. T. Kondratieva, "The role of the germinal vesicle in maturation of *Pleurodeles waltlii* oocytes induced by steroids," *Cell Differ.* **14**, 153–157 (1984).

234. H. Spring and W. W. Franke, "Transcriptionally active chromatin in loops of lampbrush chromosomes at physiological salt concentrations as revealed by electron microscopy of sections," *Eur. J. Cell Biol.* **24**, 298–308 (1981).

235. H. Steiner, "Bastardstudien bei *Pleurodeles* Molchem. Letale Fehlentwicklung in der F_2-Generation bein artispezifischer Kreuzung," *Arch. Z. Klaus.-Stift* **17**, 428–432 (1942).

236. H. Steiner, "Uber letale Fehlentwicklung des zweiten Nachkommenschaftsgeneration bei tierischen Artbastarden," *Arch. Z. Klaus.-Stift* **20**, 236–251 (1945).

237. G. Székely, "Functional specificity of cranial sensory neuroblasts in Urodela," *Acta Biol. Acad. Sci. Hung.* **10**, 107–116 (1959).

238. G. Székely, "Functional specificity of spinal cord segments in the control of limb movements," *J. Embryol. Exp. Morphol.* **11**, 431–444 (1963).

239. G. Székely and G. Czéh, "Activity of spinal cord fragments and limbs deplanted in the dorsal fin of Urodele larvae," *Acta Physiol. Acad. Sci. Hung.* **40**, 303–312 (1971).

240. G. Székely and J. Szentagothai, "Experiments with 'model nervous systems'," *Acta Biol. Acad. Sci. Hung.* **12**, 253–269 (1962).

241. A. Tournefier, J. Charlemagne, and C. Houillon, "Evolution des homo-greffes cutanées chez l'Amphibien Urodèle *Pleurodeles waltlii* Michah.: réponses immunitaires primaire et secondaire," *C. R. Acad. Sci. Paris* **D268**, 1456–1459 (1968).
242. J.-P. Vilain, "Maturation *in vitro* des ovocytes de *Pleurodeles waltlii* (Amphibien, Urodèle)," *Mém. Soc. Zool. Fr.* **41**, 93–101 (1977).
243. J.-P. Vilain, "Contributions des cations divalents au maintien du potentiel de membrane des ovocytes de *Pleurodeles waltlii* Michah. (Amphibien, Urodèle)," *C. R. Acad. Sci. Paris* **286D**, 1319–1322 (1978).
244. J.-P. Vilain, M. Moreau, and P. Guerrier, "Uncoupling of oocyte-follicle cells triggers reinitiation of meiosis in amphibian oocytes," *Dev., Growth Differ.* **22**, 687–691 (1980).
245. W. Vogt, "Gestaltungsanalyse am Amphibienkeim mit örtlicher Vitalfärbung. II. Gastrulation und Mesodermbildung bei Urodelen und Anuren," *Wilhelm Roux's Arch. Entwicklungsmech. Org.* **120**, 382–706 (1929).
246. M. A. Vorontsova, L. D. Liozner, I. V. Markelova, and E. Ch. Pukhalsky, *The Newt and the Axolotl* [in Russian], Nauka, Moscow (1952), pp. 12, 31.
247. T. Wickbom, "Cytological studies in Dipnoi, Urodela, Anura, and Emys," *Hereditas* **31**, 241–346 (1945).
248. A. J. Zamora, "The ependymal and glial configuration in the spinal cord of urodeles," *Anat. Embryol.* **154**, 67–82 (1978).
249. A. J. Zamora, "Pansegmental primordial glycogen body in the spinal cord of postmetamorphic *Pleurodeles waltlii* (Urodela)," *Anat. Embryol.* **154**, 83–94 (1978).
250. A. J. Zamora and D. Thiesson, "Tight junctions in the ependyma of the spinal cord of the urodele *Pleurodeles waltlii*," *Anat. Embryol.* **160**, 263–274 (1980).

Chapter 8

THE AXOLOTL *Ambystoma mexicanum*

N. P. Bordzilovskaya and T. A. Dettlaff

The axolotl is one of the classical subjects of developmental biology. Its advantages are the ease of keeping breeding stock in the laboratory, long season of reproduction, rapid sexual maturation, and the relatively large size of the eggs and their resistance to various experimental influences. In addition, the adult animals have a high ability of regeneration.

The occurrence of black and white strains, as well as of mutant ones [4, 7, 9, 11, 40, 42–45, 47, 62] makes it possible to have genetically marked material and to study the realization of mutation-dependent features during development [1, 6, 8, 10, 13, 24, 48, 57, 58, 60, 94].

Axolotl eggs have been successfully used for nuclear transplantations [11, 70, 85] and for many other studies, in particular those relating to the following: primary embryonic induction and ectoderm competence [21–23, 32, 34, 55, 56], formation of mesoderm, inducing effect of endoderm and somitogenesis [63, 66–69, 98, 99], determination of the material of the sense organs [28–30, 78–82], and localization and ultrastructure of primordial germ cells [52]. Using dimensionless (relative) criteria of developmental duration [17, 18], the following problems have been studied: dynamics of nuclear transformation in the process of fertilization [74], duration of different mitotic phases of the first cleavage divisions [88], cell cycle rearrangement during cleavage [76, 85] and onset of RNA synthesis [50], patterns of cytokinesis and its relationships to nuclear division [36, 37, 75, 86, 100], localization of grey crescent [33, 64], etc.

Maps of presumptive rudiments (Fig. 8.1) have been constructed for axolotl embryos at the early gastrula stages [61, 71], and the dynamics of morphogenetic movements during gastrulation have been studied [49, 53, 54].

Tables of normal development of the axolotl were published almost simultaneously by Bordzilovskaya and Dettlaff [3] and Schreckenberg and Jacobson [83].

The axolotl occupies an important place in studies of regeneration [15, 91–93, 96, 97].

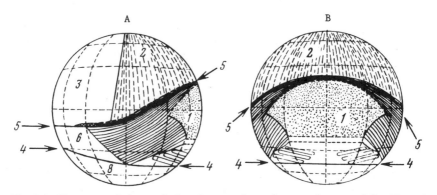

Fig. 8.1. Fate map of the axolotl embryo at the early gastrula stage (after Pasteels [71] and Lovtrup [61]). A) Lateral view; B) dorsal view. 1) Notochord; 2) neural plate; 3) epidermis; 4) site of beginning of invagination; 5) boundary of invaginating material; 6) mesoderm of lateral plates; 7) somites (solid shading); 8) endoderm.

8.1. TAXONOMY, DISTRIBUTION, AND KARYOTYPE

The axolotl *Ambystoma mexicanum* (Cope) (= *Siredon mexicanum*) is a neotenic larva of *Ambystoma* and belongs to the family Ambystomidae, order Urodela, class Amphibia.

The axolotl was initially erroneously included in the nonmetamorphosing group Perennibranchiates and given the name *Siredon pisciformis* [65]. However, after some axolotls in the Paris Botanical Gardens had spontaneously metamorphosed in 1856, this species was renamed *Ambystoma mexicanum*. Still some scientists [14, 26] continued to call it *Siredon pisciformis* rather than *S. mexicanum*. At one time an erroneous spelling of its name with an "l" was widespread, i.e., *Amblystoma* [19, 65].

In its native habitats, in Mexico, the axolotl is found throughout the year in cold mountain lakes and attains sexual maturity in the larval state [90]. It metamorphoses spontaneously but very rarely and in unusual laboratory conditions. The metamorphosed form is called Ambystoma. *Ambystoma tigrinum* is closely related to *A. mexicanum* and they are easily hybridized (see [26, 44]).

The karyotype of the axolotl has been elucidated by Signoret [84] and Hauschka and Brunst [39]. Out of 28 chromosomes, 7 pairs are metacentrics and 7 pairs are acrocentrics. The pairs of chromosomes are numbered 1–14 in decreasing order of their lengths (see Fig. 8.2).

8.2. MANAGEMENT OF REPRODUCTION IN THE LABORATORY

8.2.1. Conditions for Keeping Adults and Larvae

The axolotls are easily bred in aquaria, and colonies are maintained in many laboratories of the world. They can live 8–10 years in captivity [96]. Unlike fe-

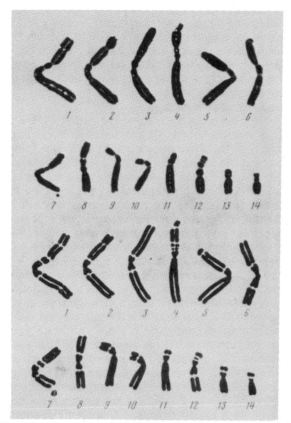

Fig. 8.2. Karyotype of axolotl blastula cells after keeping them for 100 h at 0°C. One chromosome is presented from each pair of homologs (1–14) (after Signoret [84]).

males, the males have quite distinct, strongly developed, cloacal swellings at the tailbase. In addition, females are stouter than males. By these features the sex of axolotls is quite easily distinguished.

Axolotls are very resistant to low temperatures. They can live under ice for some time, but tolerate heat poorly. The tolerance temperature range of axolotls is 8–24°C [12]. At 18—21°C they live quite well and spawn. Detailed recommendations for keeping sexually mature axolotls and larvae in the laboratory are given by Vorontzova et al. [96], Humphrey [41, 42, 44–46], Brunst [14], New [65], Fankhauser [25, 26], and Billett and Wild [2].

Very useful information on axolotl colonies in different laboratories, conditions of keeping and reproduction, diseases and mutant strains, and abstracts and bibliography of relevant investigations, can be found in the *Axolotl Newsletter* (from 1976), published by the Axolotl Colony of Indiana University (G. Malacinski, ed.). We are grateful to Profs. G. Malacinski, A. J. Brothers, and R. V. Fremery for their kind permission to use their data, published in the *Axolotl Newsletter*, in the present paper.

The conditions of keeping and reproduction of axolotls in different laboratories greatly vary. At Indiana University [12, 47], adult axolotls are kept singly in small glass aquaria or large basins in 50% modified Holtfreter solution (102 mg $MgSO_4 \cdot 7H_2O$ per liter). Young animals (from 2 months on) are kept in the same solution. All axolotls are fed by hand with thin pieces of beef or calf liver, which are kept in the frozen state, as well as earthworms. Each animal receives as many meat pieces as it actively takes. The juveniles are fed six times a week, and adult three times. The water is changed completely within 1.5 h after every feed.

Brunst [14] recommends keeping 1–2 adults or 3–4 six-month-old animals or more numerous younger animals in a 4- to 5-liter vessel two-thirds filled with dechlorinated water. Water should be changed frequently, preferably once a day, and, in addition, within several hours after feeding. If the water is replaced by colder water, which induces defecation, it is necessary to remove the food residue and feces from the vessel bottom the next day, using a large pipette. The addition of cold water prior to feeding decreases the appetite of the animals.

Brunst recommends feeding axolotls with liver, as well as calf and beef meat (without bones, fat, and tendons). The meat is cut in pieces necessary for one feeding, and frozen. Before feeding every frozen piece is cut into thin strips (4–5 mm wide and 30–40 mm long) for the adult axolotl and still smaller for larvae. The axolotls do not take meat from the bottom, and one should patiently feed them by waving a piece of meat held by long forceps in front of the head. To heighten their appetite, they are sometimes fed with earthworms, fish, tadpoles, or newborn mice instead of meat.

It is recommended that the axolotls should be fed two or three times a week, and on the day of feeding to give them food twice or at more frequent intervals. It is important that the pieces of meat are not too large, since the animals may regurgitate them afterward, and not too small since hungry animals kept together can bite off legs from each other. If this takes place, the wounded animals are placed for some time in ice-cold water in the refrigerator: under these conditions the wounds heal more rapidly and do not become infected.

It is necessary to isolate diseased animals. They are difficult to treat since their diseases are not well known [12, 20, 27, 65, 72, 73].

In the Hubrecht Laboratory [27] adult axolotls are kept in asbestos cement containers (50 × 100 × 50 cm), 10 in each. The inner surface of such container is coated completely with a waterproof chemically inert varnish. The water level is kept at about 35–40 cm. Some algal growth on the bottom and side walls keeps the water clear. Each container has its own drain. The axolotls are kept in very slowly running tap water of good quality, with pH 7.5–8.5 and at 14–18°C. The containers are illuminated 12 h a day by fluorescent lamps. Half of each container is covered with a lid to provide shelter for the animals. The containers are placed in an air-conditioned room and gaseous exchange is promoted by a mechanical water aerator. The animals are fed three times a week with beef meat cleaned of fat and tough fibers, which is cut into strips of 30 × 4 × 4 mm, rinsed with tap water, and mixed with a multivitamin and mineral preparation. Animals kept in groups seek the food spontaneously. Before and 1.5 h after feeding the containers are siphoned out to remove feces and food remains. If an animal falls ill and dies, it is siphoned out and the siphon is afterwards carefully disinfected in 1% Tego. The other animals remain in the container, and if during 6 weeks no more animals die, the container is considered safe. Males and females are kept separately. During the season of reproduction each group is used twice for spawning at an interval of 3 months. Every 2 or 3 years a new generation is introduced.

8.2.2. Incubating Embryos and Rearing Larvae

At Indiana University the embryos and hatched larvae up to 1 month of age are kept in 20% Steinberg solution at pH 7.0–7.4, and larvae from 1–2 months in 20–25% Holtfreter solution with addition of 51 mg $MgSO_4 \cdot 7H_2O$ per liter. In many laboratories the embryos are incubated until hatching, and the larvae are kept in dechlorinated water. The nondeveloping eggs and dead embryos are siphoned out and the water is changed once a day. At 18–20°C incubation to hatching lasts about 2 weeks.

The hatched larvae are transferred to glass vessels. The food should be given just before the stage when the mouth opens, prior to the complete digestion of yolk. Small crustaceans, just-hatched shrimp (Indiana University), and small *Cyclops* and water fleas (Institute of Developmental Biology, Moscow) serve as food. The food is given 3–6 times a day, and the water is changed after each feeding. *Tubifex* cut into pieces can be given within 1 week after the beginning of active feeding. At the age of 1 month, young axolotls (about 2 cm long) already eat large water fleas, *Tubifex*, and small midge larvae. At Indiana University colonies at this age are given calf liver or heart and also *Xenopus* larvae and cut earthworms. Five- to six-month-old individuals switch to a diet of chow mixture. The water is changed within 1.5 h after each feed [12].

At this time, in addition to regular feeding, removal of dead larvae and food residues, changes of water, and sorting out the animals by size is of great importance, otherwise larger larvae and young eat the younger ones. As the young grow, the density of animals is gradually decreased.

8.2.3. Reproductive Behavior and Collection of Eggs

Axolotls attain sexual maturity within 10–12 months at 20–21°C, but the maximum fecundity is reached by the age of 2–3 years; the intensity of reproduction decreases markedly after the age of 6 years (see [26, 65]). In the laboratory they spawn usually from November until May, but sometimes in summer as well. One female may be induced to lay eggs 2–3 times a year with an interval of not less than 2 months. The males can be used more frequently (Gudkov, personal communication). Readiness for reproduction can be judged by the increase of the cloacal swellings in males and of the abdomen in females.

If females and males are kept separately, their transfer to the same aquarium can alone stimulate their reproduction ([2]; Gudkov, personal communication). In some cases mating behavior is stimulated by change of temperature. Billett and Wild [2] recommend lowering the temperature in the aquarium by adding ice, but by no more than 5°C and not below 10°C.

Humphrey and Brunst [14] usually put a male with a female in an aquarium (30–45 cm) or a round vessel (37 cm in diameter), with a layer of small-sized gravel or large-sized sand on the bottom so that spermatophores can be attached.

In those cases when the origin of fertilized eggs is of no importance, one male is placed with two or three females, or two males with four or five females (Gudkov, personal communication). A male and a female are placed together early in the evening and left for the night in a quiet place. When the spawners are chosen successfully, nuptial behavior is soon manifested and, after a certain period of time, the male releases spermatophores which almost look like transparent pyramids with a denser apex consisting of a nontransparent mass of spermatozoa. The male lays 1–20 spermatophores on the ground and the female swims over them and draws a spermatophore into the cloaca. Egg laying begins after insemination, usually within 18–30 h after the animals are placed together, and lasts 24–48 h.

A female which has begun egg laying can be transferred into an aquarium with a clear bottom, without gravel and sand. The eggs will fall down and are easily collected. It is, however, more convenient to place in the aquarium some objects on which the eggs can be laid, such as aquatic plants (e.g., *Elodea*), rubber plants, glass rods, or pieces of rubber tubing. The female clamps the tube between her hindlegs and lays eggs or groups of eggs on it.

The plants or tubing can be taken out periodically and the eggs laid during a definite time interval can be collected. If this interval does not need to be very short, it is better not to disturb the female too often since it can stop laying eggs for some time. During 24 h a female can lay 300–600 and, sometimes during the whole period of laying, up to 1100 eggs (see [65]). The eggs are usually laid in groups of 5–15 eggs each.

8.2.4. Stimulated Ovulation and *in vitro* Oocyte Maturation

Egg maturation and ovulation can be artificially stimulated by gonadotropic hormones in those cases when unfertilized eggs are needed. With this aim, Fankhauser [25] injected FSH (180–200 units) or implanted frog pituitaries in females.

In vitro oocyte maturation was obtained by Brachet [5], who used oocytes from 3-year-old females which had not yet spawned; he placed fully grown oocytes for 30–60 min in a saline Barth's solution with progesterone (3 μg/ml), and then transferred them into a solution without hormone. Brachet followed oocyte maturation until the extrusion of the 1st polar body. *In vitro* oocyte maturation in a modified Barth's solution under the effect of progesterone was also obtained by Grinfeld and Beetschen [33]. Oocyte maturation in their experiments proceeded for 14 h until extrusion of the 1st polar body, i.e., $8.4\tau_0$.

8.3. GAMETES, FERTILIZATION, AND ARTIFICIAL INSEMINATION

8.3.1. Structure of Egg

The mature egg of the axolotl is at metaphase II. It is, as in all amphibians, telolecithal, holoblastic, and polylecithal. The animal region of the egg is pigmented and contains a smaller amount of relatively smaller yolk granules than the vegetal region. The latter is of yellowish-white color. An electron microscopic study revealed cortical bodies in the egg cortex [31]. The breakdown of these bodies begins when the eggs are still being transported in the lower part of oviducts. The egg is covered by a low-elastic transparent vitelline membrane, a multilayered transparent capsule envelope, and a jelly layer which often coats several eggs into one clump. The space between the vitelline membrane and the capsule is filled by capsule fluid which exerts a great pressure on the capsule wall (Fig. 8.4, 1). The egg size (without capsule) varies in different females from 1.85–2.00 mm.

8.3.2.   Fertilization

In the axolotl fertilization is internal and, as a rule, polyspermic. An egg is penetrated by 1–15 spermatozoa [74, 89], 4–5 on the average. In cases of polyspermic fertilization (about 70% of the eggs) the spermatozoa penetrate the egg from different sides, and in cases of monospermic fertilization, usually in the animal half [74]. The eggs are normally fertilized when they are transported through the cloaca past spermatheca which contains the spermatozoa. The eggs isolated from oviducts are always unfertilized [65]. Fertilization occurs just prior to egg laying. Eggs from which the jelly coat is removed within 5 min after their emergence from the oviducts are already fertilized [74]. Within 10–15 min black spots with radial margins are visible with the naked eye, where the sperm have penetrated (Fig. 8.4, 1*an*), and within 25–30 min a large perivitelline space forms in the fertilized eggs (Fig. 8.4, 1). Unfertilized eggs are found among the laid eggs and their percentage varies in different females and even in different portions of the eggs from the same female.

8.3.3.   Artificial Insemination

Artificial insemination of the eggs is possible, but both a male and a female have to be killed. The method of artificial insemination suggested by Brunst [14] consists of the following steps:

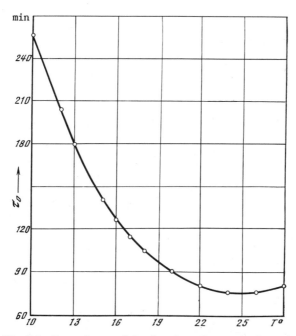

Fig. 8.3. Dependence of the duration of one mitotic cycle during the synchronous cleavage divisions (τ_0) on temperature in the axolotl (after Skoblina [88]).

1. A female is injected with 180–220 units FSH (Fankhauser [25]) and then continuously stimulated during several hours to prevent it from laying eggs. A large quantity of unfertilized eggs accumulates in the lower parts of the uterus and oviducts.

2. The female is decapitated and spined. The body wall is cut through, the oviducts are removed and opened up, beginning from the caudal part, and the eggs are removed with forceps.

3. The removed eggs are placed in a row on pieces of filter paper moistened with dechlorinated water and transferred to the bottom of a flat dish which is closed tightly with a lid to keep the eggs from drying up. In the moist chamber the eggs remain fertilizable for half an hour or more.

4. A male is sacrificed in the same way; the body wall is cut through, vasa deferentia are cut out at the caudal end and clipped with forceps so that the contents do not leak out, pulled out as much as possible, taken out, and transferred to a vessel for a suspension to be prepared. With this aim the vasa deferentia are cut into small pieces and the sperm is squeezed out with forceps.

5. The suspension of sperm is prepared either in 10% Ringer's solution or in dechlorinated water. The contents of a vas deferens are suspended in 10 ml of the solution or water. The suspension is stirred, added in a thin layer with a pipette to the eggs, and the dish is closed with a lid. The suspension should be prepared just before fertilization, since the spermatozoa lose their activity very rapidly.

6. Within 15–20 min water is added and a few minutes later the fertilized eggs are taken with needles from the paper and placed in a large vessel with water. Unfertilized eggs are discarded.

8.4. NORMAL DEVELOPMENT

Despite the large number of studies on axolotl embryos, tables of normal development were for a long time absent. Such tables were composed by Harrison [38] for *Ambystoma maculatum* (= *A. punctatum*). Illustrations from this table were published long ago [35, 77, 96], but the tables with descriptions of structure of the embryos were published posthumously [38]. These tables can be used for determination of axolotl developmental stages until the end of neurulation only, since at the later stages the embryos of *A. mexicanum* and *A. maculatum* differ markedly from each other. We have therefore undertaken a special study to construct such tables for axolotl embryos and to determine the time of the beginning of successive developmental stages [3].

The eggs were taken soon after laying, the time of appearance of the 1st cleavage furrow on the animal pole was determined for each egg, and the eggs were then placed individually in small Petri dishes in an ultrathermostat at 20°C.

During cleavage, gastrulation, and neurulation the developmental stages for the axolotl were distinguished according to Harrison's tables for *A. maculatum* [38]. In the complex of features characteristic for every stage, those features were chosen

---

Figs. 8.4–8.9. Stages of normal development of the axolotl *Ambystoma mexicanum* (drawings by N. P. Bordzilovskaya). Drawing numbers (1–44) correspond to stage numbers. an) Animal view; veg) vegetal view; d) dorsal view; v) ventral view; t) tail view. Drawings without letter designations are lateral views.

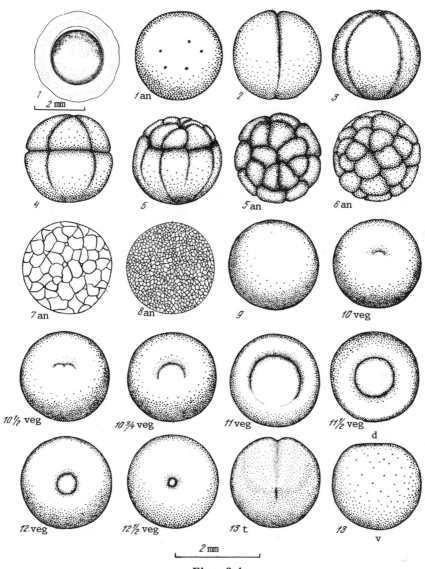

Fig. 8.4.

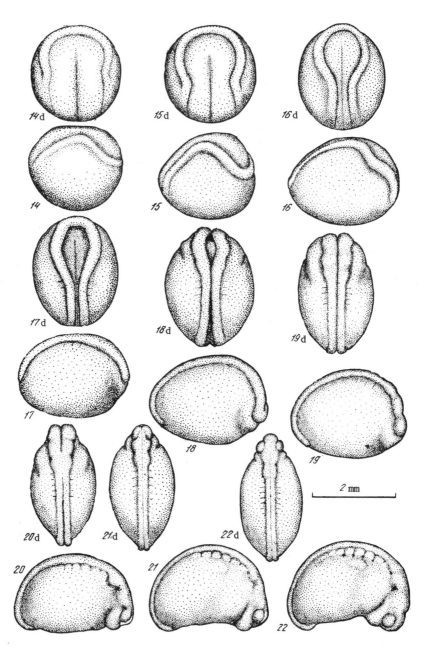

Fig. 8.5.

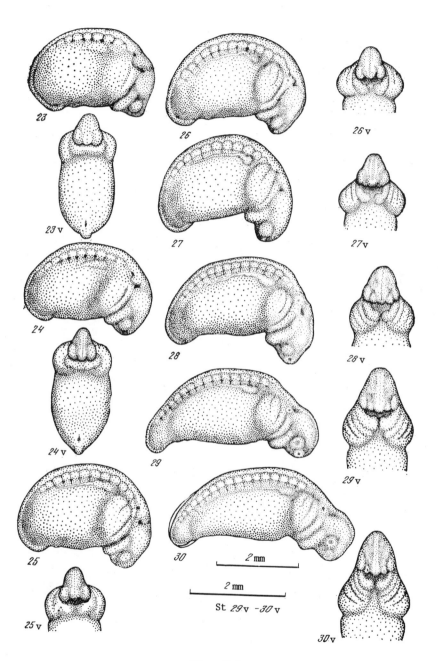

Fig. 8.6.

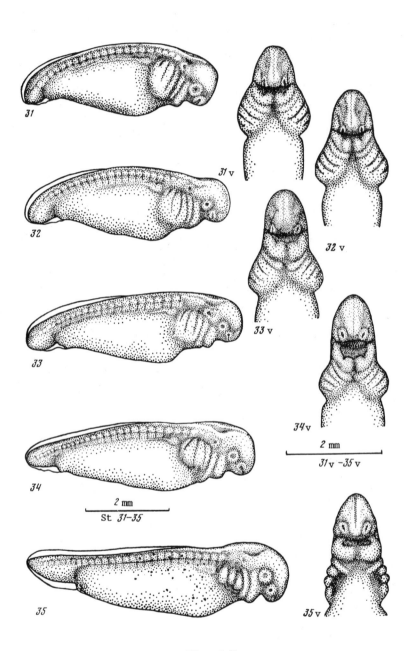

Fig. 8.7.

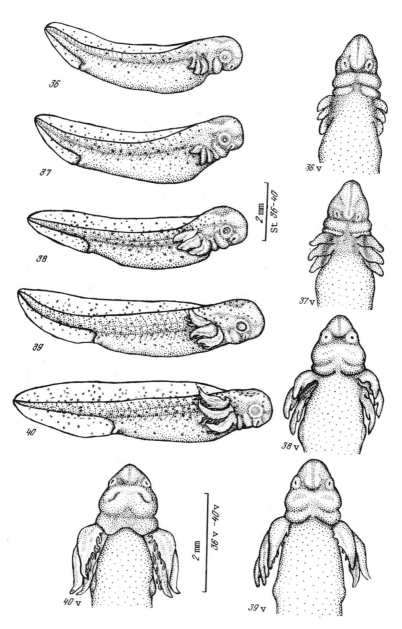

Fig. 8.8.

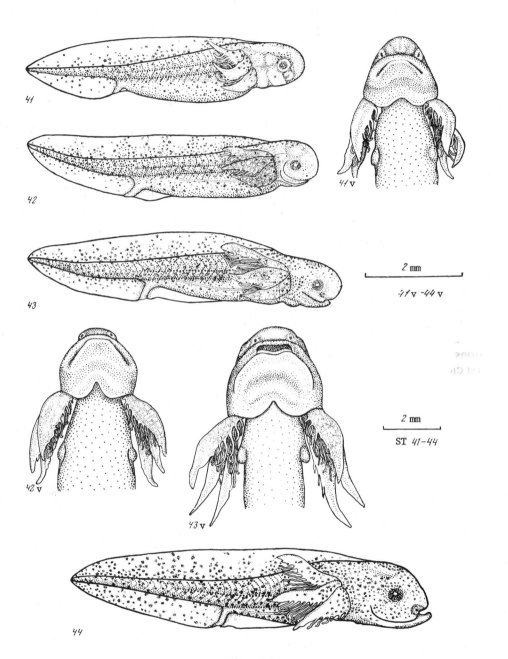

Fig. 8.9.

which are common for embryos of both species and which, in addition, allow one to distinguish the given stage from the preceding one and to determine the precise time at which it begins. The criteria of later stages, and their timing, have been defined more accurately on fixed material; this was carried out up to stage 35, at first with an interval of 0.5 and $1.0\tau_0$, then of $5\tau_0$, and the last stages with a still greater interval of about 10 and $20\tau_0$. The stages were oriented to some precisely definable features. Three to six groups of eggs that began to cleave at different times were fixed for every interval. The beginning of stages which are transitional between different periods of development, such as onset of gastrulation, slitlike blastopore, closure of neural folds, was determined on a larger number of eggs and with an increased frequency of observations. The eggs of five females were used.

Drawings of external morphological features at successive developmental stages were made using a drawing device. The drawings and the descriptions of diagnostic features for every stage were prepared by N. P. Bordzilovskaya.

Times at which different stages begin were determined on more advanced embryos for every group of eggs, thus providing, as experience has shown [18, 51, 101], more precise and reproducible results. However, since among naturally laid axolotl eggs it is difficult to have many simultaneously fertilized eggs, or to provide great frequency of observations, we can say only conditionally that the values in our table are indeed obtained on the more advanced embryos.

To obtain relative characteristics of the duration of different periods of development (τ_n/τ_0), the time from the appearance of the 1st cleavage furrow on the egg surface until the onset of the corresponding developmental stage (τ_n in minutes) is divided by the value of τ_0 (in minutes) at the same temperature. At 20°C this parameter varies from 87–93 (Skoblina, unpublished; see Fig. 8.3). In our calculations we take τ_0 as 92 min. Table 8.1 shows the time from the appearance of the 1st cleavage furrow to the corresponding stage in hours and minutes and in number of τ_0 units (τ_n/τ_0). For a number of stages the variability of the value τ_n/τ_0 is shown for embryos from different groups of eggs of the same female.

For the periods of cleavage and gastrulation our data agree fairly well with those obtained earlier by Skoblina [87] at 18 and 22°C and ten Cate [16] at 16, 18, 22, and 23°C (recalculated by Skoblina in τ_0 units), as well as with those of Valouch et al. [95] for the period of cleavage at 18–23°C. Beyond these temperatures, the ratio τ_n/τ_0 begins to change. In the interval 26–28°C, τ_0 ceases to shorten at first with a rise in temperature and then even lengthens, thus suggesting the damaging effect of these temperatures on early cleavage (see Fig. 8.3). The duration of the later and less-sensitive developmental periods continues to shorten at these temperatures and, as a result, the values of τ_n/τ_0 decrease as compared with those at optimum temperatures [18, 87]. At low temperatures (10°C; see ten Cate [16]), on the contrary, the process of gastrulation lengthens disproportionately and the closure of the neural folds is delayed. The values of τ_0 at different temperatures (Fig. 8.3) and the ratio τ_n/τ_0 can thus be used for predicting the timing of different stages of development within the middle range of optimum temperatures only where the duration of different developmental periods changes proportionally with the temperature. One should take into account that variations of the time of transition to the next stage in different embryos are usually within 10% of the total duration of the given stage.

With a small time interval between successive stages, individual differences in the stage duration for different simultaneously fertilized embryos usually exceed

TABLE 8.1. Normal Developmental Stages of the Axolotl (incubation at 20°C)

Stage No.	Time from the 1st cleavage furrow		Size, mm (D, diameter; L, length, W, width, H, height)	Diagnostic features of stages
	hours, minutes	τ_n/τ_0		
1	–	–	D 1.85–2.00	Freshly laid fertilized egg, in membranes
$1^1/_2$	–	–	D 2.00	Activated egg, a wide perivitelline space formed
2–	0	0		1st cleavage furrow on the animal pole – reference point for reading the time (no drawing)
2	1.05	0.7	D 2.0 (average)	2 blastomeres
3	2.40	1.7	D 2.0	4 blastomeres
4	4.12	2.7	D 2.0	8 blastomeres
5	5.22	3.5	D 2.0	16 blastomeres. Divisions of vegetal blastomeres lag behind those of animal ones
6	7.06	4.5	D 2.0	32 blastomeres
7	8.26	5.5	D 2.0	64 blastomeres
8	16.00	10.5	D 2.0	Early blastula (fall of mitotic index [76])
9	21.28	14	D 2.0	Late (epithelial) blastula. Surface smooth
$9^1/_2$	24.32	16	D 2.0	Onset of morphogenetic (nuclear) function [50]
10	26.00	17	D 2.0	Early gastrula I. First signs of dorsal blastopore lip formation
$10^1/_2$	32.10	21	D 2.0	Early gastrula II. Invagination continues.Blastopore is almost a horizontal slit (one quadrant)
$10^3/_4$	37.00	24	D 2.0	Middle gastrula I. Dorsal lip of blastopore forms a semicircle
11	38.30	25–26	D 2.1	Middle gastrula II. Blastopore occupies 3 quadrants of circle. Lateral lips formed, ventral lip marked by pigment accumulation only. Yolk plug reaches a maximum diameter of 1.2 mm
$11^1/_2$	41–43	27–28	D 2.1 (average)	Late gastrula I. Blastopore forms a closed circle. Invagination continues. Yolk plug decreases twice (D = 0.6 mm)

TABLE 8.1 (continued)

| Stage No. | Time from the 1st cleavage furrow | | | Diagnostic features of stages |
	hours, minutes	τ_n/τ_0	Size, mm (D, diameter; L, length, W, width, H, height)	
12	47.30	31	D 2.1	Late gastrula II. Blastopore has an oval or circular form. Size of yolk plug 0.4 × 0.5 mm
12$\frac{1}{2}$	49–52	23–34	D 2.1	Late gastrula III. Closing oval blastopore. Size of yolk plug 0.15 × 0.2 mm
13–	50.30–54	33–35		Stage of slitlike blastopore. Boundaries of neural plate are still indistinct
13	55–56	36–37	D 2.1	Early neurula I. Blastopore is a narrow vertical slit. Groove in the midline of neural plate. Boundaries of neural plate are outlined but neural folds are not yet elevated. Dorsal side is slightly flattened
14	58.15	38	D 2.2	Early neurula II. Neural plate is wide. Neural folds outlined and slightly elevated in the head region. Embryo slightly elongated
15	59.50	39	L 2.25 W 2.1	Early neurula III. Neural plate the shape of a shield. Neural folds elevated and now confine all the area of neural plate
16	63	41	L 2.15 W 2.1	Middle neurula. Neural folds become higher. The dorsal region of neural plate narrows. Neural plate sinks inside
17	64.30	42	L 2.35 W 2.0	Late neurula I. Neural folds become higher, especially in the head region. Further narrowing and sinking of neural plate both in the head and dorsal regions. Hyomandibular groove limiting the mandibular arch is outlined (yet very slightly). Segmentation of mesodermal material begins. 2 pairs of somites delineated

TABLE 8.1 (continued)

Stage No.	hours, minutes	τ_n/τ_0	Size, mm (D, diameter; L, length, W, width, H, height)	Diagnostic features of stages
	Time from the 1st cleavage furrow			
18	66– 67.30	43–44	L 2.4 W 1.9	Late neurula II. Neural plate is deeply sunken. Neural folds approach each other; they are especially high in the head region where 3 expansions corresponding to prosen-, mesen-, and rhombencephalen are delineated (yet very slightly). Neural folds in dorsal region are about to touch. Hyomandibular groove becomes more pronounced. 2 pairs of somites (when counting the number of somite pairs under SEM at stage 22 it was found to be somewhat higher; see [63])
19	69.00	45	L 2.7 W 1.7 H 2.1	Late neurula III. Neural folds touching all over but do not yet fuse. Brain curvature is quite distinct in profile. 3 brain vesicles are also distinct. Eye protrusions are outlined. Hyomandibular groove becomes deeper. 3 pairs of somites
20	70.30	46	L 2.7 W 1.5 H 1.7	Late neurula IV. Neural folds fused in dorsal region, but still touching in head region. Eye vesicles are distinct and increasing. Grooves in ectoderm appear at the level of rhombencephalon. Future gill area slightly outlined. Mandibular arch becomes prominent. 4 pairs of somites
21	72.00	47	L 2.7 W 1.5 H 1.8	Late neurula V. Neural folds completely fused. Posterior boundary of gill region more distinct. Pronephros outlined (a very small expansion). Ventral side of embryo slightly concave. Head region (from the level of mandibular arch) somewhat bent downward. Dorsal side semicircular. Occipital and parietal brain curvatures seen. 4 pairs of somites
22	73.00	48	L 2.8 W 1.4 H 2.3	Gill area and pronephros are now distinct. Tailbud is delineated (yet very slightly). Body elongates. Ventral side more concave due to greater bending of the head region downward. 5–6 pairs of somites

TABLE 8.1 (continued)

Stage No.	Time from the 1st cleavage furrow		Size, mm (D, diameter; L, length, W, width, H, height)	Diagnostic features of stages
	hours, minutes	τ_n/τ_0		
23*	~74	~48.5	L 3.5 W 1.35	Ear rudiment as a shallow depression in epithelium above future hyoid arch. In dorsal part of gill area hyobranchial groove is delineated and separates hyoid arch from 1st branchial arch I. Tailbud still small. 6–7 pairs of somites
24	80	52	L 3.0 W 1.35 H 1.85	Ear pit becomes more distinct. Hyobranchial groove elongates ventrally. Pronephros is distinct: not only pronephros itself is seen but beginning of pronephric duct as well. 8–9 pairs of somites
25	83	54	L 3.25 W 1.45 H 2.0	Gill area continues to increase. Hyobranchial groove lengthens. 1st gill cleft delineated (yet very slightly) also in dorsal region. Tailbud still small. Body elongates. Ventral side more concave; head bends more downward. 9–10 pairs of somites
26	84.30	55	L 3.25 W 1.6 H 1.9	Ear pit quite distinct. 1st gill cleft becomes more pronounced and longer. Gill area is a considerable distinct swelling, its height accentuated by a deep groove separating developing pronephros from gill area. Pronephric duct runs from pronephros along 6 somites. Olfactory rudiment appears as a protrusion in anterior head part. Tailbud gradually increases. Body elongated. Head curved downward. 10–11 pairs of somites
27*	~86	~56 (54–59)	L 3.35 W 1.6 H 1.95	2nd gill cleft appears in dorsal part of gill area (yet slightly outlined). 12 pairs of somites
28	~92	~60 (57–63)	L 3.55 W 1.6 H 1.75	Further body elongation. Head maximally curved downward. Olfactory pit distinct in front of eye. 14 pairs of somites
29	97	63.5–64	L 4.2 H 1.75	Body begins to straighten: head somewhat less curved. Tailbud increases. 16 pairs of somites

TABLE 8.1 (continued)

Stage No.	Time from the 1st cleavage furrow		Size, mm (D, diameter; L, length, W, width, H, height)	Diagnostic features of stages
	hours, minutes	τ_n/τ_0		
30*	~102	~67 (65–68)	L 4.5 H 1.65	Straightening of head and dorsal body curvature continues. Body elongates. Tailbud increases. Fin fold first seen. Dorsal fin begins at the level of 14th somite. 18 pairs of somites
31*	~109	~71 (69–74)	L 4.7 H 1.7	A groove appears in regions of lens rudiment. 3rd gill cleft outlined in dorsal part of gill area. Dorsal fin begins at the level of 12th somite. 19 pairs of somites
32*	~113	~74 (73–75)	L 5.0 H 1.7	Dorsal fin begins at level of 10th somite. 20 pairs of somites
33*		74	L 5.25 H 1.6	Dorsal fin begins at level of 8th somite. 21–22 pairs of somites
34*	~115	~75	L 5.5 H 1.6	Dorsal fin begins at level of 7th somite. 24–25 pairs of somites
35	122	80	L 6.25 H 1.6	From this stage on the body axis from the rhombencephalon to tail base is quite straight. 3 rudiments of external gills appear as nodules on surface of gill area. Lateral line is seen as far as 6th somite. Dorsal fin begins at level of 5th somite. 1st chromatophores appear and heartbeat begins. Somites are now difficult to count
36	130	85	L 7.1 H 1.7	External gills as short processes directed laterally from gill area
37*	177	115	L 7.5 H 1.7	Gills elongate ventroposteriorly. No limb buds are seen
38*	178	114–132	L 7.9 H 1.8	Gill filament rudiments appear as nodules, 2 in each gill. Opercular rudiment is delineated as a fold over hyoid arch. Opercular rudiments do not yet reach the midline. Limb buds are still only slightly outlined
39	220	144	L 9.0 H 1.9	1st gill has 2 pairs of filament rudiments. 2nd and 3rd gills have 3 pairs each. Gills cover limb buds. Both opercular rudiments almost reach the midline. Mouth angles are delineated (yet very slightly)

TABLE 8.1 (continued)

Stage No.	hours, minutes	τ_n/τ_0	Size, mm (D, diameter; L, length, W, width, H, height)	Diagnostic features of stages
	Time from the 1st cleavage furrow			
40	240	157	L 9.3 H 2.1	Gills become longer. Number of filaments increases. 1st gill has 4 pairs of filaments, 2nd and 3rd gills have 6–7 pairs each. Both opercular rudiments join at midline. Mouth angles are outlined. Limb buds as small protrusions
41	265	173–177	L 10.0 H 2.2	Gills continue to elongate. Number of filaments increases; they also become longer. Mouth becomes more distinct. 2nd lateral line runs on the flank toward limb bud and bypasses it from ventral side. Forelimb buds are still small protrusions. Hatching begins
42	296	193	L 10.5 H 2.1	Gills reach far beyond level of forelimb rudiments. Mouth is completely outlined but has not yet broken through
43	342	223	L 11.3 H 2.3	Mouth breaking through or already open

Note: For the significance of the asterisks after stage numbers, see the text.

this interval. For this reason we failed to show, for several stages defined on fixed material, the duration of time intervals between them, because they are found simultaneously in samples fixed at close intervals (in Table 8.1 they are marked by an asterisk).

Drawings of *A. mexicanum* embryos at successive developmental stages are given in Figs. 8.5–8.9, in which the numbers of the drawings correspond to stages. The diagnostic features of the embryos at successive developmental stages are given in Table 8.1. Table 8.2 presents adequate developmental stages for *A. mexicanum* [3, 83] and *A. maculatum* [38].

8.5. COMPARISON OF DIFFERENT TABLES OF NORMAL DEVELOPMENT

Since the drawings and descriptions of *A. maculatum* embryos [38] served to distinguish the stages of axolotl development, it is necessary to examine the extent to which the stages of development of *A. maculatum* and *A. mexicanum* correspond

TABLE 8.2. Comparison of Developmental Stages of the Embryos of *A. maculatum* and *A. mexicanum* from the Tables of Harrison [38], Bordzilovskaya and Dettlaff [3], and Schreckenberg and Jacobson [83]

A. maculatum	A. mexicanum		A. maculatum	A. mexicanum	
Harrison	Bordzilovskaya and Dettlaff	Schreckenberg and Jacobson	Harrison	Bordzilovskaya and Dettlaff	Schreckenberg and Jacobson
1	1	1	19	19	19
–	2	–	20	20	20
2	2	2	21	21	21
3	3	3	22	22	22
4	4	4	23	23	23
5	5	5	24	24	24
6	6	6	25	25	–
7	7	7	26	26	–
8	–	8	27	27	25
–	8	–	28	28	26
9	9–9½	8+	29	29	27
10/10–	10	9	–	–	28?
–	10½	10	–	–	29?
11–	10¾	11	30	30	30
11	11	–	31	31	31
–	11½	–	32	32	32
12	12	12	33	33	33
–	12½	13	34	34	
13	13–	–	35	35	34
					35
13+	13	14	36	36	36
					37
14	14	14+	37	37	38
15	15	16	39	39	–
17	17	17	40	40	–
18	18	18		41	40

to each other and where essential structural differences lie. Some diagnostic features indicated in Table 8.1 are common for both *A. maculatum* and *A. mexicanum*. The principal differences between specific structural features of the embryos are as follows. In axolotl embryos the gill area, hyomandibular, hyobranchial, and branchial grooves appear at earlier stages and the mouth angles also form earlier. Conversely, the limb buds develop much later than in *A. maculatum*. In the two species the size and shape of the head region also differ at similar stages. Finally, in the axolotl rudiments of balancers are absent.

In our tables there is no coincidence in the structure of embryos of *A. maculatum* and *A. mexicanum* at stage 8 (the embryo of *A. maculatum* appears to be less developed), and it is difficult to identify stage 9 by the features described in the Harrison tables, without a time parameter. For the axolotl we distinguish a greater number of stages during gastrulation (see Table 8.2). The onset of gastrulation in the axolotl (stage 10) corresponds to stage 10 in *A. maculatum*, if we examine the drawings, and to stage 10–, if we take into account their description. No stages corresponding to our stages $10^1/_2$ and 12 are found in the Harrison tables. Our stage $10^3/_4$ corresponds to stage 11 in *A. maculatum*, and at stage $12^1/_2$ the embryo of *A. mexicanum* appears somewhat less developed than that of *A. maculatum*. Our stage 13– corresponds to Harrison's stage 13 and our stage 13 to Harrison's stage 13+. All other stages, as already stated, generally coincide. Unfortunately, Harrison's tables lack an indication of the hatching stage. Thus, stage 41 in our tables cannot be compared.

It is also worthwhile to compare the stages described by us with those of Schreckenberg and Jacobson [83]. Since all authors oriented their data to Harrison's tables, no great differences between these tables should exist. This is not, however, always the case because it is sometimes difficult to see in Schreckenberg and Jacobson's photographs the features that appear diagnostic to us, and these authors did not mention some of these features at all in their descriptions, or indicated them at later stages only. In certain cases Schreckenberg and Jacobson deviate from the stage numbering of Harrison: thus the stage of gastrulation onset is marked as stage 9 rather than 10 or 10–, and the stage of slitlike blastopore as stage 14 rather than 13. As far as we can judge, the stages indicated by Schreckenberg and Jacobson and by us correspond to each other as to stages 1–7, 15–24, and 30–33. The number of the other stages in the table of Schreckenberg and Jacobson differs not only from ours but also from that in Harrison's tables (see Table 8.2). A more detailed comparison is given by Bordzilovskaya and Dettlaff [3].

With recalculation of the time of beginning of the successive developmental stages by the data of Schreckenberg and Jacobson [83] obtained at $18 \pm 0.5°C$ in the value of τ_0 (at 18°C, $\tau_0 = 105$ min; see Fig. 8.2), a satisfactory coincidence of the relative characteristics was obtained for some more distinct and short-term stages, namely onset of gastrulation, slitlike blastopore, some stages of neurulation – 18 and 20 – as well as stages 23, 30, 37, and 38 (after Schreckenberg and Jacobson). The stages of hatching (40 after Schreckenberg and Jacobson and our 41) proved, on the contrary, to be separated by a very large time interval (about $85\tau_0$!). The embryo at stage 40 after Schreckenberg and Jacobson is thus much older than the embryo at the stage of onset of hatching.

REFERENCES

1. J. B. Armstrong and M. F. Ortiz, "Alkylation treatment of the Mexican axolotl: An approach to the isolation of new mutants," *Am. Zool.* **18**, 359–368 (1978).
2. F. S. Billett and A. E. Wild, *Practical Studies of Animal Development*, Chapman and Hall, London (1975).
3. N. P. Bordzilovskaya and T. A. Dettlaff, "Table of stages of the normal development of axolotl embryos," *Axolotl Newsletter* **7**, 2–22 (1979).
4. E. C. Boterenhood, *The Mexican Axolotl. UFAW Handbook on the Care and Management of Laboratory Animals*, 3rd edn., Edinburgh & Livingstone, Ltd., London (1967).
5. J. Brachet, "Le contrôle de la maturation chez les oocytes d'Amphibiens," *Ann. Biol.* **13**, 403–434 (1974).

6. R. Briggs, "Genetic control of early embryonic development in the Mexican axolotl *Ambystoma mexicanum*," *Ann. Embryol. Morphogen., Suppl.* **1**, 105–113 (1969).

7. R. Briggs, "Developmental genetics of the axolotl," in *Genetic Mechanisms of Development*, F. H. Ruddle, ed., Academic Press, New York (1973), p. 169.

8. R. Briggs and J. Cassens, "Accumulation in the oocyte nucleus of a gene product essential for embryonic development beyond gastrulation," *Proc. Natl. Acad. Sci. USA* **55**, 1103–1109 (1966).

9. R. Briggs and R. R. Humphrey, "Studies on the material effect of the semilethal gene, *v*, in the Mexican axolotl," *Dev. Biol.* **5**, 127–146 (1962).

10. R. Briggs and J. T. Justus, "Partial characterization of the component from normal eggs which corrects the maternal effect of gene *O* in the Mexican axolotl (*Ambystoma mexicanum*)," *J. Exp. Zool.* **147**, 105–116 (1968).

11. R. Briggs, J. Signoret, and R. R. Humphrey, "Transplantation of nuclei of various cell types from neurulae of the Mexican axolotl (*Ambystoma mexicanum*)," *Dev. Biol.* **10**, 233–246 (1969).

12. A. I. Brothers, "Instructions for the care and feeding of axolotl," *Axolotl Newsletter* **3**, 9–17 (1977).

13. R. B. Brun, "Experimental analysis of the eyeless mutant in Mexican axolotl (*Ambystoma mexicanum*)," *Am. Zool.* **18**, 273–280 (1978).

14. U. V. Brunst, "The axolotl (*Siredon mexicanum*) as material for scientific research," *Lab. Invest.* **4**, 45–64 (1955).

15. B. M. Carlson, "Morphogenesis of the regenerating limb," *Ontogenez* **13**, 339–359 (1982).

16. G. ten Cate, "The intrinsic embryonic development," *Verhandel. Koninkl. Ned. Akad. Wetenschap. Afdel. Natuurk.* **51**, 257 (1956).

17. T. A. Dettlaff, "Cell divisions, duration of interkinetic states, and differentiation in early stages of embryonic development," in *Advances in Morphogenesis*, Vol. 3, Academic Press, New York (1964), pp. 323–362.

18. T. A. Dettlaff and A. A. Dettlaff, "On relative dimensionless characteristics of the development duration in embryology," *Arch. Biol. (Liége)* **72**, 1–16 (1961).

19. E. M. Deuchar, "Famous animals. 8. The axolotl," *New Biol.* **23**, 102–122 (1957).

20. E. Elkan, "Pathology in the Amphibia," in *Physiology of the Amphibia*, Vol. 3, B. Lofts, ed., Academic Press, New York (1976), pp. 273–312.

21. H. Engländer, "Die Induktionsleistungen eines heterogenen Induktors in Abhängigkeit von der Dauer seiner Einwirkungszeit," *Wilhelm Roux's Arch. Entwicklungsmech. Org.* **154**, 124–142 (1962a).

22. H. Engländer, "Die Differenzierungsleistungen des *Triturus*- und *Ambystoma*-ektoderms unter der Einwirkung von Knochenmark," *Wilhelm Roux's Arch. Entwicklungsmech. Org.* **154**, 143–159 (1962b).

23. H. Engländer and A. G. Johnen, "Die morphogenetische Wirkung von Li-Ionen auf Gastrula-Ektoderm von *Ambystoma* und *Triturus*," *Wilhelm Roux's Arch. Entwicklungsmech. Org.* **159**, 346–356 (1967).

24. L. G. Epp, "A review of the eyeless mutant in the Mexican axolotl," *Am. Zool.* **18**, 267–272 (1978).

25. G. Fankhauser, "Amphibia," in *Animals for Research*, W. Lane-Petter, ed., Academic Press, London (1963).

26. G. Fankhauser, "Urodeles," in *Methods of Developmental Biology*, F. H. Wilt and N. K. Wessels, eds., Tome's Crowel, New York (1967), pp. 85–99.

27. R. V. Fremery, "Diseases of axolotl in the amphibian colony of the Hubrecht Laboratory, Utrecht, The Netherlands, in the period 1974–1980," *Axolotl Newsletter* **9**, 7–8 (1980).

28. A. S. Ginsburg, "Transplantation of the ear ectoderm in axolotl," *C. R. Acad. Sci. USSR* **30**, 546–549 (1941).

29. A. S. Ginsburg, "On the determination of the labyrinth in axolotl and triton," *Izv. Akad. Nauk SSSR, Ser. Biol.* **3**, 215–228 (1942).

30. A. S. Ginsburg, "Changes in the properties of the ear material in the process of determination," *C. R. Acad. Sci. USSR* **54**, 185–188 (1946).

31. A. S. Ginsburg, "Cortical reaction in axolotl eggs," *Sov. J. Dev. Biol.* **2**, 515–518 (1972).

32. L. Goetters, "Differenzierungsleistungen von explantiertem Urodelenectoderm (*Ambystoma mexicanum* Cope und *Triturus alpestris* Laur) nach verschieden langer Unterlagerungszeit," *Wilhelm Roux's Arch. Entwicklungsmech. Org.* **157**, 75–100 (1966).

33. S. Grinfeld and J. C. Beetschen, "Early grey crescent formation experimentally induced by cycloheximide in the axolotl oocyte," *Wilhelm Roux's Arch. Dev. Biol.* **191**, 215–221 (1982).

34. H. Grunz, "Experimentelle Untersuchungen über die Kompetenzverhaltnisse früher Entwicklungsstadien des Amphibien-Ektoderm," *Wilhelm Roux's Arch. Entwicklungsmech. Org.* **160**, 344–374 (1968).

35. V. Hamburger, *A Manual of Experimental Embryology*, University of Chicago Press, Chicago (1947).

36. K. Hara, "The cleavage pattern of the axolotl egg studied by cinematography and cell counting," *Wilhelm Roux's Arch. Dev. Biol.* **181**, 73–87 (1977).

37. K. Hara and E. C. Boterenbrood, "Refinement of Harrison's normal table for the morula and blastula of axolotl," *Wilhelm Roux's Arch. Dev. Biol.* **181**, 89–93 (1977).

38. R. G. Harrison, *Organization and Development of the Embryo*, Yale University Press, New Haven (1969).

39. T. S. Hauschka and V. V. Brunst, "Sexual dimorphism in the nucleolar autosome of the axolotl (*Siredon mexicanum*)," *Hereditas* **52**, 345 (1965).

40. R. R. Humphrey, "A linked gene determining the lethality usually accompanying a hereditary fluid imbalance in the Mexican axolotl," *J. Hered.* **50**, 279–286 (1959).

41. R. R. Humphrey, "A chromosomal deletion in the Mexican axolotl (*Siredon mexicanum*) involving the nucleolar organizer and the gene for dark color," *Am. Zool.* **1**, Abstract 222 (1961).

42. R. R. Humphrey, "Mexican axolotls, dark and white mutant strains: care of experimental animals," *Bull. Philadelphia Herpetol. Soc.*, April–Sept., **21** (cited by New, 1966).

43. R. R. Humphrey, "A recessive factor (*o* for ova-deficient) determining a complex of abnormalities in the Mexican axolotl (*Ambystoma mexicanum*)," *Dev. Biol.* **13**, 57–76 (1966).

44. R. R. Humphrey, "Albino axolotls from an albino tiger salamander through hybridization," *J. Hered.* **58**, 95–101 (1967).

45. R. R. Humphrey, "The axolotl *Ambystoma mexicanum*," in *Handbook of Genetics*, Vol. 4, R. C. Kind, ed., Plenum Press, London (1975), pp. 3–17.

46. R. R. Humphrey, "Factors influencing ovulation in the Mexican axolotl as revealed by induced spawnings," *J. Exp. Zool.* **199**, 209–214 (1977).

47. R. R. Humphrey, "Phenotypes recognizable in the progeny of axolotl parents both heterozygous for the same two mutant genes," *Am. Zool.* **18**, 207–214 (1978).

48. C. Ide, "Genetic dissection of cerebellar development: mutations affecting cell position," *Am. Zool.* **18**, 282–288 (1978).

49. G. M. Ignatieva, "The dynamics of morphogenetic movements at the period of gastrulation in axolotl embryos," *Dokl. Akad. Nauk SSSR* **179**, 1005–1008 (1968).

50. G. M. Ignatieva, "Time relationships between the onset of RNA synthesis and the manifestation of the morphogenetic function in axolotl nuclei," *Sov. J. Dev. Biol.* **3**, 531–533 (1972).

51. G. M. Ignatieva, "Regularities of early embryogenesis in teleosts as revealed by studies of the temporal pattern of development. II. Relative duration of corresponding periods of development in different species," *Wilhelm Roux's Arch. Dev. Biol.* **179**, 313–325 (1976).

52. K. Ikeneshi and P. D. Nieuwkoop, "Location and ultrastructure of primordial germ cells (PGCs) in *Ambystoma mexicanum*," *Dev., Growth Differ.* **20**, 1–9 (1978).

53. C. O. Jacobson, "Selective affinity as a working force in neurulation movements," *J. Exp. Zool.* **168**, 125–136 (1968).

54. C. O. Jacobson, "Mesoderm movements in the amphibian gastrula," *Zool. Bidr., Uppsala* **28**, 233–239 (1969).

55. A. G. Johnen and H. Engländer, "Untersuchungen zur entodermalen Differenzierungsleistung des *Ambystoma*-Ektoderms," *Wilhelm Roux's Arch. Entwicklungsmech. Org.* **159**, 357-364 (1967).

56. A. G. Johnen, "Der Einfluss von Li- und SCN-Ionen auf die Differenzierungsleistungen des *Ambystoma*-Ektoderms und ihre Veränderung bei kombinierter Einwiklung beider Ionen," *Wilhelm Roux's Arch. Entwicklungsmech. Org.* **165**, 150–162 (1970).

57. J. T. Justus, "The cardiac mutant: An overview," *Am. Zool.* **18**, 321–326 (1978).

58. R. R. Kulikowski and F. J. Manasek, "The cardiac lethal mutant of *Ambystoma mexicanum*: a reexamination," *Am. Zool.* **18**, 349–358 (1978).

59. M. Lawrence, "Infections of the axolotl: summary of data," *Axolotl Newsletter* **9**, 2–6 (1980).

60. L. F. Lemanski, "Morphological, biochemical, and immunohistochemical studies on heart development in cardiac mutant axolotls, *Ambystoma mexicanum*," *Am. Zool.* **18**, 327–348 (1978).

61. S. Løvtrup, "Morphogenesis in the amphibian embryo, cell type distribution, germ layers, and fate maps," *Acta Zool.* **47**, 209–276 (1966).

62. G. M. Malacinski, "The Mexican axolotl, *Ambystoma mexicanum*: its biology and developmental genetics, and its autonomous cell-lethal genes," *Am. Zool.* **18**, 195–206 (1978).

63. G. M. Malacinski, B. W. Young, and A. Jurrand, "Tissue interaction during axial structure pattern formation in Amphibia," *Scanning Electron Microsc.* **11**, 207–318 (1981).

64. G. M. Malacinski and H. M. Chung, "Establishment of the site of involution at novel locations on the amphibian embryo," *J. Morphol.* **169**, 149–159 (1981).

65. D. A. T. New, "The axolotl (*Ambystoma mexicanum*)," in *The Culture of Vertebrate Embryos*, Academic Press (1966), pp. 153–160.

66. P. D. Nieuwkoop, "The formation of the mesoderm in urodelan amphibians. I. Induction by the endoderm," *Wilhelm Roux's Arch. Entwicklungsmech. Org.* **162**, 341–373 (1969a).

67. P. D. Nieuwkoop, "The formation of the mesoderm in urodelan amphibians. II. The origin of the dorsoventral polarity of the mesoderm," *Wilhelm Roux's Arch. Entwicklungsmech. Org.* **163**, 298–315 (1969b).

68. P. D. Nieuwkoop, "The formation of the mesoderm in urodelan amphibians. III. The vegetalizing action of the Li ion," *Wilhelm Roux's Arch. Entwicklungsmech. Org.* **166**, 105–123 (1970).

69. P. D. Nieuwkoop and G. A. Ubbels, "The formation of the mesoderm in urodelan amphibians. IV. Qualitative evidence for the purely ectodermal origin of the entire mesoderm and of the pharyngeal endoderm." *Wilhelm Roux's Arch. Entwicklungsmech. Org.* **169**, 185–199 (1972).

70. L. A. Nikitina, "Behavior of the nuclei of growing oocytes in the mature egg cytoplasm," *Dokl. Akad. Nauk SSSR* **267**, 463–465 (1983).

71. J. Pasteels, "New observations concerning the maps of presumptive areas of the young amphibian gastrula (*Ambystoma* and *Discoglossus*), *J. Exp. Zool.* **89**, 255–281 (1942).

72. H. H. Reichenbach-Klinke, *Krankheiten der Amphibien*, Gustav Fischer, Stuttgart (1961).

73. H. H. Reichenbach-Klinke and E. Elkan, *The Principal Diseases of Lower Vertebrates*, Academic Press, London (1965).

74. N. N. Rott, "Cytology of fertilization in the axolotl," *Sov. J. Dev. Biol.* **1**, 150–156 (1970).

75. N. N. Rott, "Correlation between karyo- and cytokinesis during the first cell divisions in the axolotl (*Ambystoma mexicanum* Cope)," *Sov. J. Dev. Biol.* **4**, 175–177 (1973).

76. N. N. Rott and D. R. Beritashvili, "Changes in the content of potassium and sodium during the early embryogenesis of axolotl," *Sov. J. Dev. Biol.* **6**, 78–80 (1975).

77. R. Rugh, *Experimental Embryology*, Burgess, Minneapolis (1962).

78. O. I. Schmalhausen, "Role of olfactory sac in the development of chondral capsule of the olfactory organ in Urodela," *Dokl. Akad. Nauk SSSR* **23**, 395–397 (1939).

79. O. I. Schmalhausen, "Development of ear vesicles in the absence of medulla oblongata in amphibians," *Dokl. Akad. Nauk SSSR* **63**, 276–279 (1940).

80. O. I. Schmalhausen, "A comparative experimental study of the early developmental stages of olfactory rudiments in amphibians," *Dokl. Akad. Nauk SSSR* **74**, 863–865 (1950).

81. O. I. Schmalhausen, "Localization and development of the olfactory organ rudiments with special reference to their origin in vertebrates," *Dokl. Akad. Nauk SSSR* **74**, 1045–1048 (1950).

82. O. I. Schmalhausen, "Conditions of formation and differentiation of the olfactory organs in embryogenesis," *Dokl. Akad. Nauk SSSR* **76**, 469–471 (1951).

83. G. M. Schreckenberg and A. G. Jacobson, "Normal stages of development of the axolotl, *Ambystoma mexicanum*," *Dev. Biol.* **42**, 391–400 (1975).

84. J. Signoret, "Étude des chromosomes de la blastula chez l'axolotl," *Chromosoma* **17**, 328–335 (1965).

85. J. Signoret, R. Briggs, and R. R. Humphrey, "Nuclear transplantation in the axolotl," *Dev. Biol.* **4**, 134–164 (1962).

86. J. Signoret and J. Lefresne, "Contribution à l'étude de la segmentation de l'oeuf d'axolotl. I. Définition de transition blastuléene," *Ann. Embryol. Morphogen.* **4**, 113–123 (1971).

87. M. N. Skoblina, "Characteristics of duration of main stages of embryogenesis in *Ambystoma mexicanum*," in *4th Embryological Conference: Abstracts* [in Russian], Leningrad University Press (1963), pp. 172–173.

88. M. N. Skoblina, "Dimensionless description of the length of mitotic phases of first cleavage divisions in axolotl," *Dokl. Akad. Nauk SSSR* **160**, 700–703 (1965).

89. F. Sládecek and J. Lanzová, "Cytology of fertilization of the eggs of axolotl," *Folia Biol. (Ceskosl.)* **5**, 372–378 (1959).

90. H. M. Smith and R. B. Smith, *Synopsis of the Herpetofauna of Mexico. I. Analysis of the Literature on the Mexican Axolotl*, E. Lundberg, N. Bennington, Vermont (1971).

91. D. L. Stocum, "Stages of forelimb regeneration in *Ambystoma maculatum*," *J. Exp. Zool.* **209**, 395–416 (1979).

92. P. W. Tank, B. M. Carlson, and T. G. Connelly, "A staging system for forelimb regeneration in the axolotl, *Ambystoma mexicanum*," *J. Morphol.* **150**, 117–128 (1976).

93. P. W. Tank and N. Holder, "Pattern regulation in the regenerating limbs of urodele amphibians," *Q. Rev. Biol.* **56**, 113–142 (1981).

94. R. Tompkins, "Genetic control of axolotl metamorphosis," *Am. Zool.* **18**, 313–320 (1978).

95. P. Valouch, J. Melichna, and F. Sládecek, "The number of cells at the beginning of gastrulation depending on the temperature in different species of amphibians," *Acta Univ. Carol. Biol.*, pp. 195–205 (1971).

96. M. A. Vorontsova, L. D. Liozner, I. V. Markelova, and E. C. Pukhalskaya, *The Newt and The Axolotl* [in Russian], Nauka, Moscow (1952).

97. H. Wallace, *Vertebrate Limb Regeneration*, Wiley New York (1981).

98. B. W. Young, R. E. Keller, and G. M. Malacinski, "An atlas of notochord and somite morphogenesis in several anuran and urodelan amphibians," *J. Embryol. Exp. Morphol.* **59**, 223–247 (1980).

99. B. W. Young and G. M. Malacinski, "Comparative analysis of amphibian somite morphogenesis: cell rearrangement patterns during rosette formation and myoblast fusion," *J. Embryol. Exp. Morphol.* **66**, 1–26 (1981).

100. A. I. Zotin and R. V. Pagnaeva, "The moment of determination of the position of primitive groove in eggs of *Acipenser* and axolotl," *Dokl. Akad. Nauk SSSR* **152**, 765–768 (1963).

101. S. E. Zubova, "Type of variation of development rate in acipenserid embryos," *Dokl. Akad. Nauk SSSR* **145**, 694–697 (1962).

Chapter 9

THE SOUTH AFRICAN CLAWED TOAD
Xenopus laevis

T. A. Dettlaff and T. B. Rudneva

The South African clawed toad *Xenopus laevis* has become one of the most popular laboratory animals in developmental biology studies (see [11, 28, 43, 51, 76]). It can be easily kept and reproduced in captivity, and one can obtain the eggs at any given time through the year by stimulation with gonadotrophic hormones; the clutch contains a sufficiently large number of eggs, and these are not readily damaged by various experimental conditions. All these features, together with the rapid rate of development and a variety of genetic markers, make this animal most suitable for experimental studies.

Studies on *X. laevis* embryos are facilitated by the fact that its development is rather well known and the normal table of its development has been elaborated {76]. Several methods have been successfully applied to *X. laevis* embryos: nuclear transplantation [31, 40], stimulation of *in vitro* oocyte maturation [44, 87], removal of the oocyte nucleus (germinal vesicle, GV) [44], transplantation of germ cells [8, 96], sex inversion [35], molecular biology investigations, and injection of labeled precursors [43, 53, 67].

Operated embryos can be reared to the state of sexual maturity, producing mature gametes [7, 52].

Studies on *X. laevis* are concerned with key problems, such as the patterns of macromolecular syntheses during gameto- and embryogenesis, cytoplasmic control of nuclear activity, gene activity during development, and the development of the fine structural organization of gametes and differentiating cells. The oocytes and eggs of *X. laevis* have been successfully used as a model system for studying the processes of translation of heterogeneous mRNAs and the interaction of mRNA and the protein-synthesizing system in different combinations ([45–47, 53, 71, 101]; see review of publications until 1971 in [28]). From 1975 onward the oocytes and eggs of *X. laevis* have been widely used for studying transcription of DNA and individual purified genes by their transplantation into the oocyte nucleus and egg. The oocytes and eggs of *X. laevis* have become a classic subject for studies known as "genetics by gene isolation" and "inverse genetics" [48–50]. The patterns of hormonal regulation of oocyte maturation and the development of the mature egg organization are also intensively studied in *X. laevis* (for review, see [21, 23, 69]).

Although there has been some previous work on other amphibian species, there was relatively little information on the time of determination of various organs and on the mechanisms of their determination. However, using *Xenopus*, a number of studied have been published which relate to these aspects. These include: the process of fertilization [36, 68, 97–99], establishment of bilateral symmetry (see [66]), karyo- and cytokinesis during cleavage [1, 9, 10, 27, 56, 70, 72, 79, 84, 86, 93], initiation and change of the cell cycle [37, 74, 75], morphogenetic movements and processes of gastrulation and neurulation [2, 62–65, 72, 77, 88, 14–17, 91], induction [59, 92, 13], and somitogenesis [16, 68, 78].

X. laevis is of special interest for comparative studies of oocyte maturation [23] and development (see [19, 20, 32, 38]), since it is a lower representative of Anura and by some features, in particular by the relative duration of the early developmental periods, is even closer to Urodela than to Anura [27, 93]. Today, however, many amphibian species are so scarce in numbers and ranges that most of the problems are being solved using *X. laevis*. As a result, there is a dangerous tendency to extend the results obtained on the *X. laevis* oocytes and embryos to the other amphibian species. The ever-increasing number of studies of *X. laevis* embryos makes it a very difficult and, in fact, an unreal task to compile the complete list of references for this chapter.

9.1. TAXONOMY, MUTATIONS, HAPLOIDS, AND POLYPLOIDS

The South African clawed toad *Xenopus laevis* Daudin (its name related to three pairs of horny claws on the hind legs) (Fig. 9.1) belongs to the subfamily Xenopinae, family Pipidae, order Anura, class Amphibia. The genus *Xenopus* includes six species (see [43]). For experimental studies *X. tropicalis* is sometimes used, in which the eggs, and hence the embryonic cells, are smaller than in *X. laevis*; this allows for their use for heteroplastic transplantations. *X. laevis* includes five subspecies which differ from each other by color [77]. The transplantations of nuclei [42] or cells [7] between the embryos of different subspecies make it easier to identify transplants.

The diploid number of chromosomes is 36 [95]. An anucleolar (0-*nu*) mutation was found [30] which is lethal in the homozygous state but quite viable in the heterozygous state. The mononucleolar strain is maintained in several laboratories (Institute of Animal Genetics in Edinburgh, Laboratory of Molecular Biology in Cambridge, U.K.). It is widely used in molecular biology studies [12] and in those cases when genetically marked nuclei are needed (experiments such as nuclear transplantation or the grafting of rudiments).

At the Institute of Developmental Biology, Academy of Sciences of the USSR (Moscow), a strain of a spontaneous mutation, temporary albinism, is maintained. The mutant eggs and embryos are fully devoid of pigment until stage 40 [58].

Different kinds of embryos have been experimentally obtained, namely: gynogenetic [83] and androgenetic [41], haploid, as well as triploid [89] and tetraploid [39] forms (see also [51]).

9.2. CONDITIONS FOR KEEPING ADULT *Xenopus*

9.2.1. Distribution and Habitats in Nature

The data on *X. laevis* given below are taken mainly from special reviews [6, 43, 51, 73, 77]. Its structure and way of life have been described by Brown [11].

All the representatives of the subfamily Xenopinae, which includes also the genera *Hymenochirus* and *Pseudohymenochirus*, occur in Central and South Africa. The South African toads live permanently in water. They are found in lakes, ponds, and swamps, live equally well in fresh or brackish water, and are very tolerant to environmental influences, living in pure and polluted, permanent and intermittent water bodies. When water bodies dry out the animals bury themselves in the mud until the next rain, or move by land over short distances into another water body. The toads can stay on land for a short time only since they dry

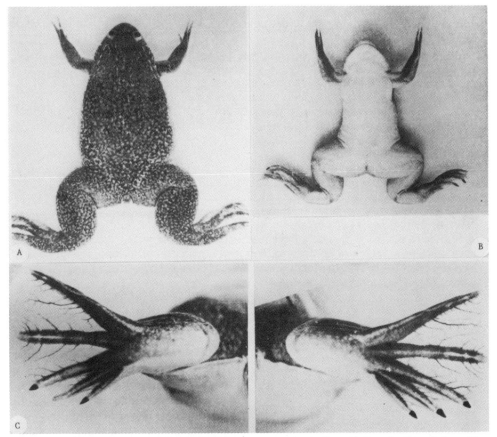

Fig. 9.1. The South African clawed toad *Xenopus laevis* Daudin. A) Female, dorsal view; B) male, ventral view; C) claw "spurs" on hind legs (on the left leg one claw is cut out to mark the animal [43]).

out easily and perish. They survive better in a humid environment, such as moistened moss, leaves, or wadding. Under such conditions (at 16–26°C) they are usually transported from Africa to European laboratories (see [43]).

9.2.2. Laboratory Conditions and Care

South African toads can live in the laboratory for up to 15 years. They are kept in aquarium tanks with dechlorinated nonrunning water. The volume should be enough for the toads to swim [according to Gurdon (1967), 10 females or 15 males may be kept in 30 liters of water]. The level of water should be much lower than the tank walls so that the toads cannot spring out. Pieces of tile or stones are put on the bottom as a shelter for the animals. The open top of the tank is covered by a net. Access to an air surface is needed, since the animals have to be able to swim to the surface for air.

Unlike Nieuwkoop and Faber [77], Gurdon and Woodland [51] recommend that running water be avoided, since it has been found to cause a condition resembling red-leg disease, which results in the death of the toads. Tanks should be emptied and refilled with clean water at least twice a week, and any excess chlorine or heavy metal ions needs to be removed by evaporation and ion-exchange filters. To prevent fungal diseases, it is recommended to add a few milliliters of saturated sea salt solution with every change of water [73].

The recommended temperatures for keeping the animals are 18–22°C [51], 18°C [77], and 20–25°C [73], although they can live normally at 10–28°C. The illumination should correspond to the day length: the animals are kept under natural illumination or given artificial light which is switched off during the night.

9.2.3. Feeding

Adult animals should be fed twice a week. The best food is chopped or coarsely minced ox or horse heart muscle or ox liver (in the latter case, the water should be changed just after feeding). Earthworms, tadpoles, enchytraeids, and tubeworms can be given together with meat. The meat should be lean, and fat should be removed. The food can be simply placed in the water; the animals take it very rapidly and there is no need to feed them specially. The food should be distributed equally all over the tank. Fifty animals should be given 100 g of meat per week [77].

9.2.4. Diseases

Fungal infections can be alleviated or cured by immersing animals in 0.002% wt/vol $KMnO_4$ aqueous solution for 30 min daily [43]. The animals sometime perish due to leg paralysis: they cannot swim to the surface for air and are suffocated. However, if the water level is lowered in time, the animals can be saved; they do not eat for a long time, up to several weeks, and then begin to take live worms and can recover fully [73]. Gurdon [43] reports that, in his colony, single animals die without any signs of disease (see [11, 29, 81, 82] for more detail about diseases of South African toads). *X. laevis* is very sensitive to antibiotics; sulfonamides are lethal, and penicillin or streptomycin are harmful or lethal if injected [51].

9.3. MANAGEMENT OF REPRODUCTION IN THE LABORATORY

9.3.1. Spawning Conditions in Nature

South African toads mate in nature from September to December. The most active spawning is observed during days when there is a rise of water temperature above 21°C. Egg laying occurs within 8 to 10 h after the temperature exceeds this critical level [73, 77]. At the beginning and end of the breeding season this critical level lies lower, but spawning is far less abundant. In nature the females lay eggs on water plants.

Spawning can be induced by a rise of temperature in the laboratory as well, but gonadotrophic hormones are much more convenient for this purpose.

9.3.2. Stimulation of *in vivo* Oocyte Maturation

The oocytes of South African toads undergo maturation during all seasons under the influence of gonadotrophic hormones of both amphibians and mammals. That is why *Xenopus* was used as a test subject for the early diagnostics of pregnancy. Although under normal conditions the females attain sexual maturity by the age of 6–8 months, it is recommended to use for breeding animals over the age of 2 years. Males are distinguished from females by the presence of nuptial calluses and the absence of cloacal lips, which are well pronounced in females. For injections of gonadotrophic hormones it is better to use stouter females with swollen cloacal lips and males with darker calluses (see Fig. 9.1A, 1B).

Maturation is stimulated by chorionic hormones: aviutrin (Parke-Davis, USA), pregnyl and cestyl (Organon Laboratories, UK), gonadotrophin (CIBA), or chorionic gonadotrophin (Richter, Hungary). The gonadotrophin is diluted in Ringer solution (for poikilotherms) or in distilled water (not the diluent supplied with the hormone): 50–300 IU in 0.5 ml and 500–600 IU of the hormone in 1 ml is taken up in a syringe and injected after passing the syringe needle through the dorsal hind-leg skin and into the dorsal lymph sac. Nieuwkoop and Faber [77] recommend piercing the skin of the thigh and the septum between the lymph sacs of the thigh and the back, rather than injecting into the dorsal sac directly. The amount of the hormone needed varies according to the condition of the animals and the hormone used.

Gurdon [44] recommends injecting 50–150 IU into a male 24–36 h, and 200 IU into a female 16 h before the desired time, and then 300 IU within 8 h (at 23°C) (the amount of hormone can be increased in May–August). Egg laying can be expected within 8 h after the second injection.

Oocyte maturation can also be induced after a single injection of choriogonin. Males are given 100–200 IU and females 300–400 IU. In this case, egg laying can begin within 12–14 h (Rudneva, personal communication; see also [51]).

After the injection the male and female are put into a small covered container, sheltered from light. In cases where embryos at later developmental stages are needed, a metal gauze with edges bent downward is put on the bottom of the container to prevent the animals from eating deposited eggs. After spawning, the animals are taken out, their excrement is removed, and the eggs are collected. When fertilized eggs or the embryos at early developmental stages are needed, it is better to put a ring of rubber tubing into the container. After the female begins to lay eggs on it, it is taken out, the eggs are removed, and the ring is placed back in. By repeating this procedure at definite time intervals, one can obtain eggs with more or less precisely known time of fertilization.

Egg laying usually begins early in the morning and lasts 12–24 h (if the temperature is lowered, spawning can be delayed). Between 2000 and 6000 eggs can be obtained during 12 h from an injected female [43]. To obtain highly fertilizable eggs from the injected females in the laboratory, it is recommended to inject the females periodically, not less than once in 4–6 months, even if eggs are not needed.

In the Hubrecht Laboratory a method has been developed for collecting naturally fertilized eggs immediately after laying [34]. A couple of animals injected with human chorionic gonadotrophin according to the usual procedure are placed in a glass tank containing water kept at 23–25°C by means of a heating unit. The tank is kept in a dark room and illuminated from above by a 6V 15W lamp provided with any green filter. The light intensity is regulated by a transformer, such as that used for a microscope lamp. In this way it is possible to watch the couple continuously, as they hardly seem to be disturbed by the constant green light. To be able to take out eggs immediately after laying, the container is divided into two compartments. The tray onto which the eggs fall through a grid can be gently drawn into the next compartment and lifted out. A few minutes after laying, the eggs are available for experiments.

9.3.3. *In vitro* Oocyte Maturation

The oocytes which have completed their growth mature quite well under the influence of gonadotrophic hormones and progesterone in a modified Barth solution but can mature in Ringer solution as well. To obtain the oocytes, a female is operated or killed and small fragments of the ovary with 2–5 large oocytes are placed in a solution with the hormones. Choriogonin in doses of 20–40 IU/ml [44] and many steroids (progesterone, pregnenolone, prednisolone, hydrocortisone, etc.) at a dose of 1 µg/ml [87] proved to be effective.

9.3.4. Rearing Embryos and Larvae

In nature fertilized eggs are deposited among water plants and hatched larvae attach themselves to these plants until the transition to active feeding. The larvae are plant-eating.

In the laboratory up to 90% of embryos can be reared under favorable conditions. In the vessel where the eggs and larvae are kept the temperature should be maintained at 20–25°C. By the end of larval development there should not be more than 2 larvae per liter of water. The water in the vessel is aerated until the larvae begin to use lungs for their respiration. This stage is reached when air bubbles appear on the water surface. If plant powder is used for feeding, there is no need to change the water, at least during the first weeks when the larvae are very fragile. The water should be completely free of chlorine and copper ions. To remove chlorine, tap water should be left in contact with air for several days.

The larvae begin to feed on the fifth day. Very good results are obtained on feeding with a powder of nettle or medick leaves, as well as dry yeasts, and a powder of dried liver [18]. The dry nettle is ground in a mortar and the fresh nettle is homogenized, treated with boiling water, and rubbed through gauze until the water becomes weakly opalescent. The larvae are fed once a day, and the amount of food should be such that the water becomes transparent by the time of the next feeding. It is not necessary to remove the excrements. Gurdon and Woodland [51] report that the larvae grow much faster on a diet of dried milk with brewer's yeast.

The larvae do not require much light. They can develop quite satisfactorily in complete darkness; indeed, bright light prevents their normal development.

During metamorphosis the larvae cease to feed, become very sluggish, and keep to the bottom. It is necessary to lower the water level so that the larvae do not drown. By the end of metamorphosis they begin to feed again; they already eat small worms, such as enchytraeids and tubeworms. After metamorphosis they should be fed to satiety since otherwise they remain small upon attaining sexual maturity. Both juvenile and adult animals should be given vitamin D and calcium to prevent the formation of skeletal defects.

One very important requirement is sufficient space. There should not be less than 3 liters of water to each young animal, and enough space for directional swimming.

The growing animals are fed gradually on midge larvae, earthworms, and, later, fully or partially, by chopped ox heart.

9.4. STRUCTURE OF GAMETES, INSEMINATION AND FERTILIZATION

9.4.1. Mature Unfertilized Egg

The diameter of the egg of *X. laevis* is 1.2–1.5 mm (the eggs obtained from females reared in the laboratory can be smaller). The animal region of the egg is intensely pigmented and has a brownish color. Its cortex contains cortical granules, and the cytoplasm contains many yolk platelets. The nucleus is at metaphase II. The fine structure of the mature egg cytoplasm and changes in the fine structure of oocytes during maturation have been described [5, 61, 94]. The egg is surrounded by the vitelline membrane and jelly coat. Upon accelerated egg laying, after hormonal stimulation, the eggs are laid in groups surrounded by a common jelly coat. The eggs are inseminated at the moment of their deposition from the female's body as they move along the channel formed by the cloacal lips.

9.4.2. Spermatozoon

As in all amphibians, the spermatozoon has an undulating membrane [80]. Its total length is about 44 μm. There is an acrosome (1 μm) in the anterior head region, then the spirally bent nucleus (1.5 whorls of spiral) 13 μm long and 1.2 μm wide. The middle part of the spermatozoon (2.5 μm) contains mitochondria and centrioles. The fibrillar complex of the axial filament has a typical (9 + 2) structure. The tail length is about 31 μm. The spermatozoa are very mobile in water and 0.05 N Ringer solution, but only for a few minutes [100].

9.4.3. Artificial Insemination

In contrast to the situation in the common frog, ovulation in *X. laevis* is extended in time, and one cannot obtain a large number of unfertilized eggs at one time. If a female is killed after the injection of hormones, one cannot even then obtain more than a few hundred mature ovulated eggs. The following method is recommended [43]. The injected male with black calluses is killed, and one of the testes is taken out through a small cut; it is cleaned of blood, cut in 2–3 parts, placed in 1 ml of Ringer solution (the spermatozoa remain immobile in this), and the male is put in a moistened wadding and kept in a refrigerator (at 4°C). Meanwhile, the injected female lays eggs. As soon as she lays 10–20 eggs, they are transferred to a small dish, the water is removed, part of the testis is placed in the same dish and brought into contact with the eggs for 1 min and then removed. One testis can thus be used for insemination of a few hundred eggs during 3–4 h; it is kept in a tightly closed watch glass. The eggs are covered with water within 1 min.

If the eggs are fertilized, their animal pole turns upward inside the envelopes within 15–20 min. When the percentage of fertilization decreases, the second testis is taken out of the male and the eggs are inseminated in the same way for 3–4 h more. Up to 100% of fertilized eggs are obtained using this method [43].

Rybak and Gustafson [74] suggested a method for sperm collection without killing the males. The males are injected with a gonadotrophic hormone (400 IU of pregnyl in 1 ml of distilled water), and are kept at 20–21°C; 10–16 min later they are injected with norepinephrine (0.5 mg in 1 ml of Ringer solution). Thereafter they are kept for 2 h in running water (13–15°C) under dim light and in quiet conditions. Semen, often mixed with urine, is then drawn off by inserting a capillary (diameters: outside 1.1 mm and inside 0.6 mm) into the cloaca. Sperm kept in a capillary retains its fertilizing capacity for at least 2–3 h, as shown by Hara et al. [51]. The latter authors used this semen for local insemination: they used a polyvinylchloride plate with a small saw and file at one and, and provided with a reservoir and a slit about 0.5 mm wide, the sperm localizer. The localizer is set against an unfertilized egg stripped onto a microscope coverslip, faced with the slit to the wanted part of the egg. If necessary, some water is introduced between the localizer and the egg-bearing coverslip to make them stick together. The sperm fluid is then introduced into the reservoir of the localizer and comes into contact with the egg jelly at the mouth of the slit.

9.4.4.   Fertilization

Fertilization in X. laevis is normally monospermic. Changes in the fine structure of the egg during fertilization have been described [4], as well as the cortical reaction [97–99], the formation of the white crescent, and the determination of the plane of bilateral symmetry [66, 90].

Upon artificial stimulation of oocyte maturation, the percentage of fertilization can vary markedly, apparently because of premature ovulation of the eggs or, conversely, their overripening in the female body. Differences in the physiological conditions of the eggs also manifest themselves after fertilization and can result in an unstable reaction to experimental influences [43].

9.5.   TIMING OF STAGES OF NORMAL DEVELOPMENT

Nieuwkoop and Faber [77] have described in detail the external and internal structure of X. laevis embryos at successive developmental stages, from the fertilized egg to the end of metamorphosis (stages 1–66), and have given original drawings of the external appearance of the embryos and larvae at all stages. These tables are used by all those who work with *X. laevis*. The authors gave us their kind permission to reproduce the drawings in this edition (Figs. 9.4–9.13) and to cite their descriptions, which we acknowledge with sincere gratitude.

The detailed description of the structure of *X. laevis* embryos and larvae after Nieuwkoop and Faber [77] is preceded by Table 9.1, in which the diagnostic features and the time of onset are indicated for all developmental stages (from fertilization to hatching, stages 0–35/36). The time of onset of every stage is given in τ_0 units: from the moment of appearance of the 1st cleavage furrow on the animal region of the egg using the improved data of Dettlaff and Rudneva [39] and from fertilization using the data of Nieuwkoop and Faber [77].

The value of τ_0, i.e., the duration of one mitotic cycle synchronous cleavage division [25, 26], in *X. laevis* is equal to the interval between the appearance on the egg surface of the 1st and 2nd cleavage furrows [84]. The dependence of the value of τ_0 on temperature is determined by this interval and is represented as a curve (Fig. 9.2). It has been shown for *X. laevis* [22, 27] that at temperatures from 17–26°C the relative time of the onset of the same developmental stages remains practically constant (Fig. 9.3). This allows prediction within this temperature range of the time of onset of different developmental stages [24]. For this purpose, one has to read from this curve the value of τ_0 at the desired temperature and multiply it by the number of τ_0 units indicated in Table 9.1 (τ_n/τ_0).

The reading of time from the moment of appearance of the 2nd cleavage furrow (from stage 2–) given in Table 9.1 is most precise when experimenting with embryos obtained by natural fertilization. In the case of artificial insemination it is easier to read the time from the moment of insemination. The recalculation of the data by Hamilton [54], Nieuwkoop and Faber [66], and Gerhart et al. [97] has shown that in *X. laevis* the 1st cleavage furrow appears within $3\tau_0$ after adding the sperm. Thus, by adding $3\tau_0$ to the values indicated in Table 9.1 one can predict the time of onset of different stages from the moment of insemination.

It is noteworthy that recalculation of τ_0 from the data on the time of onset of a few distinct developmental stages obtained by different authors at different temperatures [27, 54, 77, 93] has given very close values for the same developmental stages.

Similar results were obtained when the data of Nieuwkoop and Faber [77] were compared with ours for most stages. In Table 9.1 (graph 3) the time is given for each developmental stage at which Nieuwkoop and Faber fixed embryos for the description of their structure recalculated by us in τ_0 units. Since the embryos in their experiments developed at 22–24°C, the recalculation of their data is given for the mean temperature 23°C. Nieuwkoop and Faber read the time of development not from the appearance of the 1st cleavage furrow, as we did, but from the moment of egg laying soon after its fertilization, that is, $\sim 3\tau_0$ earlier. For a number of stages the data on the time of their onset given in graphs 2 and 3 (Table 9.1) differ indeed by $3\tau_0$ or a similar value.

When describing the diagnostic features (Table 9.1), we introduced some changes in the description of stage 2– and, in addition, gave intermediate stages $7^1/_2$, 8+ (after 33) and $8^1/_2$ (after 3). These changes are indicated in the text; they are shown in italics enclosed in parentheses. All these changes were approved by Nieuwkoop and Faber.

TABLE 9.1. Time of Onset of Successive Stages in the Development of *Xenopus laevis* in τ_0 Units from the Moment of Appearance of the 1st Cleavage Furrow on the Egg Surface (improved data of Dettlaff and Rudneva [27]) and from Fertilization (data of Nieuwkoop and Faber [77])

Stage number	Time in τ_0 units from stage (τ_n/τ_0)		Diagnostic features of stages
	2–	0	
0	–	0	Mature unfertilized egg at the moment of insemination
1	–	–	Fertilized egg; perivitelline space formed
2–	0	3.0	Appearance of 1st cleavage furrow on the animal pole
2	0.7	3.5	2 blastomeres; 1st cleavage furrow completed on the vegetal pole
3	1.7	4.6	4 blastomeres; 2nd cleavage furrow completed on the vegetal pole
4	2.7	–	8 blastomeres
5	2.7	6.3	16 blastomeres
6	4.9	7.0	32 blastomeres
$6^1/_2$	5.7	8.1	Morula; 48–64 blastomeres
7	6.5	9.2	Early blastula I; animal blastomeres still divided synchronously
$7^1/_2$	8.0	–	Early blastula II; desynchronization of divisions of the animal blastomeres
8	8.5	11.5	Middle blastula I; active divisions of the animal blastomeres
8+	11	–	Middle blastula II; fall of mitotic index in the animal blastomeres (see also [74, 75])
$8^1/_2$	11?	–	Sudden activation of genes accompanied by a sharp increase in RNA synthesis (after [3])
9	13.0	16.1	Late epithelial blastula
10	17–19	20.8	Onset of gastrulation; accumulation of pigment in the region of dorsal blastopore lip
$10^1/_4$	20	–	Onset of invagination
$10^1/_2$	22	25.4	Crescent-shaped blastopore
11	24–25	27.0	Middle gastrula, blastopore in the form of a horseshoe, the accumulation of pigment marks the position of lateral lips

TABLE 9.1 (continued)

Stage number	Time in τ_0 units from stage (τ_n/τ_0)		Diagnostic features of stages
	2–	0	
$11^1/_2$	26–27	29.0	Formation of ventral blastopore lip; large yolk plug
12	28–29	30.5	Diameter of yolk plug decreased to half
$12^1/_2$	30–32	33.0	Very small yolk plug
13	33–35	34.0	Slitlike blastopore; distinct contours of neural plate not yet marked
$13^1/_2$	34–36	36.0	Neural plate clearly delimited
14	35–36	37.5	Neural plate; onset of elevation of neural folds
15	36–37	40.4	Stage of neural folds (early)
16	37–39	42.3	Stage of neural folds (middle)
17	39–41	43.4	Stage of neural folds (late)
18	41–44	45.4	Stage of neural groove; neural folds in trunk region very close to each other but not yet touching
19	43–46	48.0	Onset of formation of neural tube; neural ridges touching each other all along
20	47–48	50.0	Fusion of neural folds and formation of neural tube; embryo begins to elongate, lateral outline flat
21	49	52.0	Lateral outline of embryo just becoming concave, ventral outline flat
22	50–53	55.4	Eye vesicles become distinct; ventral outline of embryo slightly concave
23	52–54	57.0	A groove appears separating the jaw and gill areas
24	57–59	60.6	Eye vesicles protrude less far laterally than gill area; tail bud discernible; initial motor reactions to external stimulation
25	60–61	63.2	Eye vesicles equally far or farther laterally than gill area; gill area grooved; onset of fin formation
26	66	68.5	Myotomes are first seen; pronophros is discernible; onset of spontaneous movements
27	67–68	72.0	Lateral flattening of eye vesicles; fin translucent, except for region just behind anus
28	70–72	75.0	Onset of secretion in the sucker. Fin extended, with outer transparent and inner translucent band

TABLE 9.1 (continued)

Stage number	Time in τ_0 units from stage (τ_n/τ_0)		Diagnostic features of stages
	2–	0	
29/30	77–79	80.8	Gray eye cup showing through; fin is transparent all along; tail bud distinct
31	80–82	86.5	Tail bud equally long and broad
32	83–85	92.3	Eye cup horseshoe-shaped, protrudes distinctly; length of tail bud is 1.5 times its breadth
33/34	95–100	102.7	Melanophores appear dorsally on the head and laterally on the body in a row extending from just below the pronephros backward; length of tail bud about twice its breadth; onset of heartbeat; stomodeum forms as a shallow vertical groove
35/36	115–125	115.0	Onset of hatching; eyes entirely black; two gill rudiments formed; melanophores appearing on back, length of tail bud is 3 times its breadth

For stages 1–19, besides the drawings made by Nieuwkoop and Faber, original drawings made by N. P. Bordzilovskaya are also given which illustrate quite distinctly their diagnostic structural features.

Data published earlier [27] on the time of onset of different stages during neurulation (stages 13–19), expressed in τ_0 units, are made more precise in the present table in accordance with the drawings of these stages. (Individual variations of times of onset of developmental stages in different embryos within the limits of stages 14–18 usually exceed the time interval between successive stages.)

9.6. STAGE CRITERIA FOR THE NORMAL DEVELOPMENT OF *Xenopus laevis*

These descriptions are cited from Nieuwkoop and Faber [76] with their permission; some modifications and comments have been introduced as a result of the authors' own experiences; these are italicized and are enclosed in parentheses. An asterisk indicates that certain criteria used by Nieuwkoop and Faber have not been included.

Stage 1. Age 0 hours; length 1.4–1.5 mm.

External criteria: 1-cell stage, shortly after fertilization. Pigmentation darker ventrally than dorsally.

Internal criteria: Nucleus at metaphase II. Well-defined cortical layer and layer of subcortical plasm, mainly located in animal half of egg. Inner plasm composed

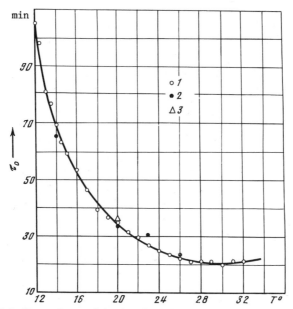

Fig. 9.2. Dependence of the duration of one mitotic cycle during the synchronous cleavage divisions (τ_0) on temperature in *Xenopus laevis* [27]. 1) Our data; 2) data of Valouch et al. [93]; 3) data of Hamilton [54]. Data of Gerhart et al. [37] also fit this curve: the period from insemination to the 1st cleavage furrow, equal to $3\tau_0$, lasts 112 min at 19.0° and 108 min at 19.5°, i.e., τ_0 equals 37 min at 19.0 and 36 min at 19.5°.

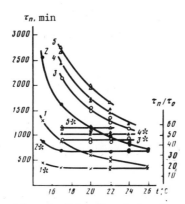

Fig. 9.3. Absolute (in min) and relative (in τ_0) duration of some developmental periods in *Xenopus laevis*. 1–5) Duration of the periods from the appearance of the 1st cleavage furrow on the egg surface to stage 10, onset of gastrulation (1); stage 13, slitlike blastopore (2); stage 19, fusion of neural folds (3); stage 22, when eye vesicles become distinct (4); and stage 24, when the motor reaction appears in response to external stimulation (5). Numerals with asterisks are absolute and those without asterisks are relative durations of corresponding periods. Stage numbers according to Nieuwkoop and Faber [76]. From [22].

of inner animal, central, and inner vegetal plasms. Special cytoplasmic inclusions near vegetal pole. Thickness of cortical and subcortical layers decreasing in animal–vegetal and dorsoventral directions: clear animal–vegetal and dorsoventral polarity.

Stage 2–. Age $1^1/_4$ hours; length 1.4–1.5 mm.

External criteria: Beginning of the first cleavage. (*The first cleavage furrow is first seen in the center of the animal region. When experimenting with embryos taken from clutches, this stage can be used as a reference point for reading the time of onset of successive developmental stages, since it can be determined with great precision (up to 1 min). In this respect we propose to make more precise the definition of this stage given by Nieuwkoop and Faber [65, p. 163]: "First cleavage groove has not yet reached vegetative pole," and consider "the very beginning of the furrow appearance in the center of the animal region" as stage 2–.*)

Stage 2. Age $1^1/_2$ hours; length 1.4–1.5 mm.

External criteria: Advanced 2-cell stage. First cleavage furrow has reached vegetal pole.

Internal criteria: Penetration of cortical and subcortical layers into interior of egg along cleavage furrows; the latter have progressed about half of egg radius into animal half of egg. Formation of very thin partition wall between the two blastomeres; blastomeres not equal in size. First plane of cleavage more or less coinciding with plane of bilateral symmetry.

Stage 3. After 2 hours; length 1.4–1.5 mm.

External criteria: Advanced 4-cell stage. Second cleavage furrow has reached vegetal pole. In animal view dorsal blastomeres usually smaller than ventral ones; the latter darker than the former. (*In the embryos we have studied we could not find distinct differences in the size of the first blastomeres, and the 1st and 2nd cleavage furrows only rarely separated the pigmented and unpigmented parts of the egg. The coincidence of the 1st cleavage furrow plane with that of bilateral symmetry was not evident in our material; see drawings by N. P. Bordzilovskaya in Figs. 9.4A , 9.5A, and 9.6A.*)

Internal criteria: Partition walls between the four blastomeres. Cleavage cavity present for the first time, mainly located in animal half of egg.

Stage 4. Age $2^1/_4$ hours; length 1.4–1.5 mm.

External criteria: Advanced 8-cell stage (3rd cleavage furrows run equatorially, at the level of about one-third of animal–vegetal axis). Dorsal micro- and macromeres usually smaller than ventral ones; dorsal micromeres less pigmented than ventral ones.

Internal criteria: Third plane of cleavage at about one third of animal–vegetal axis. Pigmented cortical layer and subcortical plasm penetrating over about one-fifth of egg radius along third cleavage furrow.

Figs. 9.4–9.13. Stages of normal development of the South African clawed toad *Xenopus laevis* Daudin. an) Animal view; veg) vegetal view; d) dorsal view; d-l) dorsal-lateral; d-t) dorsal-tail ; h) head view; v) ventral view; t) tail view. The drawings are reproduced from Nieuwkoop and Faber [76]. Drawing numbers 1–66 correspond to stage numbers; drawings without letter designations are lateral views. If the scale is changed, the embryo is drawn twice, i.e., in the previous and the new scale. The drawing of stage 13 by Nieuwkoop and Faber is renamed as $13^1/_2$, since it corresponds to a greater extent to the description of this stage given by the authors themselves. Figures 9.4A, 9.5A, and 9.6A are original drawings of stages 1–19 by N. P. Bordzilovskaya.

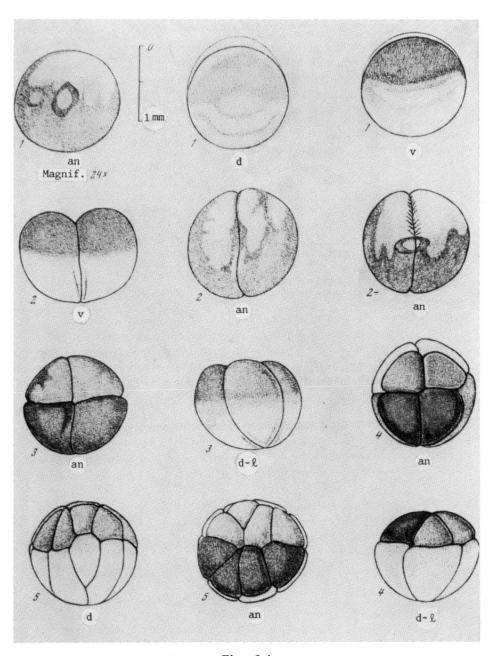

Fig. 9.4.

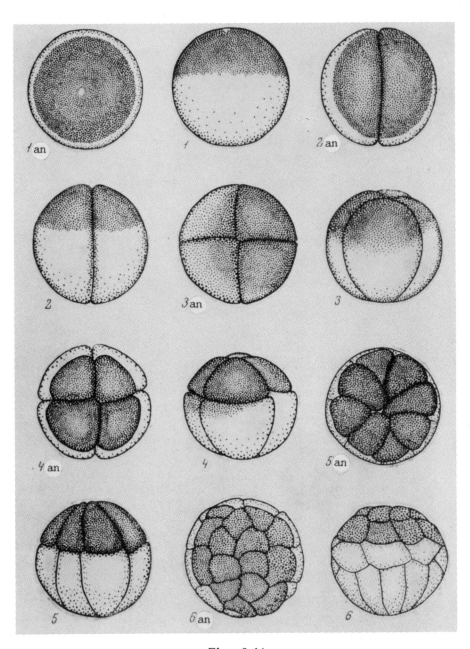

Fig. 9.4A.

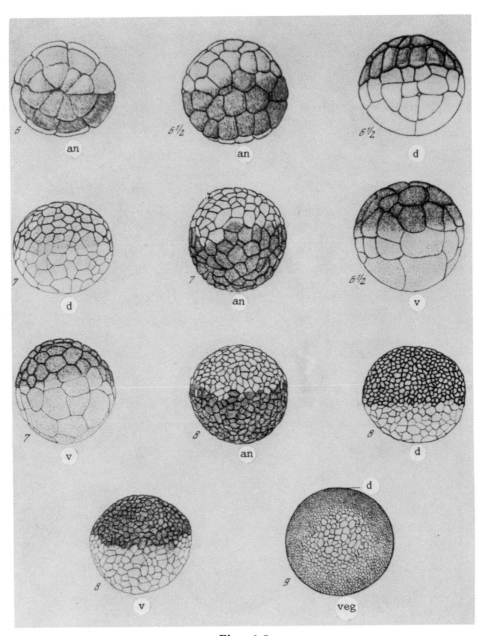

Fig. 9.5.

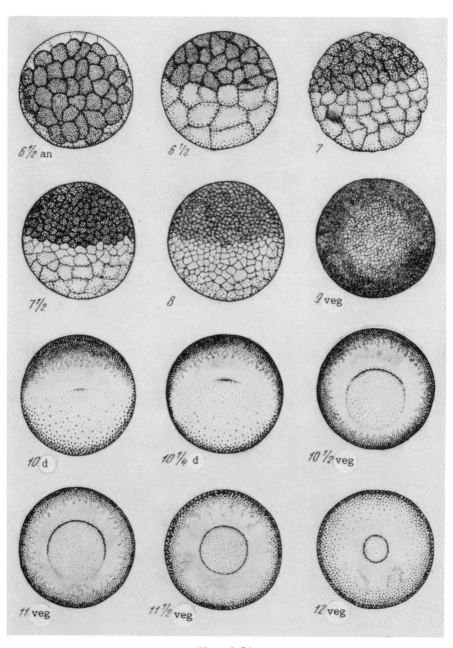

Fig.   9.5A.

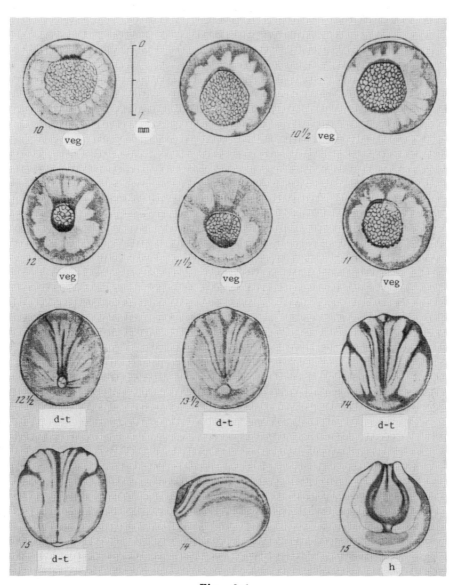

Fig. 9.6.

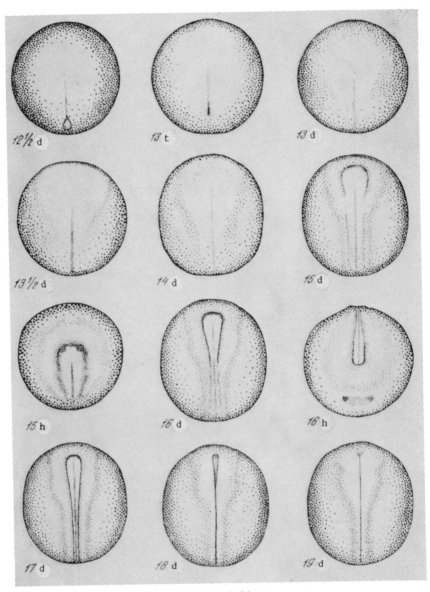

Fig. 9.6A.

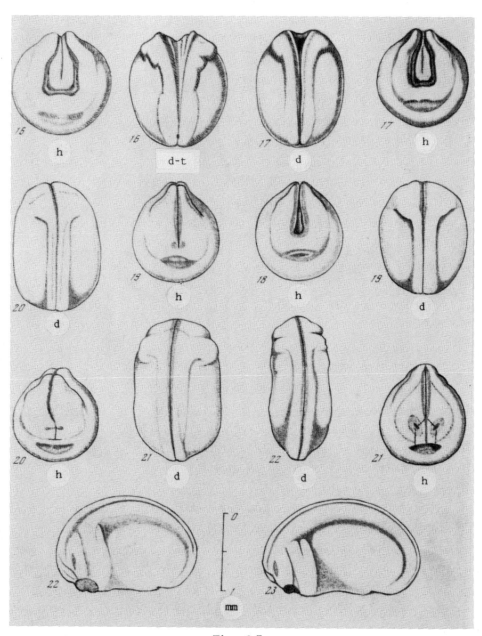

Fig. 9.7.

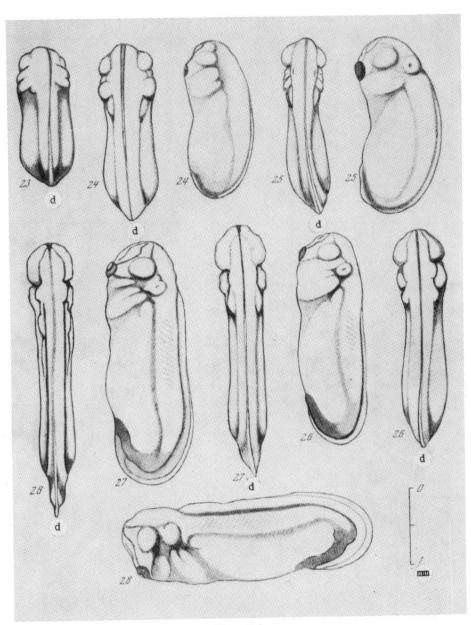

Fig. 9.8.

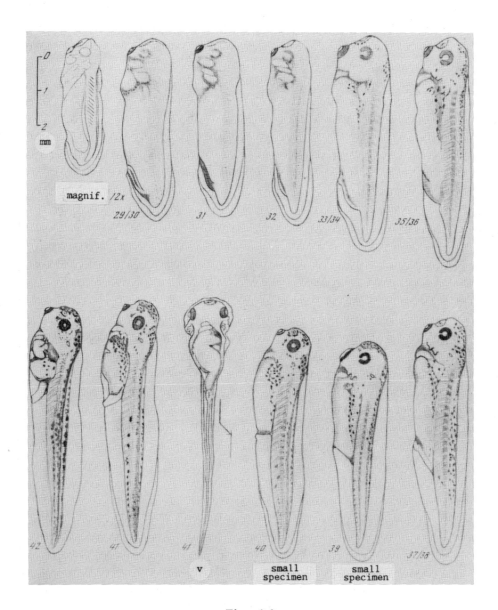

Fig. 9.9.

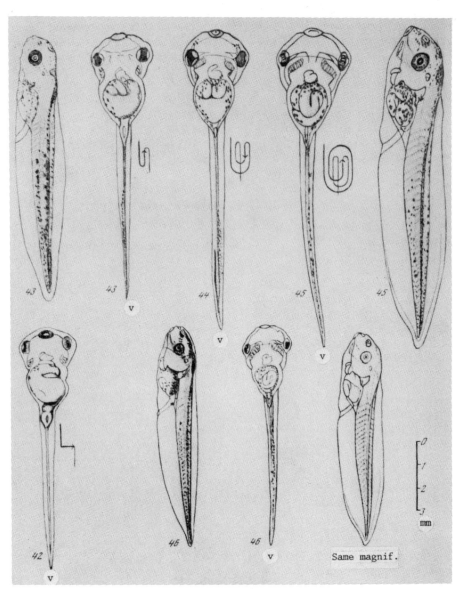

Fig. 9.10.

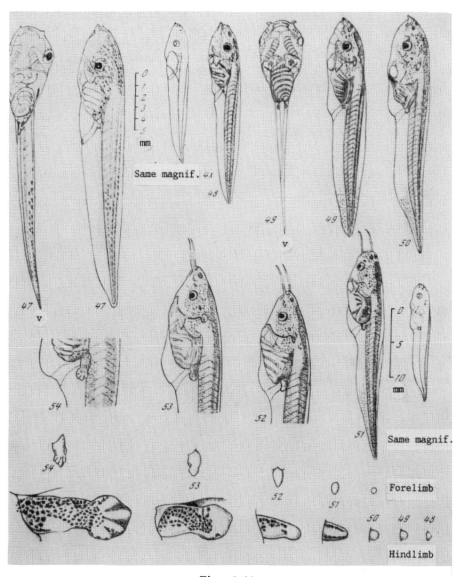

Fig. 9.11.

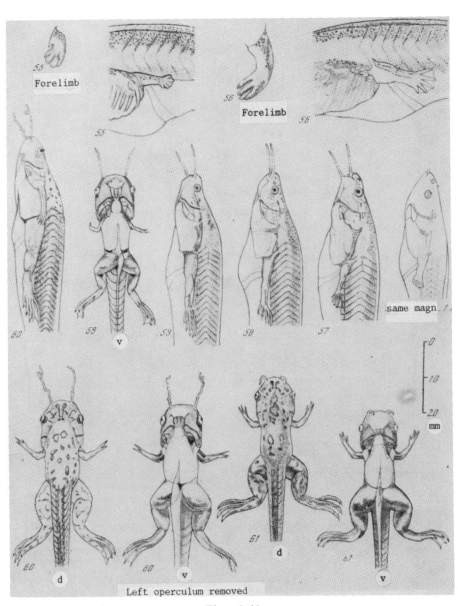

Fig. 9.12.

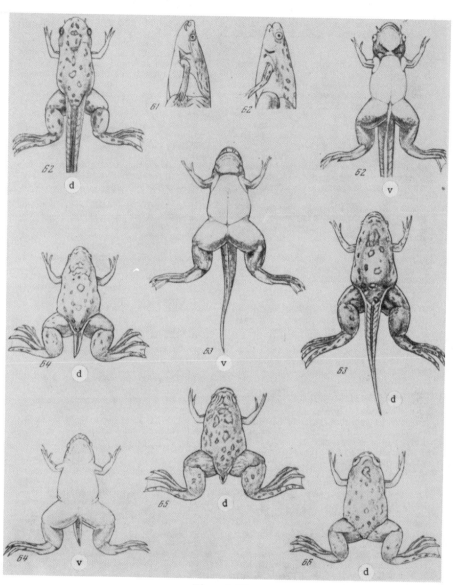

Fig.   9.13.

Stage 5. Age 2³/₄ hours; length 1.4–1.5 mm.

External criteria: Advanced 16-cell stage. Dorsal micromeres distinctly smaller and less pigmented than ventral ones. Macromeres entirely separated by cleavage grooves.

Internal criteria: Individual blastomeres varying in size and form. Cleavage cavity about as large as a micromere, located slightly eccentrically toward dorsal side.

Stage 6. Age 3 hours; length 1.4–1.5 mm.

External criteria: Advanced 32-cell stage. Distinction between dorsal and ventral micromeres as in preceding stage.

Internal criteria: Blastomeres roughly arranged in 4 rows of 8. Cleavages still nearly synchronous.

Stage 6¹/₂. Age 3¹/₂ hours; length 1.4–1.5 mm.

External criteria: Morula stage. About 48–64 blastomeres. In animal view about 6 micromeres along meridian.

Internal criteria: Blastomeres still arranged in single layer around cleavage cavity. Synchronism of cleavages gradually lost; animal blastomeres in advance compared with vegetal ones. Distinction possible between animal, equatorial, and vegetal blastomeres; transitions between cell types more gradual at dorsal than at lateral and ventral sides.

Stage 7. Age 4 hours; length 1.4–1.5 mm.

External criteria: Large-cell blastula stage. In animal view about 10 micromeres along meridian.

Internal criteria: Tangential cleavage; formation of double-layered embryo. Except for rather sharp boundary between outer equatorial and vegetal blastomeres, transition between various regions still very gradual. Pregastrulation movements noticeable for the first time: slight epibolic extension of animal and equatorial areas and beginning of ascent of plasm along cleavage furrows near vegetal pole.

Stage 8. Age 5 hours; length 1.41.5 mm.

External criteria: Medium-cell blastula stage (*early blastula*). Surface not yet entirely smooth. Border of animal pigment cap more diffuse at dorsal than at lateral and ventral sides. Gradual transition in cell size from animal to vegetal pole.

Internal criteria: Outer cell layer single, inner cell material 1–4 cells thick; first appearance of intercellular spaces between outer and inner cell material. First indication of distinction between animal, marginal, and vegetal areas in inner cell material. Continuation of pregastrulation movements.

Stage 9. Age 7 hours; length 1.4–1.5 mm.

External criteria: Fine-cell blastula stage (*late epithelial blastula*). Animal cells smaller at dorsal than at ventral side. Border between marginal zone and vegetal field distinct, particularly dorsally, owing to difference in cell size.

Internal criteria: Blastocoel has attained its full size; inner surface smooth.

Stage 10. Age 9 hours; length 1.4–1.5 mm.

External criteria: Initial gastrula stage. First indication of blastopore only by pigment concentration. No formation of groove.

Internal criteria: Formation of bottle-necked cells in outer cell layer in area of future blastopore groove. (Beginning of invagination of the internal material of marginal zone.)

Stage 10¹/₄. Age 10 hours; length 1.4–1.5 mm.

External criteria: Early gastrula stage. First formation of dorsal blastopore groove; groove still straight.

Internal criteria: Initial invagination of archenteron. Extension of inner blastopore lip from dorsal to lateral, and even partially to ventral side of inner marginal zone. Gradual delamination of prechordal portion of definitive mesodermal mantle from central endoderm mass.

Stage 10^1/$_2$. Age 11 hours; length 1.4–1.5 mm.

External criteria: Crescent-shaped blastopore stage. Blastopore groove angular. Epibolic extension of presumptive ectoderm and reduction of vegetal area, ventral border of future yolk plug indicated by pigment concentration.

Internal criteria: Marked progress in formation of definitive mesodermal mantle, which has extended up to equator at dorsal side and is forming all around vegetal area. First contact between prechordal part of archenteron roof and most caudal portion of sensorial layer of ectoderm. Invagination of short slit-shaped archenteron over 10–15°. Clear topographical and temporal distinction between rolling-in of inner marginal zone and invagination of archenteron.

Stage 11. Age 11^3/$_4$ hours; length 1.4–1.5 mm.

External criteria: Horseshoe-shaped blastopore stage. Blastopore groove surrounding about half of future yolk plug and indicated at its ventral side; future yolk plug often rounded rectangular, slightly elongated in dorsoventral direction, diameter more than two-fifths of diameter of egg (±50° of circumference).

Internal criteria: Mesodermal mantle extending to 30–40° above equator at dorsal side, to 20—30° above equator at lateral side and approximately to equator at ventral side. Invagination of archenteron extended to lateral side; archenteron still slit-shaped, extending dorsally over 40–50°. Blastocoel beginning to be displaced toward ventral side.

Stage 11^1/$_2$. Age 12^1/$_2$ hours; length 1.4–1.5 mm.

External criteria: Large yolk plug stage. Blastopore groove closed ventrally; yolk plug not yet quite circular. At ventral side concentrated superficial pigment still visible. Diameter of yolk plug about one-third of diameter of egg (~40°).

Internal criteria: First indication of neural anlage in sensorial layer of ectoderm; epithelial layer now a flattened epithelium. Mesodermal mantle extending dorsally to 20–30° from animal pole, ventrally not beyond equator. Archenteron extending over 80–90°, still slit-shaped. Endodermal mass displaced toward ventral side.

Stage 12. Age 13^1/$_4$ hours; length 1.4–1.5 mm.

External criteria: Medium yolk plug stage. Yolk plug circular; diameter somewhat less than one-quarter of diameter of egg (~25°). More and less pigmented fields radiating from yolk plug.

Internal criteria: Neural anlage extended craniad; neural and epidermal areas clearly distinguishable anteriorly. Mesodermal mantle reaching some distance from animal pole at dorsal side; beginning of segregation of prechordal and chordal areas. Archenteron extending over more than 90°; first indications of widening.

Stage 12^1/$_2$. Age 14^1/$_4$ hours; length 1.4–1.5 mm.

External criteria: Small yolk plug stage. Future position of neural plate and median groove indicated by darker pigment lines. Yolk plug usually ovoid, variable in size.

Internal criteria: Clear delimitation of ecto-, meso-, and endodermal germ layers. Neural anlage in sensorial layer of ectoderm reaching nearly up to animal pole of egg. Mesodermal mantle has reached its definitive extension, at dorsal side close to animal pole, at lateral and ventral sides to 40–50° from animal pole. Archenteron extending up to animal pole and markedly widened.

Stage 13. Age 14^3/$_4$ hours; length 1.5–1.6 mm.

External criteria: Slit-blastopore stage. Neural plate faintly delimited; slight elevation of its rostral part and slight flattening of its caudal part. Caudal part of median groove formed.

Internal criteria: Formation of neural anlage still restricted to sensorial layer of ectoderm; more intimate attachment in dorsal midline between epithelial and sensorial layer of ectoderm, and between the latter and the underlying archenteron roof. Presumptive somite mesoderm double-layered. Archenteron extended to about 10° ventral to animal pole; caudal archenteron ring-shaped around endodermal protuberance. Blastocoel rapidly decreasing in size.

Stage 13^1/$_2$. Age 15^1/$_2$ hours; length 1.5–1.6 mm.

External criteria: Initial neural plate stage. Neural plate clearly delimited.

Internal criteria: First indication of formation of neural crest zone in margin of anterior half of neural anlage.* Archenteron widened and extended to definitive position at about 45° ventral to animal pole; first formation of liver diverticulum.

Stage 14. Age 16^1/$_4$ hours; length 1.5–1.6 mm.

External criteria: Neural plate stage. Cerebral part of neural plate bent downward, with median elevation at rostral end of median groove. Initial elevation of neural folds, most pronounced in future nuchal region. Blastopore always slit-shaped.

Internal criteria: Neural folds fading out in caudal direction. Presumptive somite area clearly demarcated against lateral mesoderm.

Stage 15. Age 17^1/$_2$ hours; length 1.5–1.6 mm.

External criteria: Early neural fold stage. Presumptive cement gland faintly circumscribed. Anterior part of neural plate roundish. Neural folds distinct, except mediorostrally; initial formation of sharp inner ridges on neural folds in rhombencephalic region.

Internal criteria: Cement gland anlage delimited. First symptoms of segregation of neural crest from thickening and narrowing neural plate (s.s.). Presumptive eye anlagen beginning to sink in. Beginning of dorsal convergence and stretching movements. Clearly discernible myocoel. Blatocoel disappeared or greatly reduced.

Stage 16. Age 18^1/$_4$ hours; length 1.5–1.6 mm.

External criteria: Middle neural fold stage. Anterior part of neural plate rectangular; darkly pigmented eye anlagen present; neural plate sharply constricted in the middle. Inner ridges on neural folds forming angle of about 90° with neural plate in rhombencephalic region.

Internal criteria: First visible participation of epithelial layer of ectoderm in neural plate. Distinct elevation of neural folds along entire length of neural anlage. Eye anlagen forming deep depressions in anterolateral edges. Left and right heart anlagen beginning to fuse.

Stage 17. Age 18^3/$_4$ hours; length 1.5–1.6 mm.

External criteria: Late neural fold stage. Anterior part of neural plate oblong triangular, angles formed by eye anlagen. Neural folds approaching each other from blastopore up to anterior trunk region.

Internal criteria: In anterior half of embryo neural crest material located at lateral edges of neural anlage; in caudal half sharp delimitation of neural crest material from neural plate (s.s.). Eye anlagen showing first signs of lateral evagination. First indications of some segregation, but still continuous myocoelic cavity.

Stage 18. Age 19^3/$_4$ hours; length 1.5–1.6 mm.

External criteria: Neural groove stage. Anterior part of neural plate narrow, more or less club-shaped, often narrower toward rostral end. Parallel neural folds in trunk region very close to each other, not yet touching.

Internal criteria: First segregation of mesencephalic and rhombencephalic neural crest. Eye evaginations with slit-shaped cavities. Anterior 3–4 somites in process of segregation.

Stage 19. Age $20^3/_4$ hours; length 1.5–1.6 mm.

External criteria: Initial neural tube stage. Neural folds touching each other, except for inconstant openings at anterior and posterior end and behind nuchal region. Considerable lateral extension of brain. Lateral outline of embryo still convex.

Internal criteria: Embryonic pigment in epithelial layer of ectoderm dispersed over entire cell body. Neural crest segregated into 4–5 cell masses; beginning of lateral migration. Segregation of anterior 4–6 somites. Presumptive heart rudiment a single median mesodermal thickening.

Stage 20. Age $21^3/_4$ hours; length 1.7–1.8 mm.

External criteria: Neural folds fused, suture still present. The two eye anlagen showing through are dumbbell-shaped; eyes are hardly protruding. Beginning of stretching of embryo. Lateral outline flat.

Internal criteria: Embryonic pigment of epithelial layer of ectoderm concentrated at apical side of cells. Stomodeal–hypophyseal anlage indicated by median thickening of sensory layer of ectoderm just in front of brain. Brain subdivided into archencephalon and deuterencephalon. Massive neural crest extending anteriorly approximately to anterior border of eye anlagen. In trunk region neural crest still in contact with suture of closing neural tube. Anterior 6–7 somites segregated. Coelomic cavity indicated in most dorsal portion of lateral plate. Anlage of hypochorda indicated. First indications of 1st and 2nd visceral pouches in pharyngeal wall; oral evagination visible.

Stage 21. Age $22^1/_2$ hours; length 1.9–2.0 mm.

External criteria: Suture of neural tube completely closed. Delimitation of frontal field by pigment lines. Primary eye vesicles showing through in the form of two separate, obliquely placed oval spots; beginning of protrusion of eyes. Lateral outline of embryo just becoming concave, ventral outline flat.

Internal criteria: *First indication of development of cephalic flexure and of rhombencephalic roof formation. Closure of neural canal. *Ear placode visible for the first time as thickening of sensory layer of ectoderm. Segregation of maxillary and mandibular portions of mesectoderm. Anterior 8–9 somites segregated. First indication of pronephros anlage.

Stage 22. Age 24 hours; length 2.0–2.2 mm.

External criteria: Distinct protrusion of eyes. Initial groove between jaw and gill areas only at laterodorsal side. Lateral and ventral outlines of embryo slightly concave. Anal opening displaced to ventral side. Vitelline membrane becoming wider.

Internal criteria: First indication of formation of dorsolateral placodes in sensorial layer of ectoderm. Segregation of brain into prosencephalon, mesencephalon, and rhombencephalon; cephalic flexure approximately 135°. In anterior spinal cord beginning of mediolateral constriction of central canal and withdrawal of anterior trunk neural crest. Primary eye vesicles in broad contact with epidermis. Segregation of hyal portion of mesectoderm. Anterior 9–10 somites distinct. Formation of blood islands indicated.

Stage 23. Age 24³/₄ hours; length 2.2–2.4 mm.

External criteria: Jaw and gill areas completely separated by groove. Ventral outline of embryo more concave.

Internal criteria: Segregation of prosencephalon into telencephalon and diencephalon. Spinal cord completed at anterior end. First appearance of olfactory placodes as thickenings of sensorial layer of ectoderm. First slight depression in ear placode underneath epithelial layer of ectoderm. Segregation of 1st branchial portion of mesectoderm. Approximately 12 somites segregated. Beginning of segregation of nephrotomes; Wolffian duct anlage indicated. First contact of ectoderm and endoderm in 1st visceral pouch (formation of mandibular arch); 3rd visceral pouch indicated.

Stage 24. Age 26¹/₄ hours; length 2.5–2.7 mm.

External criteria: Eyes protruding less far laterally than gill area. Gill area more prominent than jaw area, gill area not yet grooved. Ventral outline of embryo nicked. Tail bud discernible. Initial motor reactions to external stimulation.

Internal criteria: Distinct nerve fibers in root of trigeminal nerve, and some fibers emerging from ganglia profundus and Gasseri. Mediolateral constriction of central canal extended over two-thirds of spinal cord, and anterior half of spinal cord completed. Fifteen somites segregated; axial mesenchyme becoming liberated from somites. Tail bud for the first time distinct. Segregation of collecting tube anlage in anterior portion of pronephros. Segregation of hypochorda starting anteriorly.

Stage 25. Age 27¹/₂ hours; length 2.8–3.0 mm.

External criteria: Eyes protruding equally far or further laterally than gill area. Gill area grooved. Invagination of ear vesicle indicated by pigment spot. Beginning of fin formation.

Internal criteria: Frontal glands indicated in epithelial layer of ectoderm of diencephalic region. Cephalic flexure of brain about 90°. First part of maxillomandibular nerve distinct. Most anterior neural crest of spinal cord has begun lateral migration; only posterior quarter of spinal cord not yet completed. Primary eye vesicles fully developed; medial wall decreasing, lateral wall increasing in thickness. Head somite I diminishing; 16 somites segregated. Fourth visceral pouch indicated.

Stage 26. Age 29¹/₂ hours; length 3.0–3.3 mm.

External criteria: Ear vesicle protruding. Pronephros distinctly visible. Myotomes showing through for the first time. Fin somewhat broadened at dorsocaudal end of body. Beginning of spontaneous movements.

Internal criteria: First indications of olfactory lobe formation in anterolateral portion of telencephalic wall; beginning of evagination of pineal body. In posterior fifth, spinal cord not yet completed and central canal still unconstricted. Lateral wall of primary eye vesicle showing local protrusions on inner surface. Head somite I disintegrated; 17 somites segregated. Wolffian duct anlage extended to level of trunk somite 3. First contact of ectoderm and endoderm in 2nd visceral pouch (formation of hyoid arch). Postanal gut indicated.

Stage 27. Age 32¹/₄ hours; length 3.4–3.7 mm.

External criteria: Lateral flattening of eyes. Fin translucent, except for region just behind anus. Tail bud formation accentuated in lateral outline.

Internal criteria: First part of ophthalmic profundus nerve distinct. Retinal layer of optic vesicle invaginating at anterodorsal margin; beginning of lens formation as thickening of sensorial layer of ectoderm. Ear vesicle closed, but still connected with ectoderm. Nineteen somites segregated. Endocardial anlage delimited. Wolffian duct extends to level of trunk somite 5. First contact of ectoderm and endoderm in 3rd visceral pouch (formation of 1st branchial arch).

Stage 28. Age $32^1/_2$ hours; length 3.8–4.0 mm.

External criteria: Fin extending up to anus. Fin broadened and distinctly divided into outer transparent and inner translucent band.

Internal criteria: Beginning of secretion of cement gland. Hypophyseal anlage penetrated beyond optic stalks; infundibulum rather well developed, ventral wall thinning out; first fibers (white matter) developing along ventrolateral portion of mesencephalon and rhombencephalon and along varying portions of anterior half of spinal cord. First epibranchial placodes segregated. Marginal invagination of retinal layer extended around eye vesicle, central portion still convex. Ear vesicle detached from epidermis. Between 20 and 33 somites segregated. First formation of endocardial tube and beginning of formation of pericardial cavity. First nephrostome funnel clearly indicated, the following two just distinguishable; coelomic filter chamber indicated.

Stage 29/30. Age 35 hours; length 4.0–4.5 mm.

External criteria: Gray eye cup showing through for the first time. Fin transparent up to the base over its whole length. Tail bud distinct.

Internal criteria: Chiasmatic ridge separated rostrally from lamina terminalis by shallow groove. Nerve fibers extending over 50–60% of length of spinal cord. Eye vesicle invaginated, but central portion still convex. Head somites II and III much reduced; 24–25 somites segregated, segregation reaching tail. Endocardial heart anlage a short tube, closed at both ends; pericardial cavity surrounding endocardial anlage on lateral and ventral sides. Minor lumen indicated in collecting tube of pronephros anlage, and first appearance of glomus. Floor of pharyngeal cavity somewhat raised in front of liver diverticulum.

Stage 31. Age $37^1/_2$ hours; length 4.2–4.8 mm.

External criteria: Tail bud equally long and broad.

Internal criteria: Mushroom-shaped epiphyseal evagination; hypophyseal cell plate reaching to caudal border of chiasmatic ridge, but still connected with ectoderm. Nerves beginning to grow out from ganglion VII; appearance of ganglion VIII. Neural crest of spinal cord entirely withdrawn from neural tube; thin layer of nerve fibers extending over 70% of length of spinal cord. Nasal pit indicated. Central portion of retinal layer concave. Segregation of 2nd branchial portion of mesectoderm. 22–23 post-otic somites; appearance of "Urwirbelfortsätze" on ventral sides of anterior myotomes. First pair of aortic arches completed, and beginning of paired dorsal aorta formed; pronephric sinus formed. First nephrostome funnel completed.

Stage 32. Age 40 hours; length 4.5–5.1 mm.

External criteria: Eye cup horseshoe-shaped, standing out distinctly. Length of tail bud about $1^1/_2$ times its breadth.

Internal criteria: Hypophyseal cell strand detached from ectoderm; roof of IVth ventricle very thin. First nerve fibers passing from nasal placode to brain; distinct root of glossopharyngeal nerve. Layer of nerve fibers along spinal cord terminating at level of cloaca. Cavity of primary optic vesicle disappeared except for region near optic stalk; margins of optic cup, forming choroid tissue, touching each other; first pigment in outer layer of optic cup. First appearance of anlage of ductus endolymphaticus. About 26 post-otic somites formed. Chordal epithelium and elastica externa formed at base of tail. Posterior end of endocardial tube broadened as anlage of sinus venosus; short medial postcardinal veins present. Second nephrostome funnel completed; small lumen in anterior portion of Wolffian duct; rectal diverticula indicated. First contact of ectoderm and endoderm in 4th visceral pouch (formation of 2nd branchial arch); 5th visceral pouch indicated; paired lung anlage clearly visible.

Stage 33/34. Age 44^1/$_2$ hours; length 4.7–5.3 mm.

External criteria: Stomodeal invagination a shallow vertical groove. Dorsal part of eye more pigmented than ventral part; distinct melanophores in dorsal part. Melanophores appearing dorsally on the head and laterally in a row extending from just below the pronephros backward. Length of tail bud about twice its breadth. Beginning of heartbeat.

Internal criteria: First chromatophores differentiating in head and trunk regions. Appearance of supraorbital, infraorbital, and trunk lateral line placodes. First optic fibers in middle of chiasmatic ridge and along optic stalks; appearance of oculomotor nerve. Appearance of Rohon–Beard cells at trunk levels of spinal cord. Nearly spherical lens anlage detached from ectoderm; distal lens epithelium being formed. Sensorial anlagen of auditory vesicle indicated. Anlage of musculus interhyoideus segregated. About 32 post-otic somites formed. Dorsal mesocardium completed and heart twisted; 3rd aortic arch completed; paired dorsal aortae reaching to pronephros; omphalomesenteric veins opening into functional ducts of Cuvier; anterior cardinal vein formed. Appearance of anlagen of anterior pair of lymph hearts. Third nephrostome completed; lumen in entire collecting tube of pronephros. Thyroid anlage clearly discernible. Anterior wall of liver rudiment bulging forward.

Stage 35/36. Age 50 hours; length 5.3–6.0 mm.

External criteria: Stomodeal invagination roundish. Eye entirely black, choroid tissue nearly closed. Formation of two gill rudiments, anterior one nipple-shaped. Melanophores appearing on back. Posterior outline of proctodeum still curves. Length of tail bud about 3 times its breadth. Beginning of hatching.

Internal criteria: Supraorbital and infraorbital lateral line anlagen extending to eye; hyomandibular lateral line placodes proliferating on both sides of cement gland; dorsal and middle trunk lateral lines extending beyond pronephros. Interretinal part of optic nerve distinct; trochlear nerve has reached musculus obliquus superior. Cavity of optic stalk obliterated; beginning of differentiation of pars optica retinae; lens cavity fully developed. About 36 post-otic somites formed. Chambers of S-shaped heart distinct; 4th, 5th, and dorsal part of 6th aortic arches completed; secondary vessels of 3rd and 4th aortic arches formed; paired dorsal aortae united as median dorsal aorta; medial postcardinals united behind cloaca; lateral head veins and musculoabdominal veins formed. Rich blood supply to pronephros from cardinal veins; coelomic filter chamber and glomus extending over entire length of pronephros. Oropharyngeal cavity separated from gastroduodenal cavity by high transverse ridge; first contact of ectoderm and endoderm in 5th visceral pouch (formation of 3rd branchial arch). Horizontal ridge separating primary liver cavity from tracheal cavity, the former opening distally into submesodermal space; dorsal pancreas anlage indicated; canalis neurentericus occluded.

Stage 37/38. Age 53^1/$_2$ hours; length 5.6–6.2 mm; length of tail with fin about 1.8 mm.

External criteria: Stomodeal invagination much deeper, opening round. Both gill rudiments nipple-shaped, a branch of the anterior one indicated. Posterior outline of proctodeum straight, forming very obtuse angle with ventral border of tail myotomes. Melanophores spreading over tail.

Internal criteria: Appearance of anlagen of occipital and ventral trunk lateral lines. Subcommisural organ differentiating; first fibers in comm. posterior. N. branchialis X^2 has reached middle of 2nd branchial arch. Rohon–Beard cells appearing in anterior tail levels of spinal cord. Beginning of aggregation of spinal ganglion cells. Organon vomero-nasale segregated from main olfactory organ. Nuclei arranged in three layers in pars optica retinae; beginning of differentiation of visual cells. Anlagen of musculus quadratohyoangularis and musculus orbitohyoideus segregated. About 40 post-otic somites formed. Paired dorsal aortae extending forward as internal carotid artery; median dorsal aorta extending backward into tail as caudal artery. Future dorsal branch of caudal vein appearing in dorsal tail fin. Wolffian ducts and rectal diverticula communicating; entire pronephros functional. Stomach and duodenum dorsally segregated; anterior wall of primary liver cavity forming numerous folds; the two ventral pancreas anlagen discernible.

Stage 39. Age $56^1/_2$ hours; length 5.9–6.5 mm; length of tail with fin maximum 2.6 mm.

External criteria: Melanophores appearing around nasal pits. Opening of stomodeal invagination transversely elongated. Melanophores on back arranged in a superficial and a deeper layer. Outlines of proctodeum and tail myotomes forming angle of about 135°. Melanophores appearing along ventral edge of tail musculature.

Internal criteria: Infraorbital lateral line anlagen segregating into sense organs. First fibers in comm. anterior. N. hyomandibularis VII connected with musculus interhyoideus; ganglion superius IX and ganglion petrosum IX separated; roots of the vagus nerve distinct. Thin layer of fibers on ventral surface of spinal cord; a few cells of central gray matter bulging into white matter in anterior levels of spinal cord. Arteria hyaloidea have reached interior of eye cup; delicate layer of mesenchyme around eye cup; eye muscle anlagen outlined. Sensorial epithelium of ear vesicle split into mediocaudal and laterocranial anlage. About 43 post-otic somites. Sixth aortic arch completed with ductus Botalli and pulmonary artery. First mesonephric cells appearing in caudal trunk region. Contact between ectoderm and endoderm obliterated in 1st and 2nd visceral pouches, representing anlage of middle ear and eustachian tube; anlage of primitive tongue formed; cavities of trachea and pharynx separated by horizontal ridge. Duodenum curving to left side as first symptom of coiling of intestinal tract; most posterior part of postanal gut disintegrating.

Stage 40. Age 66 hours; length 6.3–6.8 mm.

External criteria: Mouth broken through. Length of gills about twice their breadth, posterior one sometimes also showing a branch. Outlines of proctodeum and tail myotomes forming angle of 90°. Beginning of blood circulation in gills.

Internal criteria: Supraorbital and occipital lateral line anlagen segregating into sense organs; dorsal and middle lateral lines of trunk extending to base of dorsal fin at base of tail, and segregating proximally into sense organs. First fibers in comm. habenularis. Spinal nerves reaching lateral plate in trunk region. Outer and inner plexiform layers forming in pars optica retinae; fasciculus opticus formed; nuclei of central lens fibers degenerating; inner corneal layer formed. Laterocranial sensorial anlage of ear vesicles split into medial and lateral (crista externa) anlagen. The various parts of cranium and visceral skeleton well distinguishable as mesenchymatous condensation. Individualization of musculares levatores mandibulae and musculus intermandibularis. Formation of musculus geniohyoideus. About 45 post-otic somites. Elastica interna forming at base of tail. Anlage of forelimb atrium formed dorsocaudal to 5th visceral furrow. Hepatic vein present; left omphalomesenteric vein, receiving subintestinal vein, has lost connection with heart and has broken into capillaries in liver; small external jugular and pulmonary veins formed. Oral plate ruptured and 3rd and 4th visceral pouches perforated; 6th visceral pouch indicated; tracheal cavity separated from gastroduodenal cavity; paired thyroid anlage detached from thyroglossal duct; anlagen of thymus gland and ultimobranchial bodies indicated. Primary liver cavity occluded; gall bladder anlage formed; the three pancreatic rudiments fused.

Stage 41. Age 76 hours; length 6.7–7.5 mm.

External criteria: Gills broader and flatter, more laterally directed. Formation of a left rostral and a right caudal furrow in yolk mass; torsion of interjacent part about 45°†; formation of conical proctodeum, forming angle of about 60° with tail myotomes. Formation of fin rostral to proctodeum; ventral outline of yolk mass and proctodeum a smooth concave line.

Internal criteria: Yolk consumed in skin and brain structures. Lateral infundibular recesses and optic ventricles formed; pia mater discernible at level of mesencephalon. First motor neurons enlarging at edge of gray matter in tail region of spinal cord. Lens cavity disappeared. Crista posterior split off from mediocaudal sensorial anlage at ear vesicle; first indications of formation of semicircular canals. Several parts of cranium and visceral skeleton procartilaginous; fen. basicranialis enclosed by procartilaginous structures. Segregation of musculus tentaculi; anlagen of musculus dilatator laryngis and musculares constrictores laryngis recognizable; further individualization of visceral muscles; eye muscles differentiating. Spiral valve present in conus arteriosus; other heart valves indicated; ductus Botalli and 1st aortic arch absent; coeliacomesenteric artery formed; right omphalomesenteric vein has lost connection with the heart and has broken into capillaries in the liver. Beginning of segregation of primordial germ cells with dorsal endoderm. Appearance of filter process anlagen along caudal margin of 3rd visceral arch. Main gastric glands forming; oblique right caudal fold in surface of intestine, corresponding to future tip of helix; post-anal gut disappeared.

Stage 42. Age 80 hours; length 7.0–7.7 mm.

External criteria: Beginning of formation of opercular folds. Torsion of intestine about 90°; proctodeum connected with yolk mass by short horizontal tube.

Internal criteria: First appearance of dura-endocranial membrane at level of mesencephalon; anlage of Stirnorgan beginning to segregate from epiphyseal anlage; first fibers in comm. cerebellaris. Nervus ophthalmicus profundus V and nervus maxillomandibularis separated by processus ascendens palatoquadrati; nervus glossopharyngeus innervating musculus levator arcuum branchialium. Rods and cones distinguishable in the optic retina. Beginning of chondrification of visceral skeleton and palatoquadratum. Head somites I and II disappeared; left and right anlagen of musculares constrictores fused dorsally and ventrally. Right omphalomesenteric vein receiving main portal vein; abdominal vein present. First

---

†The torsion of the intestine is indicated by diagrams added to the drawings of stages 41–45 (Figs. 9.9 and 9.10).

appearance of interrenal cells. 5th visceral pouch perforated; filter process anlagen appearing along caudal margin of 4th and 5th visceral arches.

Stage 43. Age 87 hours; length 7.5–8.3 mm.

External criteria: Lateral line system becoming visible externally. Cement gland losing its pigment. Torsion of intestine about 180°; proctodeum narrower, arched or S-shaped.

Internal criteria: Ventral lateral line has reached pre-anal fin. Beginning of development of cerebral hemispheres; first signs of choroid plexus formation in prosencephalic and rhombencephalic roof. Narrow cell strand connecting lateral part of nasal organ with pharyngeal roof. Beginning of formation of pectinate membrane of eye. Laterocranial sensorial anlage of ear vesicle split into anlagen of crista anterior and macula utriculi. Beginning of chondrification of cranium; Meckel's cartilages, ceratohyal and basihyobranchials chondrified. Hindlimb region indicated. Anterior and posterior valves capable of closing atrioventricular aperture of heart. Accumulation of mesenchyme indicating future spleen anlage. Continuous mesonephric anlage formed. Primordial germ cells beginning to form median, unpaired genital ridge. Hypochorda disappeared. Epithelial bodies indicated on ventral side of 3rd and 4th visceral pouches. Long intestinal loop formed, reaching anteriorly as far as liver.

Stage 44. Age 92 hours; length 7.8–8.5 mm.

External criteria: Appearance of tentacle rudiments. Opercular folds protruding further. Coiling part of intestine showing S-shaped loop; torsion about 360°. Blood circulation in gills usually ceased (gills smaller).

Internal criteria: Caudal lateral line appearing on pre-anal fin. Beginning of invagination of plexus formations. First appearance of duraendocranial membrane around anterior portion of spinal cord; motor neurons standing out in tail region of spinal cord. Beginning of segregation of inner choroid and outer scleral coat of eye. Ear vesicle divided into pars superior and pars inferior. Beginning of chondrification of comm. quadratocranialis anterior and arcus subocularis; plana branchiala chondrified. Head somite III disappeared. Urwirbelfortsätze splitting into anlagen of musculus obliquus abdominis and musculus rectus abdominis. Valves in conus fully developed; circulation in filter apparatus established; lateral postcardinal veins indicated; posterior vena cava established. Aditus laryngis partially formed; anlagen of filter apparatus developing along cranial margins of visceral arches.

Stage 45. Age 98 hours; length 8–10 mm.

External criteria: Operculum partly covering gills, edge still straight. Intestine spiralized in ventral aspect, showing $1^1/_2$ revolutions. Beginning of feeding.

Internal criteria: Parietal sense organs present. Nasal organs beginning to withdraw from hemispheres. Motor neurons standing out in posterior trunk region of spinal cord. Choroid tissue closed except for passage of hyaloid artery. Outgrowth of parachordalia and formation of hypochordal commissure. Mesenchymal condensations of hindlimb anlagen connected by future anlage of pelvic girdle dorsal to anal tube; first thickening of ectoderm over hindlimb bud area. Partition wall between atria completed. Anlage of spleen well defined. Three to four pairs of mesonephric units discernible in mesonephric anlage. Paired genital ridges formed. Interrenal blastema being formed. Rakes on 3rd visceral arch branched; anlagen of filter processes on inner surface of operculum.

Stage 46. Age 106 hours; length 9–12 mm.

External criteria: Edge of operculum becoming convex. Xanthophores appearing on eye and abdomen. Intestine showing 2–2$^{1}/_{2}$ revolutions. Hindlimb bud visible for the first time.

Internal criteria: Duraendocranial membrane surrounding entire brain; cells of Purkinje differentiating in corpora cerebelli. Some spinal ganglion cells enlarging and equaling Rohon–Beard cells in tail region. Distal end of ductus endolymphaticus enlarged as saccus endolymphaticus; canalis lateralis partially separated from utriculus by septum formation. Beginning of chondrification of ear capsule with cupula anterior and posterior; complete chondrification of visceral skeleton; formation of fenestra subocularis and foramen caroticum. Anterior tip of notochord degenerating. Anlage of forelimb atrium detached from ectoderm; first mesenchymal accumulation of forelimb anlage. Double-layered ectoderm over clearly defined mesenchymal masses of hindlimb rudiments. Occipitovertebral artery formed. Four to five mesonephric units distinguishable. Esophagus free of yolk. Food appearing in intestinal tract.

Stage 47. Age 132 hours; length 2–15 mm.

External criteria: Tentacles larger. Edge of operculum forming quarter of a circle. Xanthophores forming opaque layer on abdomen. Intestine showing 2$^{1}/_{2}$–3$^{1}/_{2}$ revolutions. Hindlimb bud more distinct.

Internal criteria: Cement gland beginning to degenerate. Prosencephalic plexus branched; pars intermedia of adenohypophysis beginning to differentiate. Two to three Jacobson's gland anlagen discernible on organon vomero-nasale. Formation of posterior, and beginning of formation of anterior semicircular canal; formation of maculae lagenae and sacculi and of papillae amphibiorum and basilaris; saccus endolymphaticus strongly lobed and apposed to choroid plexus of IVth ventricle; beginning of formation of perilymphatic system. Chondrification of crista otica and processus muscularis capsulae auditivae; lateral wall of ear capsule only pierced by foramen ovale; foramina jugulare and prooticum completed by procartilaginous connections. Head somite IV disappeared and trunk somite I reduced. Forelimb atrium formed. Secondary vessels of 3rd and 4th aortic arches disappeared. Appearance of blood corpuscles in spleen anlage. Pronephros free of yolk; caudal end of pronephros and cranial end of mesonephric anlage have reached each other; 6–8 mesonephric units segregated, central lumen in oldest units; anlagen of first glomeruli appearing. Genital ridges beginning to protrude into coelomic cavity. External gills degenerated to great extent. First indication of pyloric part of stomach; intestinal coils arranged in outer and inner spiral.

Stage 48. Age 180 hours; length 14–17 mm.

External criteria: Beginning of pigmentation around nervus acusticus. Forelimb bud visible for the first time. Shining gold-colored abdomen. Hindlimb bud semicircular in lateral aspect.

Internal criteria: Peripheral nuclear layer well differentiated in rostral part of optic tectum; ecto- and endomeninx separated by interjacent membrane at level of mesencephalon. Spinal nerves II, III, and IV reaching forelimb anlage; spinal nerves VIII, IX, and X reaching hindlimb anlage. Beginning of formation of saccus perilymphaticus. Connection between ear capsule and parachordalia cartilaginous; anlagen of arytenoids form well-defined mesenchymatous condensations. Basal portions of atlas arches and 3rd and 4th vertebral arches procartilaginous. Trunk somite I disappeared. Elastica externa and interna beginning to thicken in tail region. Forelimb rudiment a well-defined mesenchymal bud covered by thickened epithelium of atrium. Medial postcardinals fused as interrenal vein as far as junction with posterior vena cava; lateral postcardinals completed, opening into interrenal vein at level of cloaca; mesonephric sinus formed. First 6–8 pairs of mesonephric tubules coiling, 2–3 pairs communicating with Wolffian ducts; first nephrostomes appearing on ventral surface of mesonephros. First taste buds appearing in roof of oropharyngeal cavity; thyroid gland consisting of 4–6 diffuse lobes. Yolk completely disappeared from alimentary canal.

Stage 49. Age about 12 days; length 17–23 mm.

External criteria: Melanophores usually appearing around thymus gland and nerves and blood vessels of head; xanthophores appearing on pericardium. Forelimb bud distinct. Hindlimb bud somewhat longer, distal outline still circular, no constriction at base. Melanophores appearing on dorsal and ventral fins.

Internal criteria: Appearance of bilateral telencephalic choroid invaginations; infundibular process of neurohypophysis beginning to develop; first anlage of tuberal part of adenohypophysis; small tubular anlage of paraphysis; at least five nuclear layers in optic tectum; first cells of molecular layer of cerebellum formed; choroid plexus formation reaching lateral border of rhombencephalon. Nervus abducens discernible. Semicircular canals with ampullae completed; macula lagenae located in separate diverticulum of sacculus; saccus perilymphaticus extending into foramen jugulare. Formation of canales semicirculares anterior and lateralis of ear capsule. Cartilaginous connection between occipital arch and ear capsule. Ten to twelve pairs of functional mesonephric units and 6–8 pairs of glomeruli formed; second generation of mesonephric tubules and glomeruli appearing. Scattered cells from mesonephric and interrenal blastemata entering genital ridges. Taste buds appearing in epithelium of primitive tongue and in lateral regions of oropharyngeal roof; follicle formation starting in thyroid gland. Colon segregating from intestine.

Stage 50. Age about 15 days; length 20–27 mm.

External criteria: Forelimb bud somewhat oval-shaped in dorsal aspect. Hindlimb bud longer than broad, constricted at base, distal outline somewhat conical.

Internal criteria: Large number of unicellular glands developing in larval skin. Olfactory bulbs beginning to fuse; paraphysis beginning to ramify; Stirnorgan shifted to position dorsal to paraphysis; segregation of tectal and tegmental parts of mesencephalon. Nervus abducens has reached musculus rectus posterior. Rohon–Beard cells beginning to decline; cells of lateral motor columns beginning to differentiate at lumbar levels of spinal cord. Brachial ganglia with a few large cells. Forelimb and hindlimb nerve entering limb anlagen; sympathetic chain ganglia formed in trunk region. Choanae perforated. Utriculus divided into pars anterior and pars posterior; ductus perilymphaticus and recessus perilymphaticus papillae amphibiorum formed. Formation of canalis semicircularis posterior of ear capsule; separation of foramina endolymphaticum and acusticum; processus ventrolateralis palatoquadrati chondrified. Beginning of differentiation of mesenchyme in hindlimb bud; paired pelvic girdle anlage indicated. Brachial vein and vena cutanea magna being formed; ischiadic vein clearly distinguishable. About 20 pairs of glomeruli, and open nephrostomes in mesonephros. Complete gonadal rudiments formed. Outpocketings of lungs protruding into dorsal coelomic sacs behind ear capsules. Beginning of colloid formation in thyroid follicles. Diffuse follicles formed in ultimobranchial bodies.

Stage 51. Age about 17 days; length 28–36 mm.

External criteria: Tentacles much longer. Forelimb bud oval-shaped in lateral aspect. Hindlimb bud conical in shape, its length about $1^{1}/_{2}$ times its breadth; melanophores appearing on it.

Internal criteria: Cement gland entirely disappeared; formation of unicellular glands in epidermis of hindlimb. Paraphysis with many tubules. Cells of lateral motor columns beginning to differentiate at brachial levels of spinal cord. Four to five Jacobson's glands on organon vomero-nasale. Ventrolateral lobe of saccus endolymphaticus extending forward up to mesencephalon. Crista parotica well developed; formation of tectum posterius; fen. basicranialis closed; separation of foramina opticum and oculomotorium; beginning of formation of parasphenoid and frontoparietalia. Chondrification of basal portions of atlas arches completed. First indication of formation of secondary muscles from primary somatic musculature. Forelimb anlage segregated into anlagen of shoulder girdle and forelimb (s.s.). First indication of femur anlage. Arteria iliaca communis extending to hindlimb anlage. Interrenal vein losing connection with caudal vein, the latter splitting backward; appearance of vena femoralis. Appearance of first posterior pair of lymph hearts. Third generation of mesonephric tubules and glomeruli; number of glomeruli augmented to 40–50. Adipose substances appearing in cells of fat bodies. Taste buds appearing anterior to primitive tongue; filter processes of branchial arches fitting in between folds of pharyngeal roof; folds appearing in caudal portions of lungs; resorption vacuoles appearing in colloid of thyroid gland. Typhlosole formed in duodenum and marginal gut.

Stage 52. Age about 21 days; length 42–56 mm.

External criteria: Forelimb bud irregularly conical. Hindlimb bud showing first indication of ankle constriction and first sign of flattening of foot.

Internal criteria: Sense organs of caudal lateral line appearing on postanal fin. Stirnorgan situated just cranial to paraphysis. Dorsal column (dorsal horn) beginning to appear at postbrachial levels of spinal cord. Lumbar ganglia with a few large cells. About 10 Jacobson's glands on organon vomero-nasale. Separation of foramina acusticum anterius and posterius. Chondrification completed in 4th–9th vertebral arches; chondrification in lateral wall of perichordal tube extending from atlas down to intervertebral region between 2nd and 3rd vertebrae; 3 pairs of ribs discernible as mesenchymatous condensations. Femur rudiment procartilaginous, tibia and fibula mesenchymatous; local condensations of mesenchyme indicating beginning of muscle development. Appearance of arteria subclavia. Beginning of sexual differentiation of gonads. Small pigment cells appearing in cortical region of thymus. Five complete revolutions in outer and in inner spiral of intestinal helix.

Stage 53. Age about 24 days; length 50–60 mm.

External criteria: Fore- and hindlimbs in paddle stage. Hindlimb (excluding foot) somewhat longer than broad; 4th and 5th toes indicated.

Internal criteria: Stirnorgan just cranial to level of lamina terminalis; first development of granular layer from ependyma in cerebellum; Purkinje layers fusing in dorsal midline. Nervus trochlearis passing through canal in crista trabeculae. Dorsal column extending into prebrachial levels of spinal cord; descent of lumbar ganglia started. Formation of canales olfactorii and tectum anterius; beginning of formation of goniale; beginning of chondrification of arytenoids and annulus cricoideus. Chondrification of occipital arches completed; chondrification completed in 3rd pair of vertebral arches; perichordal tube beginning to chondrify anteriorly underneath vertebral arches; ventral perichordal tube chondrified down to region of 4th vertebra. Ten pairs of mm. interarcuales distinguishable; pars externus superficialis of musculus obliquus abdominis and musculus transversus abdominis developing. Shoulder girdle divided into scapular and coracoidal portions. Beginning of chondrification of pelvic girdle. Femur chondrifying, tibiale and fibulare procartilaginous, skeleton of foot mesenchymatous. Beginning of histogenesis of muscle fibers in thigh and leg. Congestion of pronephros. Typhlosole extending into beginning of second revolution of outer intestinal spiral.

Stage 54. Age about 26 days; length 58–65 mm.

External criteria: All 4 fingers indicated; edge of hand slightly scalloped between fingers; melanophores appearing on forelimb. Length of hindlimb (excluding foot) nearly twice its breadth; all 5 toes indicated, the 2nd only very slightly.

Internal criteria: Dorsal column extending throughout trunk levels and over short distance into postsacral levels of spinal cord. Ventrolateral lobe of saccus endolymphaticus extending from diencephalon to 1st spinal nerve. Chondrification completed in 1st and 2nd pairs and caudad to 10th pair of vertebral arches; chondrification of ventral perichordal tube reaching to 6th vertebra; perichondral ossification as well as destruction and calcification of cartilage starting in vertebral arches. Shallow acetabulum formed and iliac process well defined. Chondrification of hindlimb has proceeded to tibiale and fibulare; metatarsalia procartilaginous. Some muscles of hindlimb with indication of both origin and insertion. First signs of atrophy in pronephros. Up to 7 revolutions in outer and inner spiral of intestinal helix.

Stage 55. Age about 32 days; length 70–80 mm.

External criteria: Hand pronated about 90°; free parts of fingers about equally long as broad. Length of hindlimb (excluding foot) about 3 times its breadth; length of cartilages of 4th and 5th toe about 4 times their breadth.

Internal criteria: Lumbar motor column nuclei of spinal cord showing twice the cross-sectional area of adjacent nonmotor nuclei. Cartilage present in medial portion of sclera of eye. Ventrolateral lobe of saccus endolymphaticus reaching from telencephalon to 2nd spinal nerve. Fusion of cristae occipitales laterales with tectum posterius. Eleventh pair of vertebral arches chondrified; chondrification of ventral perichordal tube extending over entire trunk region; dorsal perichordal tube chondrified vertebrally down to 6th pair, and intervertebrally down to 3rd pair of vertebral arches. Mesenchymatous perichordal tube of tail thickening. Scapula, procoracoid, coracoid, and humerus procartilaginous. Shoulder girdle musculature and upper arm musculature indicated. Perichondral ossification of femur started. All major muscles of thigh and leg present with both origins and insertions; foot musculature well defined. Signs of local degeneration in middle region of mesonephros; anlage of urinary bladder discernible. Small tubular cavities appearing in compact testis; medullary tissue of ovary largely excavated; first meiotic division in ovary. Single row of maxillary tooth germs developed.

Stage 56. Age about 38 days; length 70–100 mm.

External criteria: Elbow and wrist clearly indicated; length of free parts of fingers 3–4 times their breadth. Length of cartilages of 4th and 5th toes about 6 times their breadth.

Internal criteria: Stirnorgan shifted to position just caudal to fused olfactory bulbs. In descent of lumbar ganglia, root of ganglion 8 overlapping part of ganglion 7, the root of ganglion 9, and all of ganglion 8. Olfactory organ divided into two portions; anlagen of glandula oralis interna and nasolachrymal duct formed. Ventrolateral lobe of saccus endolymphaticus extending caudad to 3rd spinal nerve, dorsolateral lobe extending craniad to diencephalon. Frontoparietalia fused. Twelfth pair of vertebral arches chondrified; chondrification of dorsal perichordal tube extending vertebrally to 9th vertebra and intervertebrally to 5th vertebra;

beginning of ossification of vertebral portions of perichordal tube; beginning of chondrification of hypochord; 2nd pair of ribs beginning to chondrify. Coracoid slightly ossified perichondrally; scapula, procoracoid, and coracoid chondrified. Chondrification of forelimb up to phalanges, which are still blastematous; ossification extended to forearm. Shoulder girdle musculature differentiated except for origins and insertions. Greater part of forelimb musculature still in myoblast stage. Perichondral ossification of iliac processes. Skeleton of hindlimb completely chondrified, ossification extended distally to tibiale and fibulare. Anlage of ostium tubae indicated. Cavity forming in middle ear anlage. Urinary bladder forming distinct pocket on ventral side of cloaca.

Stage 57. Age about 41 days; length 75–105 mm.

External criteria: Pigment-free spot appearing above Stirnorgan. Angle of elbow more than 90°; fingers stretched out in forelimb atrium, their length about 7 times their breadth.

Internal criteria: First signs of cornification of epidermis at tips of 1st, 2nd, and 3rd toes; stratum spongiosum and stratum compactum beginning to develop underneath stratum germinativum of skin, the latter beginning to form new epidermis and numerous glands under remainder of larval epidermis in metamorphosing areas of skin; glands sinking into stratum spongiosum. Cross-sectional area of largest lumbar motor column nuclei of spinal cord 3–4 times that of average adjacent nonmotor nuclei. Dorsalateral lobe of saccus endolymphaticus reaching to front end of telencephalon. Tenth and eleventh pairs of vertebral arches fused; chondrification of dorsal perichordal tube reaching vertebrally to 10th and intervertebrally to 8th vertebra. Musculi intertransversarii developing. Suprascapula chondrified. Epicondyles of humerus formed; patella ulnaris chondrified; proximal phalanges of 1st and 2nd fingers, and proximal and middle phalanges of 3rd and 4th fingers procartilaginous. Musculature of radio-ulna section indicated. Left and right halves of pelvic girdle in contact with each other. Beginning of ossification of hindlimb phalanges. Oocytes beginning their major growth period. Lip folds formed along upper and lower jaw; anlagen of intermaxillary glands present, each with two glandular ducts; closure of branchial clefts beginning. Beginning of metamorphosis in alimentary canal; esophagus mucosa peeling off and first signs of histolysis in duodenum.

(*Stage 58–. Forelimbs in process of eruption, elbow piercing skin; angle of elbow about 90° and fingers folded in forelimb atrium.*)

Stage 58. Age about 44 days; length 80–110 mm (ultimate length of larva).

External criteria: Forelimbs broken through. Guanophores appearing on abdomen and thighs (adult skin area). All three claws present on hindlimb.

Internal criteria: Metamorphosing skin of forelimb area caudally continuous with metamorphosing skin of trunk. Saccus endolymphaticus extending to 4th spinal nerve. Appearance of maxillae; formation of septum and tectum nasi, crista intermedia, cartilago alaris and nasale. Appearance of musculus intermandibularis anterior. Occipito-atlantal joint formed; 12th pair of vertebral arches fused with 11th and 10th pairs; vertebrally and intervertebrally, chondrification of dorsal perichordal tube extends to 10th vertebra; hypochord chondrified from 9th to 12th vertebra. Appearance of musculus rectus abdominis superficialis and paired musculares sterno-hyoideus. Melanin beginning to be deposited intercellularly in swollen epidermis of tail; tip of tail beginning to atrophy. Processus epicoracoideus completely chondrified; scapula ossified perichondrally; suprascapula already much larger; the membrane bones cleithrum and clavicula are formed. Last two carpalia chondrified, metacarpalia and phalanges ossified perichondrally. Arm and hand musculature differentiated. Left and right halves of pelvic girdle fused ventrally; ossification of ischia started. Only tarsalia and hallux purely cartilaginous, rest of hindlimb skeleton ossified perichondrally; single bony sheath of tibia–fibula. In region of terminal phalanges muscles still in premuscular stage. Appearance of vena facialis. Pronephros no longer functional. Goblet cells appearing in oropharyngeal cavity; 5–6 glandular ducts formed in intermaxillary glands. Extensive histolysis in duodenum.

Stage 59. Age about 45 days.

External criteria: Tentacles beginning to shrivel. Stretched forelimb reaching to base of hindlimb. Guanophores appearing near base of forelimb (adult skin area); anterior border of adult skin area on abdomen not yet distinct; appearance of irregular dark spots on back.

Internal criteria: Areas of metamorphosing skin well circumscribed, that of trunk extending partly over telencephalon; lymph sacs beginning to develop underneath metamorphosing skin areas; first symptoms of degeneration in larval skin between trunk and hindlimb areas of metamorphosing skin. Cross-sectional area of largest lumbar motor column nuclei of spinal cord 5 times and of brachial motor column nuclei up to 4 times that of average adjacent nonmotor nuclei. Appearance of operculum, fenestra ovalis, and pars interna plectri as mesenchymatous condensations. Intercellular melanin deposits underneath epidermis, around blood vessels, and between muscle fibers of tail. Paired anlage of sternum blastematous. Ilia connected to last vertebra by connective tissue. New generation of mesonephric tubules formed in reorganization process. First meiotic division in testis. Second row of tooth germs formed; epithelial body detached from visceral pouches. Histolysis in nonpyloric part of stomach, and pancreas.

Stage 60. Age about 46 days.

External criteria: Nervus olfactorius still longer than bulbus olfactorius. Guanophores appearing on lower jaw (adult skin area). Openings of gill chambers still wide. Distal half of fingers of stretched forelimb extending beyond base of hindlimb; forelimb still situated behind level of heart. Adult skin area of base of forelimb covered with guanophores; anterior border of adult skin area on abdomen more distinct, reaching up to heart.

Internal criteria: Well-defined stratum corneum formed on upper jaw; appearance of horn papillae in the other adult skin areas; strip of larval skin between trunk and hindlimb areas of metamorphosing skin disappeared for greater part. Cross-sectional area of largest lumbar motor column nuclei of spinal cord 6 times that of average adjacent nonmotor nuclei; neurocoel circular in cross section. Nasolachrymal duct opening in skin cranial to eye. Rudiment of gland of Harder formed anterior to eye. Appearance of carilago obliqua and processus praenasalis superior; formation of premaxillae and septomaxillae; appearance of dentale and outer lobe of goniale; beginning of reduction of ethmoid plate and flanges; arytenoids chondrified. Musculus geniohyoideus segregated into pars medialis and pars lateralis. First indications of occipito-vertebral joint formation; tissue connecting 1st and 2nd pairs of ribs with corresponding vertebrae chondrifying; cranial portion of 13th pair of vertebral arches fused with urostyle; primary marrow cavities in basal portions of vertebral arches; chondrification of dorsal perichordal tube extending vertebrally to 12th pair and intervertebrally to 11th pair of vertebral arches. Paired anlage of sternum partially procartilaginous. Ossification starting in pubis; paired epipubes procartilaginous. Complete differentiation of hindlimb musculature. First and second nephrostome of pronephros fused; Wolffian duct between pro- and mesonephros showing signs of degeneration. Third row of tooth germs formed. Middle ear anlage a bilobed vesicle.

Stage 61. Age about 48 days.

External criteria: Head narrower. Tentacles considerably shortened, mostly curved backward. Length of nervus olfactorius equal to diameter of bulbus olfactorius. Fourth arterial arch seen just in front of adult skin area of forelimb. Openings of gill chambers considerably narrowed. Forelimb at level of posterior half of heart. Adult skin area on abdomen covering posterior half of heart. Fins considerably reduced.

Internal criteria: Metamorphosis of adult skin areas completed; adult skin area of trunk covering entire brain cavity and caudal portion of heart. Spinal ganglion VIII descended to level well posterior to root of ganglion IX, and nearly all of ganglion IX posterior to lumbar level of spinal cord. Fingerlike protrusions developed from lateral cavity of olfactory organ. Appearance of pars media and externa plectri and annulus tympanicus as mesenchymatous condensations; erosion phenomena in floor and side walls of cranium and in palatoquadratum and hyobranchial apparatus; angle between palatoquadratum and anterior floor of cranium 50°. Reduction of pars lateralis of musculus levator mandibulae anterior; musculus interhyoideus attached to ceratohyale and palatoquadratum. Eleventh pair of vertebral arches partially ossified perichondrally. Musculare sternohyoidei fused to unpaired anlage. Degenerative hypertrophy of skin at end of tail; atrophy of fins and posterior tail notochord. Paired sternal anlage partially cartilaginous. Epipubis anlagen fused to Y-shaped cartilage. First nephron of pronephros completely disappeared. First signs of degeneration of filter apparatus; middle ear cavity shifted to position somewhat cranial to eye, tuba pharyngotympanica opening into oro-pharyngeal cavity at level of caudal part of eye. Histolysis in pyloric part of stomach and in ileum.

Stage 62. Age about 49 days.

External criteria: Head still somewhat broader than cranial part of trunk. Tentacles short and straight. Nervus olfactorius shorter than diameter of bulbus olfactorius. Corner of mouth still in front of eye. Thymus gland somewhat protruding. Third arterial arch (larval aorta) seen at distance of its own diameter in front of adult skin area of forelimb. Opening of operculum reduced to curved slit. Forelimb at level of middle of heart. Anterior border of adult skin area on abdomen entirely distinct, mostly nicked medially. Ventral fin disappeared from abdomen.

Internal criteria: Areas of larval skin on head beginning to shrivel. Tiny Stirnorgan situated in front of rostral end of brain cavity. Tear furrow leading from opening of nasolachrymal duct to eye. Lower eyelid forming. Dorsal outlets of foramina prooticum confluent; processus muscularis palatoquadrati and cartilago tentaculi disappeared; angle between palatoquadratum and anterior floor of cranium 65°; appearance of squamosum and fusion of inner and outer lobes of goniale. Appearance of musculus petrohyoideus. Intervertebral cartilage calcifying in preurostyle region; cartilaginous connection of 3rd pair of ribs with vertebral column. Pars lateralis of musculus ileolumbalis, and musculus coccygeosacralis formed. Signs of atrophy in notochord over entire length of tail; degenerative hypertrophy of mesenchymatous perichordal tube and vertebral arches of tail. Medial ends of cartilaginous sternal anlagen forming sternal pouches. Appearance of arteria cutanea magna; disappearance of arteries of filter apparatus. Appearance of musculothyroid veins; complete disappearance of lateral veins; disappearance of pronephric sinuses and paired portions of medial postcardinal veins; disappearance of pharyngeal vein. Of 2nd and 3rd nephrons of pronephros, only nephrostomes remaining; 2nd nephrostome fused with ostium tubae; portion of Wolffian duct between pro- and mesonephros disappeared. Middle ear situated below caudal margin of eye. Number of intestinal revolutions reduced to about 2 in outer and 2 in inner spiral; epithelium of esophagus, nonpyloric part of stomach, duodenum, and ileum reconstituted to more or less continuous layer, that of nonpyloric part of stomach already differentiating.

Stage 63. Age about 51 days.

External criteria: Head narrower than trunk. Tentacles mostly disappeared. Corner of mouth at level of caudal border of eye. Larval aorta and thymus gland no longer externally visible. Operculum closed. Adult skin areas on abdomen and lower jaw separated by narrow band of larval skin. Forelimb at level of anterior half of heart. Fin mostly perforated near anus. Tail still slightly longer than body.

Internal criteria: Area of larval skin between upper jaw and trunk areas of adult skin disappeared for greater part. Root of spinal ganglion IX just posterior to ganglion VII. Nasolachrymal duct opening on tip of papilla. Disappearance of processus muscularis capsulae auditivae; foramen metopticum confluent with foramen caroticum; disappearance of processus cornuquadratus lateralis, processus ascendens, larval processus oticus, and medial portion of comm. quadratocranialis anterior; pars interna and media plectri cartilaginous; angle between palatoquadratum and anterior floor of cranium 85°; appearance of postpalatine commissure and pterygoid; formation of processus pterygoideus; beginning of reduction of branchial chambers. Musculus orbitohyoideus fused with musculus suspensoriohyoideus. Cartilage of 13th pair of vertebral arches and cartilage of anterior ventral perichordal tube degenerating; hypochord reaching to 13th vertebra. In tail, chordal epithelium degenerating and entire notochord shrivelling; beginning of actual degeneration of muscle segments (still 49 post-otic muscle segments distinguishable). Alae sterni forming. Ductus caroticus disappeared. Third nephrostome of pronephros fused with ostium tubae. Operculum a shapeless mass; gill slits closed. Middle ear lying against ear capsule; anlage of tympanic membrane formed. Intestinal helix making only $1^1/_2$ irregular revolutions; typhlosole extending nearly to beginning of colon; histolysis in colon.

Stage 64. Age about 53 days.

External criteria: Corner of mouth well behind eye. Various adult skin areas joined almost everywhere, borderlines still clearly visible. Length of tail (from anus) one-third of body length.

Internal criteria: Root of spinal ganglion VII anterior to ganglion VI. Appearance of adult processus oticus; pars media and externa plectri confluent; angle between palatoquadratum and anterior floor of cranium 105°; branchial chambers disappeared. Formation of adult musculus levator mandibulae posterior; musculare constrictores branchiales, musculares subarcuales recti, and musculus transversus ventralis disappeared. Entire cartilage formation of ventral perichordal tube converted into loose connective tissue and cartilage of lateral perichordal tube in process of resorption. Degeneration of notochord started in urostyle region; only about 14 post-otic muscle segments distinguishable. Paired sternal anlage fused. Fourth aortic arch (adult aorta) larger than 3rd (larval aorta). Disappearance of branchial veins and branchial branches of external jugulars. Pronephros completely disappeared. Primordia of Müllerian ducts developing. Cranial margin of mouth slit situated caudal to level of nostrils; thymus gland moved to position ventrolateral to ear capsule. Epithelium of pyloric portion of stomach and of colon reconstituted; new glandular elements in esophagus, duodenum, and ileum.

Stage 65. Age about 54 days.

External criteria: Borderlines between adult skin areas partly disappeared. Tail oblong triangular in dorsal aspect, length about one-tenth of body length.

Internal criteria: Only a small area of strongly shrivelled larval skin present rostral to forelimb. Spinal ganglion VIII situated at posterior end of lumbar area of spinal cord. Fusion of adult processus oticus with ear capsule; beginning of ossification in pars media, and beginning of chondrification in pars externa plectri. Intervertebral joints formed; 13th and 14th pair of vertebral arches in process of resorption. Notochord and somites disappeared in tail region. Posterior end of epipubis fused with pelvic girdle. Adult abdominal vein established. Primitive tongue mushroom-shaped in cross section; cranial part forming free tip; openings and proximal portions of pharyngotympanic ducts fused. Epithelium of colon differentiating.

Stage 66. Age about 58 days.

External criteria: Borderlines between adult skin areas disappeared. Tail only a very small triangle, no longer visible from ventral side.

Internal criteria: Dorsal root of spinal ganglion IX at level of ganglion VII. Conjunctival sac formed. Plectral apparatus completed and operculum fenestra ovalis confluent with wall of ear capsule; angle between palatoquadratum and anterior floor of cranium 115°; formation of hyocricoid connection. Formation of adult musculus levator mandibulae anterior and musculus depressor mandibulae; musculus interhyoideus attached to palatoquadratum only. Middle portions of ribs ossified. Tail only represented by small dorsal swelling of loose connective tissue, covered with degenerating larval skin. Roofs of sternal pouches reduced. Distal tarsalia ossifying. Reorganization of mesonephros not yet completed. Dorsal coelomic sacs filled with lymphatic tissue; dorsal horns of lungs greatly reduced or absent. Typhlosole disappeared; intestine making single loop.

REFERENCES

1. C. Aimar, M. Delarue, and C. Vilain, "Cytoplasmic regulation of the duration of cleavage in amphibian eggs," *J. Embryol. Exp. Morphol.* **64**, 259–274 (1981).
2. P. C. Baker and T. E. Schroeder, "Cytoplasmic filaments and morphogenetic movements in the amphibian neural tube," *Dev. Biol.* **15**, 432–450 (1967).
3. R. Bachvarova, E. H. Davidson, V. G. Alfrey, and A. E. Mirsky, "Activation of RNA synthesis associated with gastrulation," *Proc. Natl. Acad. Sci. USA* **55**, 358–365 (1966).
4. B. I. Balinsky, "Changes in the ultrastructure of amphibian eggs following fertilization," *Acta Embryol. Morphol. Exp.* **9**, 132–154 (1966).
5. B. I. Balinsky and R. J. Devis, "Origin and differentiation of cytoplasmic structures in the oocytes of *Xenopus laevis*," *Acta Embryol. Morphol. Exp.* **6**, 55–108 (1963).
6. F. S. Billett and A. E. Wild, *Practical Studies of Animal Development*, Chapman and Hall, London (1975).
7. A. W. Blackler, "Transfer of primordial germ cells between two subspecies of *Xenopus laevis*," *J. Embryol. Exp. Morphol.* **10**, 641–651 (1961).

8. A. W. Blackler and M. Fischberg, "Transfer of primordial germ cells in *Xenopus laevis*," *J. Embryol. Exp. Morphol.* **9**, 634–641 (1961).

9. J. G. Bluemink, "The first cleavage of the amphibian egg. An electron microscope study of the onset of cytokinesis in the egg of *Ambystoma mexicanum*," *Ultrastruct. Res.* **32**, 142–166 (1970).

10. J. G. Bluemink, "Cytokinesis and cytochalasin-induced furrow regression in the first cleavage zygote of *Xenopus laevis*," *Z. Zellforsch.* **121**, 102–126 (1971).

11. A. L. Brown, *The African Clawed Toad Xenopus laevis. A Guide for Laboratory Practical Work*, Butterworths, London (1970).

12. D. D. Brown and J. B. Gurdon, "Absence of ribosomal RNA synthesis in the anucleolate mutant of *Xenopus laevis*," *Proc. Natl. Acad. Sci. USA* **51**, 139–146 (1964).

13. J. Cooke, "Properties of the primary organization field in the embryo of *Xenopus laevis*. I–V," *J. Embryol. Exp. Morphol.* **28**, 13–56 (1971); **30**, 49–62, 283–300 (1973).

14. J. Cooke, "Morphogenesis and regulation in spite of continuous mitotic inhibition in *Xenopus* embryos," *Nature (London)* **242**, 55–57 (1973).

15. J. Cooke, "Local autonomy of gastrulation movements after dorsal lip removal in two anuran amphibians," *J. Embryol. Exp. Morphol.* **33**, 147–157 (1975).

16. J. Cooke, "The control of somite number during amphibian development. Models and experiments," in *Vertebrate Limb and Somite Morphogenesis*, Cambridge University Press (1977), pp. 433–448.

17. J. Cooke, "Scale of body pattern adjusts to available cell number in amphibian embryos," *Nature (London)* **290**, 775–778 (1981).

18. R. Deanesly and A. S. Parkes, "The preparation and biological effects of iodinated proteins. VIII. Use of *Xenopus* tadpoles for the assay of thyroidal activity," *J. Endocrinol.* **4**, 324–355 (1945) (cited by Nieuwkoop and Faber [65]).

19. T. A. Dettlaff, "Cell divisions, duration of interkinetic stages, and differentiation in early stages of embryonic development," *Adv. Morphogen.* **3**, 323–362 (1964).

20. T. A. Dettlaff, "Some thermal-temporal patterns of embryogenesis," in *Problems of Experimental Biology* [in Russian], Nauka, Moscow (1977), pp. 269–287.

21. T. A. Dettlaff, "Development of the mature egg organization in amphibians and fish during the concluding stages of oogenesis, in the period of maturation," in *Modern Problems of Oogenesis* [in Russian], Nauka, Moscow (1977), pp. 99–144.

22. T. A. Dettlaff, "Adaptation of poikilothermic animals to development under varying temperatures and problem of integrity of developing organism," *Ontogenez* **12**, 227–242 (1981).

23. T. A. Dettlaff, "Development of the mature egg organization in amphibians, fish, and starfish during the concluding stages of oogenesis, in the period of maturation," in *Oocyte Growth and Maturation*, T. A. Dettlaff and S. G. Vassetzky, eds., Consultants Bureau, New York (1988).

24. T. A. Dettlaff, "The rate of development in poikilothermic animals calculated in astronomical and relative time units," *J. Therm. Biol.* **11**, 1–7 (1986).

25. T. A. Dettlaff and A. A. Dettlaff, "On relative dimensionless characteristics of the development duration in embryology," *Arch. Biol.* **72**, 1–16 (1961).

26. T. A. Dettlaff and A. A. Dettlaff, "Dimensionless criteria as a method of quantitative characterization of animal development," in *Mathematical Biology of Development* [in Russian], Nauka, Moscow (1982), pp. 25–39.

27. T. A. Dettlaff and T. B. Rudneva, "Dimensionless characteristics of the duration of embryonic development of the spur-toed frog," *Sov. J. Dev. Biol.* **4**, 423–432 (1973).

28. E. M. Deuchar, "*Xenopus laevis* and developmental biology," *Biol. Rev.* **47**, 37–112 (1972).

29. E. Elkan, "Amphibia IV-*Xenopus laevis* Daudin," in *The UFAW Handbook on the Care and Management of Laboratory Animals*, A. N. Worden and W. Lane-Peter, eds., Universities Federation Animal Welfare, London (1957) (cited by New [64]).

30. T. Elsdale, M. Fischberg, and S. Smith, "A mutation which reduces nucleolar number in *Xenopus laevis*," *Exp. Cell Res.* **14**, 642–643 (1958).

31. T. R. Elsdale, J. B. Gurdon, and M. Fischberg, "A description of the technique for nuclear transplantation in *Xenopus laevis*," *J. Embryol. Exp. Morphol.* **8**, 437–444 (1960).

32. D. P. Filatov, *A Comparative-Morphological Direction in Developmental Mechanics, Its Object, Aims, and Ways* [in Russian], Izd. Akad. Nauk SSSR, Moscow (1939).

33. S. Filoni, L. Bosco, C. Cioni, and G. Venturini, "Lens forming transformation in larval *Xenopus laevis* induced by denatured eye cup or its whole protein complement," *Experientia* **39**, 315-317 (1983).

34. R. V. de Fremery, C. A. W. Gorlee, and P. Tydeman, "A method for collecting naturally fertilized eggs of *Xenopus laevis* immediately after laying," *Lab. Anim.* **11**, 263–264 (1977).

35. L. Gallien, "Inversion expérimentale du sexe chez un Anoure inférieur *Xenopus laevis* (Daudin). Analyse des conséquences génétiques," *Bull. Biol. Fr. Belg.* **90**, 163–183 (1956).

36. J. Gerhart, G. Ubbels, S. Black, K. Mara, and M. Kirschner, "A reinvestigation of the role of the gray crescent in axis formation in *Xenopus laevis*," *Nature (London)* **292**, 511–516 (1981).

37. J. Gerhart, M. Wu, and M. Kirshner, "Cell cycle dynamics of an M-phase-specific cytoplasmic factor in *Xenopus laevis* oocytes and eggs," *J. Cell Biol.* **98**, 1247–1255 (1984).

38. A. S. Ginsburg, "Species peculiarities of the initial stages of labyrinth development in amphibians," *Dokl. Akad. Nauk SSSR* **73**, 229–232 (1950).

39. J. B. Gurdon, "Tetraploid frogs," *J. Exp. Zool.* **141**, 519–544 (1959).

40. J. B. Gurdon, "The developmental capacity of nuclei taken from differentiating endoderm cells of *Xenopus laevis*," *J. Embryol. Exp. Morphol.* **8**, 505–526 (1960).

41. J. B. Gurdon, "The effects of ultraviolet irradiation on uncleaved eggs of *Xenopus laevis*," *Q. J. Microsc. Sci.* **101**, 299–311 (1960).

42. J. B. Gurdon, "The transplantation of nuclei between two subspecies of *Xenopus laevis*," *Heredity* **16**, 305–314 (1961).

43. J. B. Gurdon, "African clawed frogs," in *Methods in Developmental Biology*, F. H. Wilt and N. K. Wessels, eds., Crowell, New York (1967), pp. 75–84.

44. J. B. Gurdon, "On the origin and persistence of a cytoplasmic state inducing nuclear DNA synthesis in frogs' eggs," *Proc. Natl. Acad. Sci. USA* **58**, 545–552 (1967).

45. J. B. Gurdon, "The translation of messenger RNA injected in living oocytes of *Xenopus laevis*," *Acta Endocrinol. Suppl.* **180**, 225–243 (1973).

46. J. B. Gurdon, *The Control of Gene Expression in Animal Development*, Oxford University Press, New York (1974).

47. J. B. Gurdon, "Molecular biology in the living cell," *Nature (London)* **248**, 772–776 (1974).

48. J. B. Gurdon, "Molecular mechanisms in the control of gene expression during development," *15th CIBA Medal Lecture, Biochemical Society Meeting*, Vol. 9 (1981), pp. 13–21.

49. J. B. Gurdon, "Concepts of gene control in development," in *Developmental Biology Using Purified Genes*, D. D. Brown and C. F. Fox, eds., Academic Press, New York (1982).

50. J. B. Gurdon and D. A. Melton, "Gene transfer in amphibian eggs and oocytes," *Annu. Rev. Genet.* **15**, 189–218 (1981).

51. J. B. Gurdon and H. R. Woodland, "*Xenopus laevis*," in *Handbook of Genetics*, R. C. King, ed., Academic Press, New York (1975).

52. J. B. Gurdon, T. Elsdale, and M. Fischberg, "Sexually mature individuals of *Xenopus laevis* from the transplantation of single somatic nuclei," *Nature (London)* **183**, 64–65 (1958).

53. J. B. Gurdon, C. D. Lane, H. R. Woodland, and G. Marbaix, "Use of frog eggs and oocytes for the study of messenger RNA and its translation in living cells," *Nature (London)* **233**, 177–182 (1971).

54. L. Hamilton, "Changes in survival after x-irradiation of *Xenopus* embryos at different phases of the cell cycle," *Radiat. Res.* **37**, 173–180 (1969).

55. K. Hara, P. Tydeman, and C. W. A. Gorlee, "A new method for local artificial insemination of *Xenopus laevis*," *Mikroskopie (Wien)* **35**, 10–12 (1979).

56. K. Hara, P. Tydeman, and M. Kirshner, "A cytoplasmic clock with the same period as the division cycle in *Xenopus* eggs," *Proc. Natl. Acad. Sci. USA* **77**, 462–466 (1980).

57. G. Hirose and M. Jacobson, "Clonal organization of the central nervous system of the frog. 1. Clones stemming from individual blastomeres of the 16-cell and earlier stages," *Dev. Biol.* **71**, 191–202 (1979).

58. O. A. Hoperskaya, "The development of animals homozygous for a mutation causing periodic albinism (a^p) in *Xenopus laevis*," *J. Embryol. Exp. Morphol.* **34**, 253–264 (1975).

59. A. Hughes, "Development of the primary sensory system in *Xenopus laevis* (Daudin), *J. Anat.* **91**, 323–328 (1957).

60. M. Jacobson and G. Hirose, "Clonal organization of the central nervous system of the frog," *J. Neurosci.* **1**, 271–284 (1981).

61. M. R. Kalt, "Ultrastructural observations on the germ line of *Xenopus laevis*," *Z. Zellforsch.* **138**, 41–62 (1973).

62. R. E. Keller, "Vital dye mapping of the gastrula and neurula of *Xenopus laevis*. I. Prospective areas and morphogenetic movements of superficial layer," *Dev. Biol.* **42**, 222–241 (1975).

63. R. E. Keller, "Vital dye mapping of the gastrula and neurula of *Xenopus laevis*. II. Prospective areas and morphogenetic movements of deep layer," *Dev. Biol.* **51**, 118–137 (1976).

64. R. E. Keller, "Time-lapse cinemicrographic analysis of superficial cell behavior during and prior to gastrulation in *Xenopus laevis*," *J. Morphol.* **157**, 223–248 (1978).

65. R. E. Keller, "An experimental analysis of the role of bottle cells and the deep marginal zone in gastrulation of *Xenopus laevis*," *J. Exp. Zool.* **216**, 81–101 (1981).

66. M. Kirschner, J. C. Gerhart, K. Hara, and G. D. Ubbels, "Initiation of the cell cycle and establishment of bilateral symmetry in *Xenopus* eggs," in *The Cell Surface; Mediator of Developmental Processes*, S. Subtelny and N. K. Wessels, eds., Academic Press, New York (1980), pp. 187–214.

67. C. D. Lane, C. M. Gregory, and C. Moral, "Dueck-hemoglobin synthesis in frog cells. The transplantation and assay of reticulocyte GS-RNA in oocytes of *Xenopus laevis*," *Eur. J. Biochem.* **34**, 219–227 (1973).

68. G. M. Malacinski and H. M. Chung, "Establishment of the site of involution at novel locations on the amphibian embryo," *J. Morphol.* **169**, 149–159 (1981).

69. Y. Masui and H. Clarke, "Oocyte maturation," *Int. Rev. Cytol.* **57**, 185–282 (1979).

70. Y. Misumi, S. Kurata, and K. Yamana, "Initiation of ribosomal RNA synthesis and cell division in *Xenopus laevis* embryos," *Dev., Growth Differ.* **22**, 773–780 (1980).

71. V. A. Moar, J. B. Gurdon, C. D. Lane, and G. Marbaix, "The translational capacity of living frog eggs and oocytes as judged by messenger RNA injection," *J. Mol. Biol.* **61**, 93–104 (1971).

72. N. Nakatsuji, "Studies on the gastrulation of amphibian embryos: numerical criteria for stage determination during blastulation and gastrulation of *Xenopus laevis*," *Dev., Growth Differ.* **16**, 257–265 (1974).

73. D. A. T. New, *The Culture of Vertebrate Embryos*, Academic Press, New York (1966).

74. J. Newport and M. Kirschner, "A major developmental transition in early *Xenopus* embryos. 1. Characterization and timing of cellular changes at the midblastula stage," *Cell* **30**, 675–686 (1982).

75. J. Newport and M. Kirschner, "A major developmental transition in early *Xenopus* embryos. 2. Control of the onset of transition," *Cell* **30**, 687–696 (1982).

76. P. D. Nieuwkoop and J. Faber, *Normal Table of Xenopus laevis (Daudin)*, North-Holland, Amsterdam (1956).

77. P. D. Nieuwkoop and P. A. Florschütz, "Quelques caractères speciaux de la gastrulation et de la neurulation de l'oeuf de *Xenopus laevis* Daudin et de quelques autres anoures. I. Étude descriptive," *Arch. Biol., Paris* **61**, 113–150 (1950).

78. M. Pearson and T. Elsdale, "Experimental evidence for an interaction between two temporal factors in the specification of somite pattern," *J. Embryol. Exp. Morphol.* **51**, 27–50 (1979).

79. M. M. Perry, "Microfilaments in the external surface layer of the early amphibian embryo," *J. Embryol. Exp. Morphol.* **35**, 127–146 (1975).

80. S. C. Reed and H. P. Stanley, "Fine structure of spermatogenesis in the South African clawed toad *Xenopus laevis* Daudin," *J. Ultrastruct. Res.* **41**, 277–295 (1972).

81. H. H. Reichenbach-Klinke, *Krankheiten der Amphibien*, Gustav Fischer, Stuttgart (1961).

82. H. H. Reichenbach-Klinke and E. Elkan, *The Principal Diseases of Lower Vertebrates*, Academic Press, London (1965).

83. J. Rostand, "Parthénogénèse expérimentale chez le Xenope (*Xenopus laevis*)," *C. R. Soc. Biol.* **145**, 1453–1454 (1951).

84. T. B. Rudneva, "Duration of karyomitosis and cell division in cleavage divisions II–V in the clawed frog, *Xenopus laevis*," *Sov. J. Dev. Biol.* **3**, 526–530 (1972).

85. B. Rybak and T. Gustafson, "Fécondation artificielle de *Xenopus laevis* sans sacrifice du géniteur mâle," *Experientia* **9**, 25–26 (1953).

86. N. Satch, "Matachronous cleavage and initiation of gastrulation in amphibian embryos," *Dev., Growth Differ.* **19**, 111–117 (1977).

87. S. Schorderet-Slatkine, "Action of progesterone and related steroids on oocyte maturation in *Xenopus laevis*. An *in vitro* study," *Cell Differ.* **1**, 179–181 (1972).

88. T. E. Schroeder, "Neurulation in *Xenopus laevis*. An analysis and model based upon light and electron microscopy," *J. Embryol. Exp. Morphol.* **23**, 427–462 (1970).

89. S. Smith, "Induction of triploidy in the South African clawed frog *Xenopus laevis* (Daudin)," *Nature (London)* **181**, 1290–291 (1958).

90. J. Stewart-Savage and R. D. Grey, "The temporal and spatial relationships between cortical contraction, sperm tail formation, and pronuclear migration in fertilized *Xenopus* eggs," *Wilhelm Roux's Arch. Dev. Biol.* **191**, 241–245 (1982).

91. S. Sudarwati and P. D. Nieuwkoop, "Mesoderm formation in the anuran *Xenopus laevis* (Daudin)," *Wilhelm Roux's Arch. Dev. Biol.* **166**, 189–204 (1971).

92. D. Tarin, "Histological features of neural induction in *Xenopus laevis*," *J. Embryol. Exp. Morphol.* **26**, 543–570 (1971).

93. P. Valouch, J. Melichna, and F. Sládeček, "The number of cells at the beginning of gastrulation depending on the temperature in different species of amphibians," *Acta Univ. Carol. Biol.*, pp. 195–205 (1971).

94. P. Van Gansen and A. Schram, "Ultrastructure et cytochimie ultrastructurale de la vesicule germinative de l'oocyte mur de *Xenopus laevis*," *J. Embryol. Exp. Morphol.* **20**, 375–389 (1968).

95. T. Wickbom, "Cytological structure in Dipnoi, Urodela, Anura, and Emys," *Hereditas* **31**, 241–346 (1945).

96. P. Whittington and K. E. Dixon, "Quantitative studies of germ cells during early embryogenesis of *Xenopus laevis*," *J. Embryol. Exp. Morphol.* **33**, 57–74 (1975).

97. D. P. Wolf, "The cortical granule reaction in living eggs of the toad *Xenopus laevis*," *Dev. Biol.* **36**, 62–71 (1974a).

98. D. P. Wolf, "On the contents of the cortical granules from *Xenopus laevis* egg," *Dev. Biol.* **38**, 14–29 (1974b).

99. D. P. Wolf, "The cortical response in *Xenopus laevis* ova," *Dev. Biol.* **40**, 102–115 (1974c).

100. D. P. Wolf and J. L. Hedrick, "A molecular approach to fertilization. II. Viability and artificial fertilization of *Xenopus laevis* gametes," *Dev. Biol.* **25**, 348–352 (1971).

101. H. R. Woodland, C. C. Ford, J. B. Gurdon, and G. D. Lane, "Some characteristics of gene expression as revealed by a living assay system," in *Molecular Genetics and Developmental Biology*, M. Sussman, ed., Prentice Hall, Englewood Cliffs, New Jersey (1972), pp. 399–423.

Chapter 10

THE COMMON FROG *Rana temporaria*

N. V. Dabagyan and L. A. Sleptsova

The oocytes, embryos, and larvae of *Rana temporaria* have long been traditional subjects of diverse experimental embryological studies. Experimental embryologists have often used the embryos of *R. temporaria* for studies concerning the mechanisms of development. In the first half of this century experiments were carried out on eggs collected from natural spawnings. Among many works, one should mention those of Dürken [11], Hamburger [14, 15], and the founder of the school of experimental embryologists in the Soviet Union, Professor D. P. Filatov [12] and his numerous associates [8, 20, 19, 27, 32].

The elaboration of methods for obtaining embryos outside the season of natural spawning and the ease with which larvae could be kept and reared further widened the possibilities of using *R. temporaria* and established it as a standard laboratory animal. At present the embryos and larvae of *R. temporaria* are widely used in studies of early development [23, 37] and organogenesis [18, 24]. The development of *R. temporaria* has also served as material for certain theoretical speculations [38].

10.1. TAXONOMY, DISTRIBUTION, AND REPRODUCTION

The common frog belongs to the genus *Rana*, family Ranidae, order Anura, class Amphibia. The common or "grass" frog is *Rana temporaria temporaria* L. (Linné, 1758) (= *R. fusca*). The species named *temporaria* takes its origin from "tempora" (there are temporal spots in this species).

Besides *R. temporaria*, experimental embryological studies involve *R. arvalis, R. esculenta, R. ridibunda,* and, in America, *R. pipiens, R. palustris,* and *R. catesbeiana. R. temporaria temporaria* L. is the most widespread of all Palearctic *Rana* species; this species occurs all over Europe, from the middle Volga flow in the south to beyond the Polar circle in the north, and in Transuralia [1, 35].

R. temporaria is a representative of typical land frogs. It occurs in forests, often in rather dry areas such as glades, and sometimes in kitchen gardens. At the end of August to the beginning of September common frogs begin to migrate to water. These frogs usually winter deep in running water, beneath undermined banks. After awakening in spring they congregate for spawning.

Beyond the breeding season, common frogs occur singly but during the period of spawning they accumulate in water in large numbers: in the Moscow District in March–April, near Leningrad at the end of April to the beginning of May, near Kiev at the end of February and beginning of May, and in the Laplandsky State Reserve in the second half of May to the beginning of June [34, 35].

The beginning of spawning is preceded by the mating of males and females in water. The male and female mate for rather a long time and then the female lays eggs over a short period and the male pours sperm on the eggs. The egg batch is a slimy (jelly) lump, 15–30 cm in diameter. It first falls to the bottom and returns to the water surface within 1.5–2 h, when the egg envelopes become swollen.

Common frogs lay eggs in shallow water. The eggs often form whole "islands" as batches of many females stick together. The spawning proceeds synchronously for 1–2 weeks. It begins in bodies of water situated on high and open places and terminates in lowland and sheltered forest waters. The collection of eggs in nature does not require any special skills; one has only to observe the time of spawning.

10.2. KEEPING EMBRYOS AND LARVAE IN THE LABORATORY

10.2.1. General Care

On the basis of our experiment of working with common frogs, the following ways of keeping embryos and larvae in the laboratory are recommended. It is important that they are kept in a sufficient volume of water: 25 embryos, or not more than 10 tadpoles, in 500 ml of water [28]. If their density is higher, the development of different individuals proceeds asynchronously and with a general delay. In shallow glass aquaria, with water level no higher than 10 cm, the embryos and tadpoles can be kept until metamorphosis.

Tap water should be dechlorinated in open vessels for not less than 2 days. Insufficiently dechlorinated water can induce rapid mass death of animals. Changes of water in the aquaria are of great value for normal development. In the first 2 days after fertilization water should be changed twice a day and then, until the transition to active feeding, once a day, and later once in 2–3 days.

In some laboratories water is not changed but is topped up with distilled water to the initial level [28]. The embryos and tadpoles do not require constant aeration, but water plants such as *Elodea* may prove useful in aquaria.

The development of embryos and larvae does not depend directly on the conditions of illumination; light is necessary for the normal life of water plants and, hence, maintenance of a favorable oxygen regime in the aquarium. It is necessary to protect the embryos from direct sunlight.

During the period of metamorphosis (see Table 10.1, stages 49–53) a very small amount of water is left in the aquaria and they are inclined so that a part of the bottom is free of water. The bottom is covered with filter paper so that the froglets do not slip too much when coming out of the water. Small pebbles and branches of houseplants are placed on the water-free part of the bottom. The aquaria are covered with nets so that the metamorphosed frogs cannot jump out. During this period water should be changed every day.

The metamorphosed frogs are placed in glass vessels in small groups. A small amount of cotton covered with filter paper and well wetted with water is placed on the bottom, together with small pebbles and branches. The vessels are covered by a net or gauze; they are cleaned every day and cotton and filter paper are changed.

10.2.2. Feeding

From the beginning of active feeding (stage 34) the tadpoles are fed with the yolk of hard-boiled eggs. Within 2–3 days the egg diet is supplemented with plant food such as homogenized fresh lettuce or nettle. The fresh green plants can be replaced successfully by dried nettle. The dried nettle leaves are homogenized, mixed with boiling water, cooled, filtered through gauze, and given as a food to tadpoles. Within a few days the yolk is replaced by meat (boiled or raw). Crushed meat is given to tadpoles in the morning, and in the evening its residues are removed from the aquarium, the water is changed, and plant food is given.

During the period of metamorphosis the tadpoles do not feed. After metamorphosis in nature the young of that year feed on many small animals and plant parts; representatives of the orders Coleoptera and Diptera predominate in their food. Such a ration cannot be reproduced in the laboratory, where the froglets are fed with *Drosophila* flies. They catch the flies quite well and are rapidly sated.

10.3. MANAGEMENT OF REPRODUCTION IN THE LABORATORY

10.3.1. Stimulation of *in vivo* Oocyte Maturation

At present one can work with the common frog embryos and oocytes during the whole autumn–winter period. To induce precocious oocyte maturation, the female is injected with pituitary suspension. As in *R. pipiens* [28], the recommended injection is a suspension of 5 pituitaries in October–December, 4 in January–February, and 3 in March. Pituitaries can be taken from both males and females since they differ slightly by their activity [13].

To obtain pituitaries, the necessary number of frogs is immobilized (by either spination or sharp blow on the head). Deep cuts are made with small scissors between the upper and lower jaws, and the lower jaw is bent back. A deep cut is then made at the junction of the first vertebra with the os parasphenoideum and exoccipitalia. The palatinum soft tissues are cut and removed and parasphenoid is exposed. Thereafter the lateral parasphenoid and exoccipital surfaces are cut on the right and on the left by introduction of scissors in the skull cavity and parasphenoid is bent forward with forceps. As a result, the pituitary, lying on the diencephalon surface, becomes exposed (Fig. 10.1). The pituitary is often torn off from the brain tissue and is found on the internal surface of the bent parasphenoid. The pituitary, of milk-pink color, is bean-shaped, 1×2 mm in size, and is surrounded by looser and lighter endolymphatic tissue. When the pituitary is excised with fine forceps, it is better not to take this tissue, or it should be removed afterwards. The isolated pituitaries are placed in a small porcelain mortar, ground with 1 ml of distilled water or

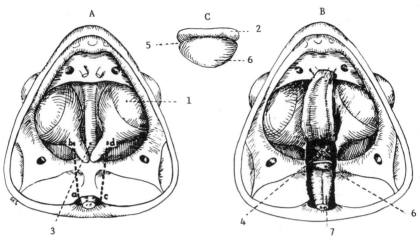

Fig. 10.1. Diagram of hypophysectomy procedure (from [15]). A) Head of frog, ventral view, lower jaw removed; B) brain and pituitary exposed; C) isolated pituitary. 1) Eye; a, b and c, d) lines of cuts; 2) anterior pituitary; 3) diencephalon; 4) optic chiasma; 5) intermediate pituitary; 6) posterior lobe; 7) medulla oblongata.

Ringer solution, well mixed, taken into a syringe, and injected into subcutaneous lymphatic sacs of the female.

To obtain mature sperm in the laboratory during the autumn–winter period, males are best stimulated with a small dose of pituitaries 24 h before the sperm is taken, although in some males the active sperm can be obtained without stimulation, since by the time of onset of hibernation the testes already contain a sufficient amount of mature sperm.

After the injection, females are placed in aquaria with a small amount of water at room temperature (preferably at 12–16°C). Males are placed in another aquarium under the same conditions. Oocyte maturation is completed within 36–48 h, depending on temperature. The passage of oocytes to the lower part of the oviduct ("uterus") after ovulation can be determined by the swelling of the abdomen. It is difficult to determine the optimal time for killing the females to obtain eggs, but one should bear in mind that both the too-early removal of eggs and their delay in the "uterus" result in the deterioration of the egg and decreased fertilizability.

Artificial stimulation of maturation is best performed on already hibernating frogs. The largest frogs are taken for injection. Before the experiment, frogs are kept in aquaria with a shallow layer of water (a layer of moist *Sphagnum* moss can be placed on the bottom) at 4°C. It is necessary to wash the frogs with cold water once every 3–5 days. If they are kept under poor conditions, the frogs rapidly lose the capacity to respond to the injection of pituitary suspension by oocyte maturation or, if they retain such capacity, produce eggs of poor quality.

10.3.2. *In vitro* Oocyte Maturation

Oocyte maturation in saline solution is obtained mainly with the aim of studying the mechanisms of maturation. Pituitary suspension is prepared in Ringer solution for poikilothermic animals (1 pituitary to 50–100 ml), and pieces of ovary with 1–6 fully grown oocytes are placed in it (up to 50 oocytes in 5 ml of suspension). They can be kept in the suspension for varying times; to obtain a greater number of ovulated oocytes they are kept in this solution until the beginning of ovulation. The highest incidence of oocyte maturation was observed in progesterone solution (1–5 μg/ml of Ringer solution), but not all oocytes ovulated. The best results were obtained when the oocytes were incubated in the solution with both progesterone and pituitary or cholesterol [13, 31].

Oocytes ovulated *in vitro* are coated by a vitelline membrane only and are not fertilizable, but they can be activated and will develop upon nuclear transfer [25].

10.4 STRUCTURE OF GAMETES

10.4.1. Oocyte

Common frog oocytes ovulate at the stage of germinal vesicle breakdown; when they are transported along the upper third of oviducts, the 1st polar body is extruded and they reach the "uterus" at metaphase of the 2nd maturation (meiotic) division (metaphase II). They remain at this stage until fertilization [28]. If they are not fertilized, the eggs die within a certain time.

10.4.2. Mature Egg

The eggs are mesolecithal, of telolecithal type. The animal pole is heavily pigmented with melanin and the vegetal pole is light colored. The boundary between the pigmented and nonpigmented parts is unclear and runs closer to the vegetal pole. The egg diameter is about 2 mm and, with the swollen envelopes, 8–10 mm. The size and color of eggs varies in different females. The egg diameter and batch size depend on the female's age. The laid eggs are covered by two envelopes, an internal primary or vitelline membrane, and an external jelly coat containing mucopolysaccharides and having several layers. By origin this is a tertiary envelope which forms when the egg is transported along the oviduct. Therefore, eggs maturing *in vitro* are devoid of this envelope. After the egg is placed or laid in water, the jelly coat swells gradually within 30–40 min. At this time it becomes sticky and causes the clumping of eggs in a characteristic batch and their attachment to substrate or aquarium walls.

10.4.3. Spermatozoon

The length of the spermatozoon tail corresponds roughly to that of its needle-shaped head. The boundary between acrosome and nucleus is indistinct; the neck is short and wide, its boundary with head and tail indistinct (Fig. 10.2).

10.5. ARTIFICIAL INSEMINATION

For artificial insemination ovulated oocytes should be taken from the "uterus." With this aim, the female is immobilized within 36–48 h after injection (depending on temperature) and a cut is made with scissors along the midabdominal line through the skin and muscle wall. If the ovulated oocytes are in the "uterus," it is

Fig. 10.2. Spermatozoa of the common frog.

not dissected, but the sperm is obtained first. The abdominal cavity is exposed in the immobilized males, and the testes are cut out; they are oval, yellowish-white, often irregularly pigmented bodies attached to the dorsal wall of the abdominal cavity. The isolated testes are minced with scissors in a small vessel, mixed with 10 ml dechlorinated water, and left for 3–5 min, during which time the physiological activation of spermatozoa takes place. The acquisition of motility by spermatozoa is monitored under the microscope. If dechlorinated water is replaced by 0.25% NaCl solution [2], the fertilizing capacity of spermatozoa, and hence the percentage of fertilized eggs, increases, as well as the percentage of polyspermic eggs.

If a large number of fertilized eggs is needed, the eggs from both the oviducts are used. To obtain them, the female is taken in the left hand by its head and hind limbs, and the walls of one and then another oviduct are cut through with scissors. The eggs fall out on the bottom of a dry vessel, where the lumps of eggs are rapidly spread with a glass spatula in a relatively thin layer. The eggs are then flooded with the prepared sperm suspension so that the pieces of minced testes do not fall on the eggs.

If several small successively inseminated portions of eggs are needed, only one oviduct is cut through and the eggs are taken out; thereafter the cut margins of the body wall are brought together so that the rest of the eggs do not dry out. The eggs taken are transferred to a Petri dish, spread on the bottom in one layer, and flooded with the sperm suspension. This procedure can be repeated several times with new portions of the eggs, which are taken first from one oviduct and then from another.

After mixing egg and sperm, the vessel is swirled lightly for 3–5 min so that the sperm suspension wets all the eggs simultaneously. Water is then added to cover the eggs, and the vessel is shaken for 20–30 min so that the swelling and sticky envelopes do not attach to the aquarium bottom or walls.

In some laboratories the eggs are allowed to attach to the vessel bottom and, after fertilization is completed, detached with a spatula or scissors. Sometimes glass is placed on the vessel bottom as a substrate for the eggs. The water with sperm is replaced with fresh dechlorinated water within 20–30 min. The layer of fertilized eggs is separated with scissors in pieces with 25–50 eggs.

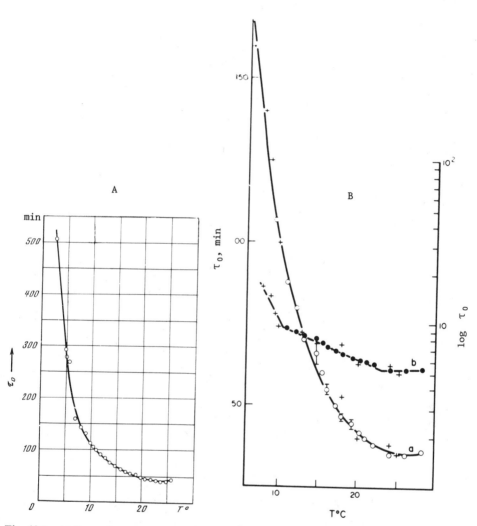

Fig. 10.3. A) Dependence of the duration of one mitotic cycle during synchronous cleavage divisions (τ_0) on temperature in *Rana temporaria* (after Chulitskaya [5]). B) Temperature dependence of the duration of one mitotic cycle during synchronous cleavage divisions (τ_0) in *Rana ridibunda* in normal (a) and semilogarithmic scales (b). +) Data of Chulitskaya [5]; ○) data of Mazin et al. [21]. ●) Data of Mazin et al. in semilogarithmic scale.

TABLE 10.1. Stages of Embryonic Development of the Common Frog *Rana temporaria*

Stage number	Time from insemination, τ_n/τ_0	Diagnostic features
0	0	Egg at the moment of insemination
1	0.5	Fertilized activated egg with swollen jelly coat
2–	3	Appearance of first cleavage furrow in the center of animal region
2	3.5	First cleavage furrow embracing 3/4 of egg circumference
3	4.5	Second cleavage furrow embracing 3/4 of egg circumference
4	5.5	8 blastomeres
5	6.5	16 blastomeres
6	7.5	32 blastomeres
7	8.5	64 blastomeres; morula
8	9.5	Early blastula
$8^1/_2$	12	Desynchronization of nuclear divisions in animal blastomeres (after Chulitskaya [6])
9	17	Middle blastula; fall of mitotic index in animal blastomeres (after Chulitskaya [6])
10	21	Late epithelial blastula
11	25.3	Onset of gastrulation; accumulation of pigment in the region of future dorsal blastopore lip
12	28	Crescent-shaped blastopore (its arch forms one-fourth of circumference)
13	32.7	Middle gastrula; formation of lateral blastopore lips; blastopore arch forms one-half of circumference
14	36.5	Lateral lips formed, blastopore forms three-fourths of circumference
15	37.0	Ventral blastopore lip formed; large yolk plug
16	42	Diameter of yolk plug decreased by half (cells of yolk plug become smaller)
17	45.8	Very small yolk plug; diameter of yolk plug decreased to one fourth
18	47.8	Residue of yolk plug elongated dorsoventrally
19	50	Neural plate boundaries delineated on the dorsal side of embryo, mainly in its head region; neural groove outlined. Slit-like blastopore

TABLE 10.1 (continued)

Stage number	Time from insemination, τ_n/τ_0	Diagnostic features
20	52	Neural groove and wide flat rudiment of neural plate limited by neural folds clearly seen
$20^1/_2$	59	Neural folds more elevated
21	55	Neural folds elevated, begin to approach, especially in the region of future brain and spinal cord
$21^1/_2$	55.5	Neural folds almost touching, especially in the trunk region
22	58.5	Neural folds almost touching all along. Embryo slightly elongated craniocaudally. Dorsal side slightly concave. Gill area prominent
23	60	Neural folds fuse in the anterior part of embryo. Three brain vesicles are forming. Neural tube not yet closed in its posterior region. The site of sucker formation and mouth pit is first outlined in the head region on the ventral side
24	64	Head region formed. Eye vesicles distinct. 2–3 pairs of somites are visible. Traces of ectodermal suture are seen above neural tube. Sucker is bordered by a fold. Division of gill area begins
25	65	A rudiment of pronephros seen, pronephric duct grows caudally. No suture is seen on the dorsal side at the site of fusion of neural folds. Dorsal curvature is distinct. Tail rudiment is outlined. A recess separates the head region from the branchial region
26	68	Tail bud is growing, protrudes on the dorsal side. Sucker distinct; onset of secretion. Eye vesicles enlarged and protrude almost like gill area. Olfactory pits are delineated. Pronephric duct reaches the middle of back
27	71.5	Tail bud markedly elongated. Back strongly concave. Posterior end of pronephric duct reaches the 11th pair of somites. Sucker folds elevated, markedly approached. A rudiment of cloaca appeared. First muscle contractions in response to external stimulation. Tail bud flattened, fin fold appears around it
28	89	Back of embryo somewhat straightened. Length of tail bud equals its width. Slit of sucker markedly deepened. Spontaneous muscle contractions appeared
29	92	Onset of hatching. Active muscle contractions. Two rudiments of external gills formed

TABLE 10.2. Stages of Larval Development of the Common Frog *Rana temporaria*

Stage number	Time from insemination, τ_n/τ_0	Diagnostic features
30	4	Mass hatching. Onset of rhythmic heart contractions. Rudiments of the first two pairs of gills begin to branch (2–3 outgrowths in each). Rudiments of the 3rd pair of external gills have appeared as a small tubercle under the first 2 pairs. Four-lobed mouth rudiment has stretched along body axis
31	4.5	Blood movement is seen in vessels of gill outgrowths. The first 2 pairs of external gills have branched. Rudiment of 3rd pair of gill enlarged, not branched. Right and left part of sucker approached all over. Mouth rudiment becomes three-lobed, stretched transversely to body axis. Fin fold markedly increased and begins to become less opaque
32	5	Two protuberances have appeared in rudiment of 3rd pair of gills. Onset of clearing of covering epithelium above the eye. Onset of sucker reduction. Common slit and external fold in the place of sucker bridge have disappeared, both parts joined by internal fold. Mouth rudiment markedly deepened, upper and lower lip forming
33	6	Mouth opening broken through. Right and left parts of sucker fully separated. Branchial arches with rudiments of internal gills have appeared under external gills. Tail fin becomes translucent.
34	7.5	Cornea transparent. Onset of active feeding. Internal gills clearly seen; three rows of digitlike outgrowths
35	8	Opercular fold closed. Onset of sucker degeneration; internal slit disappearing, folds still elevated. Lower external mouth lip consists of 2 folds; internal fold with a shallow constriction
36	9	Anal opening broken through. Sucker folds flattened, remnants of internal slit still seen. Onset of hornification of internal lips
37	9.5	Internal slit in sucker has disappeared. Onset of overgrowing of opercular fold (from the right side). Right gill as a small cord, left gill also shortening
38	9.5	Opercular fold overgrown from the right side. Left gill seen from opercular opening
39	10	Mouth apparatus: one continuous row of teeth on upper lip and one continuous and one interrupted row on lower one. Mouth papillae formed on the external margin of lower lip. Rudiments of hind limbs as a small round tubercle. Gills covered by opercular fold, their small fragment sometimes seen through opercular opening

TABLE 10.2 (continued)

Stage number	Time from insemination, τ_n/τ_0	Diagnostic features
40	17–20	Mouth apparatus same as previous stage. Sucker remnants as platelets. Limb rudiments as elongated tubercles
41	30–35	Mouth apparatus: upper lip, one continuous and one interrupted row of teeth; lower lip, one interrupted and two continuous rows. Sucker remnants as accumulation of melanin in skin. Limb rudiments begin to curve slightly
42	40–41	Mouth apparatus same as previous stage. Rudiment of knee joint
43	43–45	Mouth apparatus: upper lip, one continuous and two interrupted rows of teeth; lower lip, three continuous rows. Limb rudiments elongating and distinctly curving. Their distal end flattened to a spatula shape
44	47–49	Mouth apparatus: upper lip, one continuous and two interrupted rows of teeth; lower lip, four continuous rows. Limb rudiment as a spatula with rudiments of three toes. Curvature of talocrural joint
45	52–54	Mouth apparatus same as previous stage. All toes are outlined in limb rudiments
46	56	Mouth apparatus same as previous stage. All five toes of hind limbs separated
47	58	Mouth apparatus: upper lip, one continuous and two interrupted rows of teeth; lower lip, three interrupted rows. Hind limb curves distinctly at the level of knee, toes elongated, phalanges outlined. 2nd toe advanced in its development. Formation of talocrural joint. Rudiments of interdigital web formed between 1st and 2nd and between 2nd and 3rd toes
48	62	Mouth apparatus same as previous stage. Interdigital web fully formed
49	64	Tail attains its maximum length (14 mm). Outlines of forelimbs seen through thin skin from the ventral side. Hind limbs bent in the knee joint at an obtuse angle. Partial disappearance of horny parts of the mouth apparatus
50	66	Left forelimb coming out through opercular opening. Mouth opening archlike; remnants of mouth papillae seen in lip angles
51	66.5	Opercular membrane broken through by right forelimb. Mouth elongating laterally. Horny dents and mouth papillae disappeared completely. Onset of tail reduction

TABLE 10.2 (continued)

Stage number	Time from insemination, τ_n / τ_0	Diagnostic features
52	67	Tail reduction; its length is half the maximum previously attained (7 mm). Hip and knee joints bent at right angle
53	68–69	Tail length one third of the maximum previously attained (2.8–3 mm). All joints of hind limbs bent at acute angles
54	70	End of metamorphosis. Complete reduction of tail

There is another, "dry" method of artificial fertilization. The eggs are spread on the vessel bottom in one layer. A testis is placed in the same vessel, cut into 2 to 4 pieces, taken with forceps by the intact end and applied to the eggs so that sperm leaks from the cut surface. The eggs are then flooded with a small amount of water so that it covers all the eggs simultaneously. The rest of the procedure is as described above. The first signs of successful development during the first hour after fertilization are the turn of the eggs' animal pole upward, followed by appearance of the first cleavage furrow (see Table 10.1).

10.6. NORMAL DEVELOPMENT

Normal development of different species of the Ranidae has been described elsewhere: *R. pipiens* [30, 33], *R. sylvatica* [26], and *R. dalmatina* [3].

The first table of normal development for *R. temporaria* was constructed by Hertwig [16] and included nine stages. Kopsch [17] provided data on the biology of this species and tables of embryonic and larval development. He distinguished 30 stages, and described external features and internal structure of the embryos and larvae at these stages. It is, however, rather difficult to use these tables since the embryonic stages are insufficiently detailed, and the size of the drawings does not allow examination of the structural details and identification of the embryos and larvae under study. For these reasons, we have constructed new tables of the normal development of *R. temporaria*.

To study the normal development of *R. temporaria* we took eggs obtained both by pituitary injection and collected in nature. The eggs were incubated from fertilization until hatching in incubators at constant temperatures, and the time of onset of successive developmental stages was determined in hours and minutes. Larvae, after hatching and until the end of metamorphosis, were reared at temperatures which varied from 17–20°C and the mean time of onset of successive developmen-

Figs. 10.4–10.11. Stages of normal development of the common frog *Rana temporaria* (drawings by L. A. Sleptsova). Drawing numbers (0–54) correspond to stage numbers. an) Animal view; veg) vegetal view; d) dorsal view; v) ventral view; h) head view; t) tail view. Drawings without letter designations are lateral views.

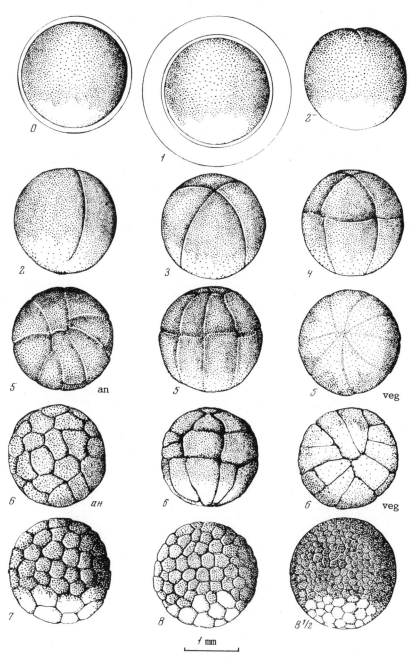

Fig. 10.4.

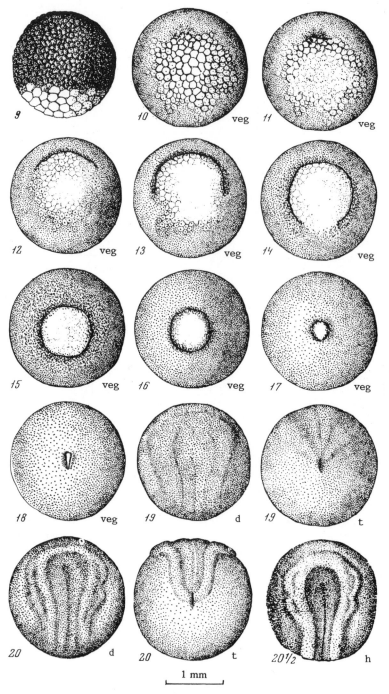

1 mm

Fig.   10.5

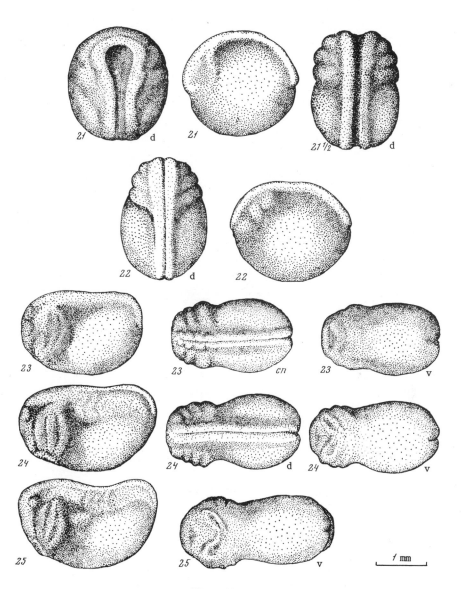

Fig. 10.6.

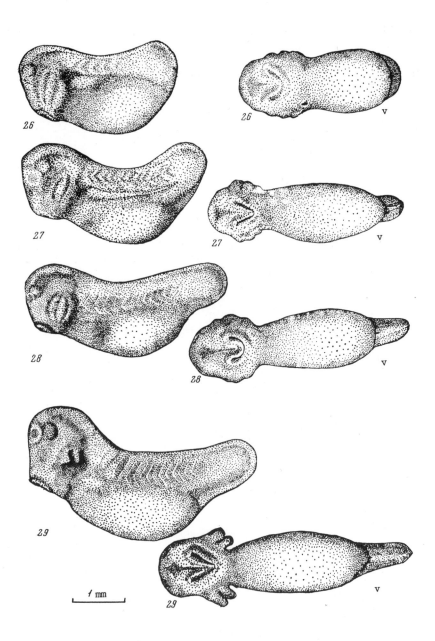

Fig. 10.7.

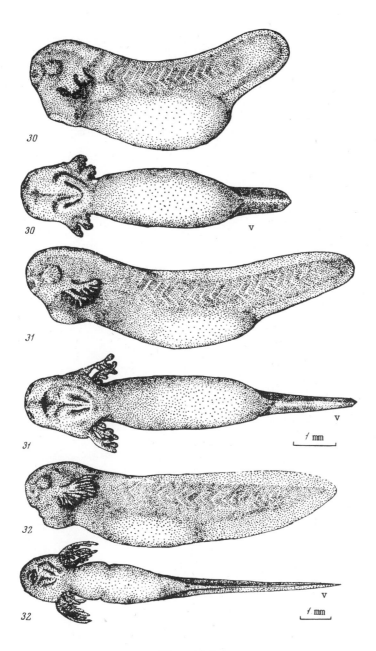

Fig. 10.8.

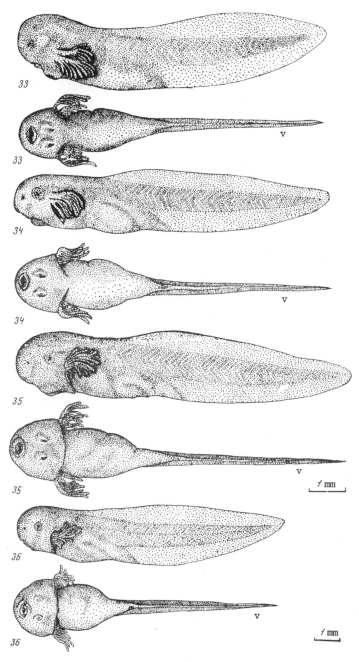

Fig. 10.9.

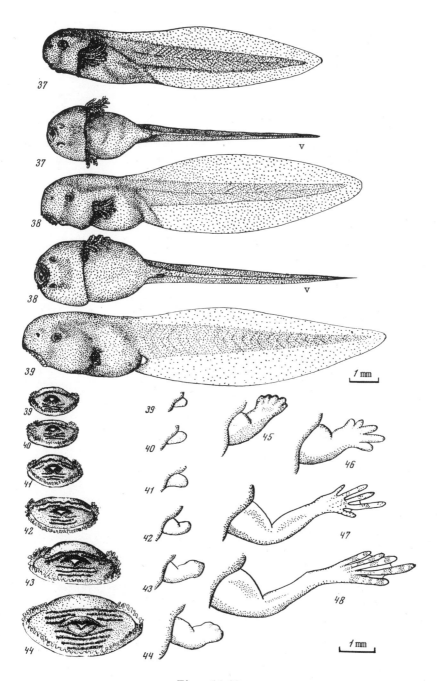

Fig. 10.10.

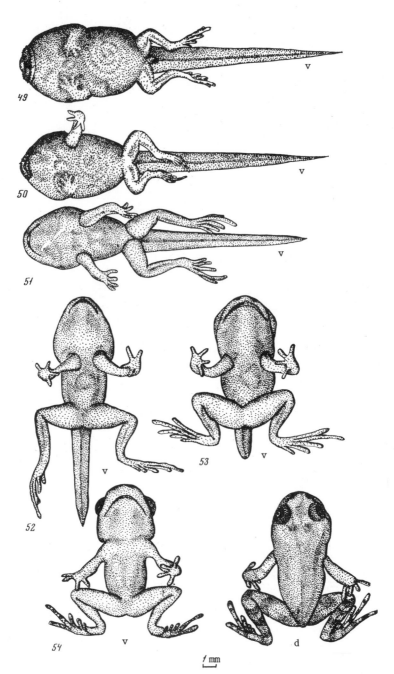

Fig.   10.11.

tal stages (by the advanced larvae in the batch) was determined in days (Table 10.2). After the transition to active feeding the asynchrony of larval development increases markedly.

The description of embryonic and larval stages is given by external features, and a few diagnostic features have been established for every stage, which allow one to distinguish the given stage and to determine precisely the time of its onset. For two stages only such features as the character of nuclear division are diagnostic: Stage $8^1/_2$, desynchronization of nuclear divisions in the animal blastomeres, and stage 9, mitotic index fall in animal blastomeres (data of Chulitskaya [6]). For stages 39–48, the structure of mouth and hind limbs is used as a diagnostic criterion.

To obtain the relative characteristics of the time of onset of every stage, the absolute time was recalculated in number of τ_0 units at the given temperature (τ_0 = duration of one mitotic cycle during the period of synchronous cleavage divisions; see Dettlaff [9]; Dettlaff and Dettlaff [10]). The value of τ_0 was determined from the τ_0 versus temperature curve for *R. temporaria* and for some other *Rana* species [21, 22] (Fig. 10.3A, B).

Table 10.1 provides the numbers of stages, time of their onset in τ_n/τ_0, and diagnostic features. The values of τ_n/τ_0 for stages 11–29 are made more precise (as compared with the first publication [7]) by Ryabova et al. [29]. For the stages of onset and end of gastrulation they agree quite well with those obtained earlier [4, 5, 36].

The drawings for all these stages are given in Figs. 10.4–10.11, where the numbers of the drawings correspond to those of the stages. The drawings were made with a drawing device by one of the authors (L. A. Sleptsova).

REFERENCES

1. A. G. Bannikov, I. S. Darevsky, and A. K. Rustamov, *Amphibians and Reptiles of the USSR* [in Russian], Mysl, Moscow (1971).
2. A. Brachet, "La polyspérmie expérimentale comme moyen d'analyse de la fécondation," *Wilhelm Roux's Arch. Entwicklungsmech. Org.* **30**, 261–303 (1910).
3. R. Cambar and B. Marrot, "Table chonologique du développement de la grenouille agile (*Rana dalmatina* Bon.)," *Bull. Biol. Fr. Belg.* **88**, 168–177 (1954).
4. G. ten Cate, "The intrinsic embryonic development," *Verhandel. Konikl. Ned. Akad. Weteschap. Afdel. Natuurk.* **51**, 1–257 (1956).
5. E. V. Chulitskaya, "Relative durations of the periods of cleavage and gastrulation and latent differentiation of the labyrinth rudiment in *Rana temporaria* embryos at different temperatures," *Dokl. Akad. Nauk SSSR* **160**, 489–492 (1965).
6. E. V. Chulitskaya, "Origin of desynchronization and changes in the rate of nuclear divisions during cleavage," *Dokl. Akad. Nauk SSSR* **173**, 1473–1476 (1967).
7. N. V. Dabagyan and L. V. Sleptsova, "The common frog *Rana temporaria* L.," in *Objects of Developmental Biology*, T. A. Dettlaff, ed. [in Russian], Nauka, Moscow (1975), pp. 442–462.

8. T. A. Dettlaff, "A study of the properties, morphogenetic potencies, and prospective fate of outer and inner layers of ectodermal and chordamesodermal regions during gastrulation, in various Anuran amphibians," *J. Embryol. Exp. Morphol.* **75**, 67–86 (1983).

9. T. A. Dettlaff, "The rate of development in poikilothermic animals calculated in astronomical and relative time units," *J. Therm. Biol.* **11**, 1–7 (1986).

10. T. A. Dettlaff and A. A. Dettlaff, "On relative dimensionless characteristics of development duration in embryology," *Arch. Biol.* **72**, 1–16 (1961).

11. B. Dürken, "Über Entwicklungskorrelationen und lokal Rassen bei *Rana fusca*," *Biol. Zbl.* **37**, 127–139 (1917).

12. D. P. Filatov, "On morphogenetic effect of the eye cup on the trunk epithelium in the common frog," *Biol. Zh.* **3**, 261–268 (1934).

13. B. F. Goncharov, "Functional relationship between length of the hormone-dependent period of follicle maturation in the common frog and the concentration of suspension. A new method for testing pituitaries," *Sov. J. Dev. Biol.* **2**, 50–55 (1971).

14. V. Hamburger, "Über den Einfluss des Nervensystems auf die Entwicklung der Extremitäten von *Rana fusca*," *Wilhelm Roux's Arch. Entwicklungsmech. Org.* **105**, 149–201 (1925).

15. V. Hamburger, *A Manual of Experimental Embryology*, University of Chicago Press, Chicago (1947).

16. O. Hertwig, "Über den Einfluss der Temperatur auf die Entwicklung von *Rana fusca* und *Rana esculenta*," *Arch. Mikrosk. Anat.* **51**, 319 (1898).

17. F. Kopsch, *Die Entwicklung des braunen Grasfrosches Rana fusca Roesel (dargestellt in der Art der Normentafeln zur Entwicklungsgeschichte der Wirbeltiere)*, Thieme-Verlag, Stuttgart (1952).

18. R. V. Latsis and N. Y. Saraeva, "Differentiation of hemopoietic cells in the explants of embryonic organs of *Rana temporaria*," *Sov. J. Dev. Biol.* **9**, 524–527 (1978).

19. G. V. Lopashov and O. G. Stroeva, *Development of the Eye. Experimental Studies*, Israel Program for Scientific Translations, Jerusalem (1964).

20. N. A. Manuilova, "Morphogenetic peculiarities of the developing eye cup and lens in amphibians," *Tr. Nauchn. Issled. Inst. Morfogen. Mosk. Gos. Univ.* **6**, 107–129 (1938).

21. A. L. Mazin, V. N. Vitvitzky, and V. Ya. Aleksandrov, "Temperature dependence of oocyte cleavage in three thermophilically different frogs of the genus *Rana*," *J. Therm. Biol.* **4**, 57–61 (1979).

22. A. L. Mazin and T. A. Dettlaff, "Dependence of the duration of one mitotic cycle during synchronous cleavage divisions (τ_0) on temperature in four species of the genus *Rana* and the limits of optimal temperatures for their reproduction and early development," *Ontogenez* **16**, 382–388 (1985).

23. A. T. Mikhailov and N. A. Gorgolyuk, "*In vitro* effect of heterogeneous inducers in early gastrula ectoderm of *Rana temporaria*. I. Inductive effect of extracts from chick embryonic retina, brain, and liver," *Sov. J. Dev. Biol.* **11**, 31–38 (1980).

24. I. V. Neklyudova and N. V. Dabagyan, "Differentiation of retina and pigment epithelium in ontogenesis of *R. temporaria*. I. Retina," *Ontogenez* **14**, 398–405 (1983).

25. L. A. Nikitina, "Transplantation of the nuclei from the ectoderm and neural rudiment of *Bufo bufo, Rana arvalis,* and *Rana temporaria* embryos in enucleated eggs of the same species," *Dokl. Akad. Nauk SSSR* **156**, 1468–1471 (1964).

26. A. W. Pollister and J. A. Morre, "Tables for the normal development of *Rana sylvatica," Anat. Rec.* **68**, 489–496.

27. V. V. Popov, "On morphogenesis of cornea in Anura," *Tr. Nauchn. Issled. Inst. Morfogen. Mosk. Gos. Univ.* **6**, 65–79 (1938).

28. R. Rugh, *Experimental Embryology,* Burgess, Minneapolis (1965).

29. L. V. Ryabova, O. B. Trubnikova, and L. V. Sleptsova, "Chronology of embryogenesis of the common frog (*Rana temporaria*) (addition to and precision of earlier data)," *Ontogenez* **15**, 91–93 (1984).

30. W. Shumway, "Stages in the normal development of *Rana pipiens.* I. External form," *Anat. Rec.* **78**, 139–144 (1940).

31. M. N. Skoblina, Z. G. Shmerling, and O. T. Kondratieva, "*In vitro* maturation of *Rana temporaria* oocytes under the effect of cholesterol," *Sov. J. Dev. Biol.* **11**, 81–84 (1980).

32. O. G. Stroeva and V. I. Mitashov, "Developmental potential of vertebrate eye tissues for regeneration of retina and lens," in *Problems of Developmental Biology,* N. G. Khrushchov, ed. [Russian translation], Mir, Moscow (1981), pp. 168–207.

33. A. C. Taylor and J. J. Kollors, "Stages in the normal development of *Rana pipiens* larvae," *Anat. Rec.* **94**, 7–23 (1946).

34. P. V. Terentiev, *A Survey of Amphibia of the Moscow District* [in Russian], Gosizdat, Moscow (1924).

35. P. V. Terentiev and S. A. Chernov, *A Key to Reptiles and Amphibians,* 3rd edn. [in Russian], Nauka, Moscow (1949).

36. P. Valouch, J. Melichna, and F. Sládecek, "The number of cells at the beginning of gastrulation depending on the temperature in different species of Amphibians," *Acta Univ. Carol. Biol.,* pp. 195–205 (1971).

37. L. E. Zavalishina and L. V. Belousov, "Mobility of membrane receptors and morphological polarization of cells in embryonic tissue explants of amphibians," *Sov. J. Dev. Biol.* **13**, 131–136 (1982).

38. A. V. Zhirmunsky and V. I. Kuzmin, *Critical Levels in the Processes of Development of Biological Systems* [in Russian], Nauka, Moscow (1982).

Chapter 11

THE DOMESTIC FOWL *Gallus domesticus*

M. N. Ragozina

The chick embryo is a classical subject of developmental biology. Chick embryos are not only widely used in descriptive and experimental embryology, but are increasingly valuable in medical research, as in work on viruses and cancer. Their development has been well studied, beginning from the last century [4, 5, 16]. Naturally, the earlier data have been expanded and somewhat changed. Comprehensive reviews of recent studies are provided by Lillie [36], Hamilton [26], Romanoff and Romanoff [59–61], Huenter [28], Ragozina [56], Novik [45], and Rolnik [58]. In the book by Rolnik [58] the problems of developmental physiology of birds are considered in detail.

Among the works of Soviet authors, one should mention the studies of the proliferation of the blastoderm cells in early embryogenesis [17–19], of development of the immune system in birds ([14, 15, 30–32]; see also [12, 41]), as well as of biochemistry of early development [2, 39, 74, 75].

Recent studies of somitogenesis should also be mentioned as well [9–11, 38, 68, 69].

In studies of avian development, various genetically determined developmental defects are widely used. A review of these data, as well as characteristics of different types of mutations, is provided by Abbott [1].

Methods of experimental studies on avian embryos are described in a number of handbooks (see [23, 33, 43, 63]). Tables of normal development were constructed by Hamburger and Hamilton [24] and, in more detail for the early developmental stages, by Vakaet ([70]; see also [21, 43, 44, 62, 71]). In this chapter the tables of Hamburger and Hamilton, with some additions, are given; I am thankful to Professor Hamburger for his kind permission to use them.

Tables of successive developmental stages have also been compiled for the turkey [57], among other domestic birds. It should be recognized, however, that quail embryos now compete favorably with chick embryos in the number of experimental studies. Quails reproduce easily in the laboratory; their eggs are much smaller than hens' eggs and the time of incubation is shorter. These features make the quail very convenient for some experimental studies, and quite a bit is known about its embryogenesis and egg incubation [35, 51, 54, 73, 76].

Obviously, chick embryonic development is of special interest for poultry farming since the industrial incubation of hen's eggs is widely used [50, 56].

11.1. TAXONOMY AND REPRODUCTION

The domestic fowl *Gallus domesticus* L. belongs to the genus *Gallus*, family Phasianidae, order Galliformes, class Aves. Domestic fowl breeds are numerous and diverse. To study the development of their embryos, all domestic fowl breeds are best classified by the productive properties of the stock. This allows us to distinguish three main groups: egg-producing, general, and meat-producing breeds. Embryogenesis in the first two groups is similar and is completed 24 h earlier than in the third group. The total duration of egg incubation amounts to 20–21 days.

The egg-producing breeds of domestic fowl lay eggs suitable for incubation throughout the whole year provided they are kept under the correct regime. Incubation of eggs by the hen (which has not lost this instinct despite factory methods) is timed to the spring–summer season. In some breeds the eggs are laid daily or within an interval of a few days. Egg laying in domestic fowl, and especially in egg-producing breeds, has no strict cyclic pattern, but some hens with moderate egg production show particular intervals between batches of laid eggs. Romanoff and Romanoff [60] called this phenomenon "the rhythm of egg laying." The total productivity of a given hen can be estimated by counting from the beginning of egg laying and the number of eggs in the clutch. The number of eggs in the cycle can vary from 2 to over 10. Some hens are known to lay eggs without interruption, i.e., 365 days a year.

The duration of the interval between two successive egg layings in different egg-laying cycles was estimated by Heywant [27]. If there were two eggs in the clutch, the interval between the egg layings amounted to 28 h, on average. The more eggs in the clutch, the less time is spent in the formation of the second egg. In a clutch of 6 eggs this time is reduced to 25.5 h, and in a clutch of 10 eggs to 24 h. The same pattern of egg formation is observed between two egg layings in the middle and the end of the cycle. Within each cycle, the shortest time in the oviducts is spent by eggs laid in the middle of the cycle, and the longest time by eggs laid last.

Differences in developmental stages of the embryos prior to egg laying is related to the duration of its transport in the oviducts. In this respect the hen's age is of great importance. It is known that in young hens the oviducts are much shorter than in hens which have laid eggs for two consecutive years. In the former, the eggs are always smaller and the time of their transport in the oviduct is shorter. The delay of the egg in the maternal organism depends, to a certain extent, on the clutch size. The rate of egg laying is influenced by environmental factors varying from season to season, such as temperature and light which affect the endocrine system of the hen.

11.2. HORMONAL REGULATION OF OOCYTE MATURATION AND OVULATION

Oocyte growth and maturation are induced by follicle-stimulating hormone and ovulation by luteinizing hormone (LH). The injection of LH can accelerate the onset of ovulation and, thus, increase the number of the eggs laid during the defined period of time. Egg size in this case is smaller.

Oocyte maturation was studied by Olsen [46, 47, 49], Borkovskaya [13], and Fraps [22]. About 24 h prior to ovulation the nuclear membrane in the oocyte is disintegrated and the first polar body forms. It is extruded 1 h prior to ovulation and remains under the zona radiata. Just prior to ovulation the oocyte attains metaphase of the 2nd maturation division (metaphase II).

Ovulation of the eggs proceeds at metaphase II. The follicle epithelium becomes gradually thinner and is then broken through in the part of the follicle which is called the stigma. The process of ovulation lasts 1–1.5 min. After ovulation the egg is released either in the body cavity or directly into the oviduct funnel. If the egg is released into the body cavity, it is transported into the oviduct funnel within 5–7 min.

Warren and his coworkers [53, 72] performed laparotomy and observed the process of egg release from the ovary. According to their data, the interval between ovulations amounts to 30 min, on average, and according to Neher et al. [42] between 20 and 130 min. The latter authors obtained *in vitro* oocyte ovulation while keeping the follicles excised from the ovary in 0.9% NaCl solution in the incubator at 41.7°C.

11.3. GAMETES, INSEMINATION, FERTILIZATION, AND EGG MEMBRANES

The ovulated hen's egg is coated with a vitelline membrane (see below). It "overripens" rapidly and loses fertilizability. Novik [45] showed that within 40 min after ovulation the eggs were not fertilizable.

The spermatozoa are capable of surviving for 23–35 days in the hen's reproductive tract. They appear in the oviduct funnel on the third day after mating or artificial insemination. Artificial insemination is performed by a single introduction into the lower part of the oviduct, the "uterus," of 0.025–0.050 ml of undiluted sperm.

After the end of the third week the vitality of the spermatozoa is apparently reduced, as eggs laid during the fourth and fifth weeks after the isolation of hens from the cock exhibit, in about 15% of the case, abnormal (parthenogenetic) cleavage which soon ceases [26]. In birds complete parthenogenesis occurs only in the turkey, and it hardly can be called natural parthenogenesis since it correlates intimately with susceptibility to fowl pox virus, which can somehow induce parthenogenetic development [48]. One can hope to obtain parthenogenesis in chickens as well, since the infested hens tend to parthenogentic activation [34].

Fertilization takes place in the oviduct funnel. Many spermatozoa always penetrate the egg (physiological polyspermy). According to Bekhtina [6, 7], the 2nd maturation division proceeds in the fertilized egg within 10–30 min after ovulation, and the pronuclei fuse within 3–5 h.

During the transport of the fertilized egg along the oviduct, isthmus, and "uterus," the "egg formation" takes place, as well as embryogenesis. The embryo is covered by tertiary envelopes and acquires the structure shown in Fig. 11.1.

The structure, which is usually termed the "vitelline membrane" in birds, is not homogeneous in its origin. The zona radiata which surrounds the oocyte prior to ovulation is both the product of the activity of the follicle epithelium cells and a derivative of the ooplasm. In the oviduct funnel the zygote is covered on top of the zona radiata by a net of protein fibers, which represent the first portion of albumen and form a part of the "vitelline membrane."

The fine histological structure of the "vitelline membrane" has been studied by a number of authors [8, 37, 40, 65] but remains as yet unclear.

The definitive "vitelline membrane" has no pores or openings; it consists of a homogeneous layer with fibers of protein origin intertwining in it. Water penetrates easily from the albumen through the vitelline membrane to the yolk.

The tertiary egg envelopes are represented by three layers of albumen which arise by secretion from the upper, longest part of the oviduct, by two layers of the outer shell membrane formed in a short oviduct part (isthmus), and by the calcareous egg shell formed in the lower part of the oviduct, the "uterus."

The albumen contains 85.7% water, 12.7% proteins, 0.3% lipids, and 0.3–0.7% mineral substances. As the embryo develops, it fully utilizes the albumen. The shell membranes and egg shell are of transient value and, except for a small quantity of calcium salts of the calcareous egg shell, are discarded upon hatching.

After the formation of the calcareous egg shell, only air from the environment is available for the embryo. The products of gas exchange are removed and an excess of water is evaporated from the egg.

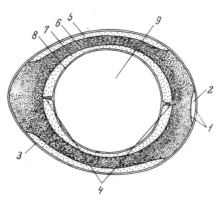

Fig. 11.1. A structural diagram of the hen's egg, after Romanoff and Romanoff [60]. View from above. 1) Outer shell membrane; 2) air chamber; 3) egg shell; 4) chalazae; 5) external liquid layer of egg albumen; 6) dense layer of egg albumen; 7) internal liquid layer of egg albumin; 8) vitelline membrane; 9) yolk.

11.4. NORMAL DEVELOPMENT DURING PASSAGE OF THE EGG ALONG THE OVIDUCT

According to Patterson [52], the first cleavage furrow is formed within about 3 h after ovulation. Within 20 min the second cleavage furrow appears at right angles to the first. Cleavage takes place simultaneously with the formation of the egg envelopes as the egg is transported along the oviduct. Cleavage is partial (i.e., a small part of the egg cleaves), discoidal, and unequal. It becomes asynchronous after the 2nd–3rd cleavage division. Radial cleavage furrows are followed by transverse ones which separate the central part of the germ disc from the peripheral part. Later on, the central cell of the germ disc divides at a greater rate than those in its periphery [29]. Recent autoradiographic studies have shown that in the chick embryo the nuclei divide synchronously during the first five cleavage divisions and the cell cycle is rearranged from the 6th cleavage division [20].

According to Olsen [46], an egg in the oviduct develops to the 8-cell stage only. Its subsequent cleavage, up to the 128-cell stage, proceeds in the isthmus and "uterus." However, Patterson [52] observed 16- and even 32-cell stages in the upper part of the oviduct. The egg stays there for 3–6 h and in the isthmus, "uterus," and cloaca for 16–24 h. The laying of the egg surrounded by all layers of tertiary envelopes, i.e., already in the cloaca, can be accelerated by intramuscular injection of Pituitrin. The eggs can also be pulled out of the oviducts by hand within 5.5 h after the laying of the preceding egg, and the hen is not damaged [21]. The developmental stages are determined by morphological features and the age of the embryos by approximate "stock age," i.e., by the interval between egg laying and extraction of the egg under study, minus 5.5 h.

Using this method, Eyal-Giladi and Kochav [21] describe six cleavage stages lasting 10–11 h. They noted a high rate of cleavage divisions (5–6 divisions during the first 2 h). Cleavage is completed by the formation of a many-layered blastodisc. At the third cleavage stage a part of yolk under the germ disc becomes liquid, and a subgerminal cavity (blastocoel) forms. Bilateral symmetry is established under the effect of gravity.

Rosenquist [62], Nicolet [44], and Volk and Eyal-Giladi [71] have shown that gastrulation in birds proceeds in two phases. The first phase consists of the formation of intestinal endoderm. It appears as a pellucid area in the center of the blastodisc. The separation of intestinal endoderm takes place by delamination and, somewhat later, by immigration of cells. Outside the pellucid area an opaque area forms, the rudiment of extraembryonic endoderm. The formation of the inner blastodisc layer (endoderm) is the essence of the first phase of gastrulation.

During the second phase of gastrulation, which proceeds mainly after egg laying (see Fig. 11.3, 1–3), the cellular material of the embryonic shield is displaced from the anterior to the posterior end. The flows of the cells from both sides meet at the posterior margin of the embryonic shield, change their direction, and thereafter move along the center of the embryo from its posterior end forward. Both cell flows form a thickened cellular strip along the midline (primitive streak).

The anterior end of the primitive streak moves forward and meets the cell masses moving in the opposite direction, and a dense accumulation of cells (Hensen's node) forms at the place of meeting. Hensen's node is the upper blastopore lip, the primitive streak margins are its lateral lips, and the blastopore itself is represented by a slit (see Figs. 11.3 and 11 4).

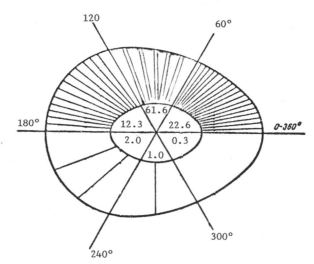

Fig. 11.2. Diagram of the primary positions of chick embryos in different sectors of the egg. The density of lines indicates the incidence (%). For explanation, see text.

In the second phase of gastrulation, the notochord, or head process, appears in front of Hensen's node (see Fig. 11.4). The yolk sac mesoderm forms at the expense of the primitive streak material. The formation of the middle gastrula layer (mesoderm) is the essence of the second phase of gastrulation.

11.5. ROTATION OF THE EMBRYO WITHIN THE EGG SHELL DURING EMBRYOGENESIS

The position of the long embryonic body axis at early developmental stages (third day of incubation) with respect to the long shell axis is easily determined by ooscopy. According to von Baer's rule [4], the long body axis at this time forms a right angle with the long shell axis, and its head is directed upward if the egg is placed with its obtuse end to the left. The angle between the positive direction of the long shell axis and the positive direction of the long body axis is determined counterclockwise. This angle is designated as *A*. Our studies carried out on intact (*in ova*) material [55], as well as the data of Ancel [3], suggest that von Baer's rule does not generally apply.

Four groups of eggs can be distinguished by the position of the embryos in them: in the first group (61.6% of all eggs), *A* varies from 60 to 120°; in the second group (22.8%) from 0 to 60°; in the third group (12.3%) from 120 to 180°; and in the fourth group (3.3%) from 180 to 360° (Fig. 11.2).

From 16 days of incubation until hatching the long body axis of the chick embryo coincides with the long shell axis. The chicks in the most favorable conditions are those which will get their beaks into the air chamber and receive the first portion of atmospheric oxygen, thus providing their transition from allantoic to lung respiration. The hatching and rotation of the embryo within the egg shell from 16 days of incubation on have been described by Hamburger and Oppenheim [25].

11.6. NORMAL DEVELOPMENT
FROM EGG LAYING UNTIL HATCHING

The descriptions and photographs of successive stages of chick embryogenesis are reproduced after Hamburger and Hamilton [24] who, in turn, used the data of Spratt [66, 67] and Saunders [64].

To study early developmental stages (up to 8 days of incubation), the eggs of an egg-producing breed, White Leghorn, were used; in addition, to study the subsequent stages, the eggs of meat-producing breeds, Barred Plymouth Rock and Rhode Island Red, were used. Up to 8.5 days of incubation the temperature of incubation was 39.4°C, and from 10–20 days, 37.5°C (the difference depended on the method of egg heating: heating from surface was 39.4°C, and heating as a whole was 37.5°C, although the rate of development was the same in both cases). For these observations 296 embryos were used.

Hamburger and Hamilton [24] distinguish 46 stages in the whole period of embryogenesis, from egg laying until hatching. The photographs and drawings of the embryos at these stages are presented in Figs. 11.1–11.23, and a brief description of external diagnostic features is provided in the text for each stage. For example, Hamburger and Hamilton [24] have chosen intervals of three pairs of somites as stages for the second day of incubation.

During the third day of incubation (up to stage 22) rapid progress in the development of limbs provides the most convenient diagnostic criterion. The developmental phase between 4 and 9 days is characterized by rapid changes in the wings, legs, and visceral arches. From the 8th to the 12th day, feather germs and eyelids provide the most useful criteria. By the end of incubation, when practically no new structures are formed, the lengths of the beak and middle toe can serve as diagnostic features. It should, however, be mentioned that enormous variations may occur in embryos. Many factors are responsible for the lack of correlation between chronological and structural age. Among these are: genetic differences in the rate of development of different breeds; differences in the stage of development when incubation is started; differences in the time and temperature between laying and incubation; temperature differences inside the incubator; type of incubator and position of the egg inside the incubator; differences in the size and form of individual eggs; thickness of egg shell, etc.

My additions to the tables by Hamburger and Hamilton [24] concern mainly the structure of extraembryonic organs and characteristics of the range of stage variations at the same time of incubation. These data have already been published [56].

Those embryos the number of which was maximal at the given stage for the given time of incubation were considered as typical. The diagnosis of the typical embryo included, besides the morphological features characterizing the degree of differentiation of organ rudiments, the state of extraembryonic organs. Extraembryonic organs provide the main life-functions of the developing embryo: respiration, feeding, and excretion of the products of nitrogen metabolism. As a result, they respond to a greater extent to changes in environmental conditions, i.e., the conditions of incubation, than the organ rudiments. For example, the degree of development of the yolk sac vascular system, and later of the allantoic vascular system, depends directly on the oxygen partial pressure in the air; the rapid growth of the allantois can be delayed by insufficiently frequent rotation of eggs during the incubation or a low degree of humidity in the incubator. A delay of the closure of the allantois margins under the albumen will adversely affect the subsequent feeding of the chick embryo, and will lead to its death by the end of incubation. Such examples are numerous. When describing the stages, I took into account the turns of the body of the embryo with respect to the egg axes, and the decrease and subsequent appearance of albumen residues from the albumen sac, measured the size of the air chamber, and considered a number of other developmental peculiarities of the chick embryo. My additions to the tables of Hamburger and Hamilton [24] are given in italics and are enclosed in parentheses.

From the 1st until the 8th day of incubation, typical embryos with rudiments of extraembryonic organs are presented in photographs or drawings. The interval between their ages is 24 h. Later, until hatching, the interval is 48 h.

From the 5th day of incubation the embryos are presented as drawings with surrounding egg envelopes. All drawings and descriptions are based on material fixed in Bouin's solution or formalin. Eggs of the Russian White breed were used (the most common breed at the Bratzevsky poultry factory in Moscow). The eggs were incubated according to the regime adopted at the factory, which gives the maximal yield of chickens.

I am grateful to S. O. Peltser for consultations concerning egg incubation under industrial conditions and for help in obtaining a sufficient amount of synchronous material. Thanks are also due to Dr. E. A. Krichinskaya, who gave some valuable advice.

Stage 1 (after egg laying). Prior to the appearance of the primitive streak. An "embryonic shield" may be visible, due to the accumulation of cells toward the posterior half of the blastoderm.

Stage 2. Initial streak ("short–broad beginning streak" of Spratt [66]). A rather transitory stage in which the primitive streak first appears as a short, conical thickening, almost as broad as long (0.3–0.5 mm in length), at the posterior border of the pellucid area. Usually obtained after 6–7 h of incubation.

Stage 3. Intermediate streak (12–13 h). The primitive streak extends from the posterior margin to approximately the center of the pellucid area. The streak is relatively broad throughout its length, and is flared out where it touches the opaque area. No primitive groove.

Stage 4. Definitive streak (18–19 h). The primitive streak has reached its maximal length (average length 1.88 mm; Spratt [66]). The primitive groove, primitive pit, and Hensen's node are present. The area pellucida has become pear-shaped and the streak extends over two-thirds to three-quarters of its length.

Stage 5. Head-process (19–22 h). The notochord or head-process is visible as a rod of condensed mesoderm extending forward from the anterior edge of Hensen's node. The head fold has not yet appeared. Since the length of the notochord increases during this stage, it is suggested that the length of the notochord in

---

Figs. 11.3–11.23. Stages of normal development of the chick embryo (*Gallus domesticus*). *W*, wing; *L*, leg; *d*, length; *w*, width. The drawings are reproduced from the tables by Hamburger and Hamilton [24]. Drawing numbers in the plates correspond to stage numbers. The drawings of the embryos with the egg envelopes and provisional organs were made by the author [56] and are designated by the letter A appended to the state number, e.g., 26A (X, boundary of allantois; 0, boundary of vascular field).

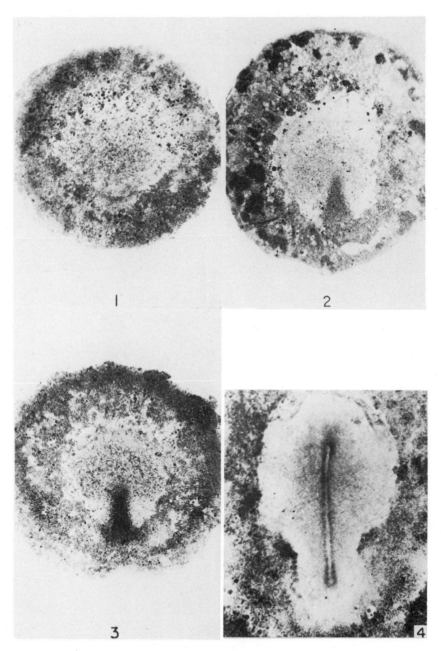

Fig. 11.3.

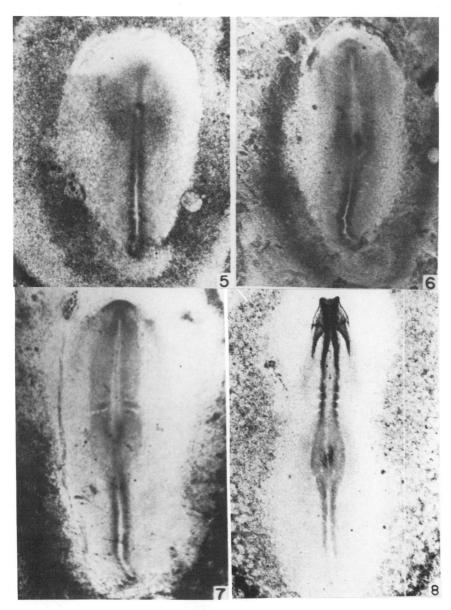

Fig. 11.4.

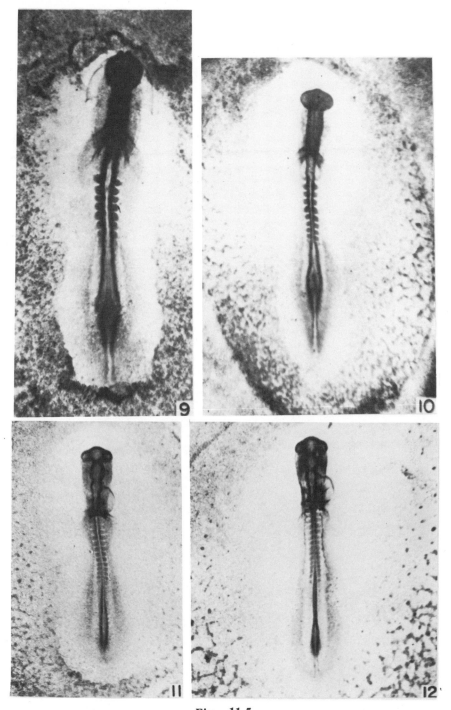

Fig. 11.5.

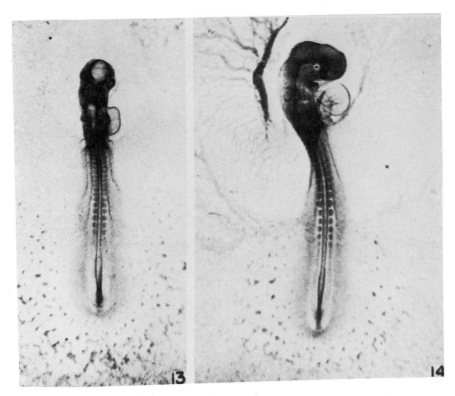

Fig. 11.6.

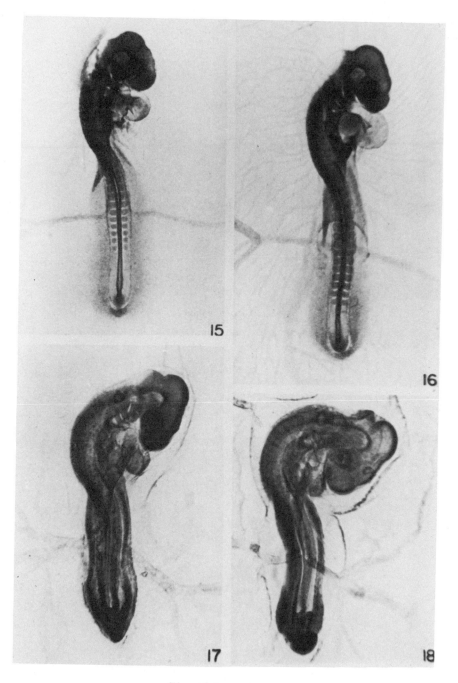

Fig. 11.6, continued.

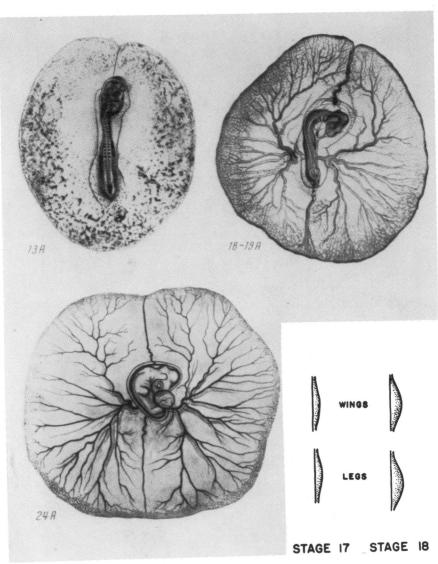

Fig. 11.7.

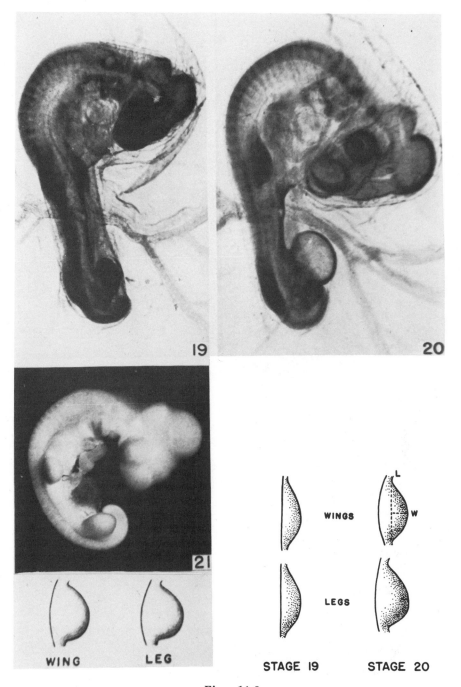

Fig. 11.8.

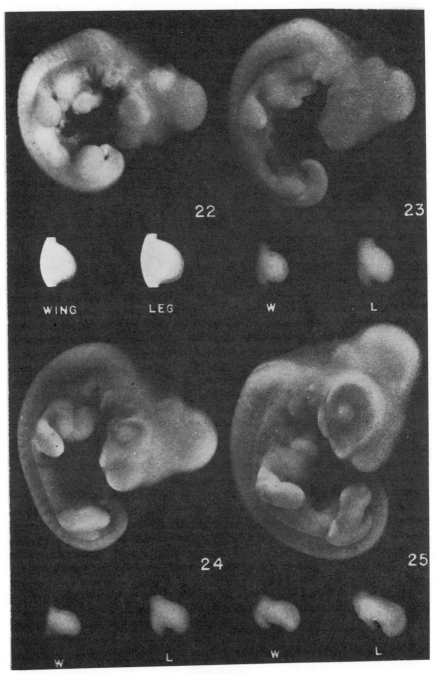

Fig. 11.9.

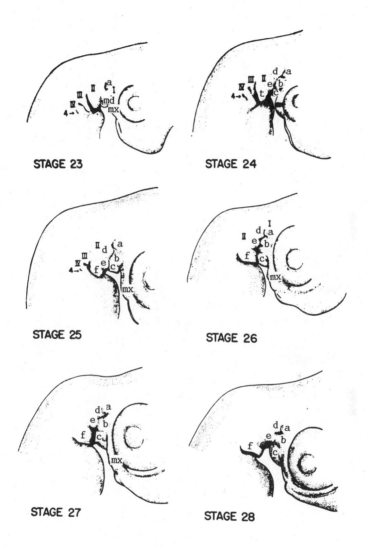

Fig. 11.10. Visceral arches of the chick embryo at stages 23–28.
I–IV) Visceral arches; mx and md) maxillary and mandibular processes
of first visceral arch; 4) visceral cleft. See text for explanation of letters
a–f. sc – serosoamniotic canal.

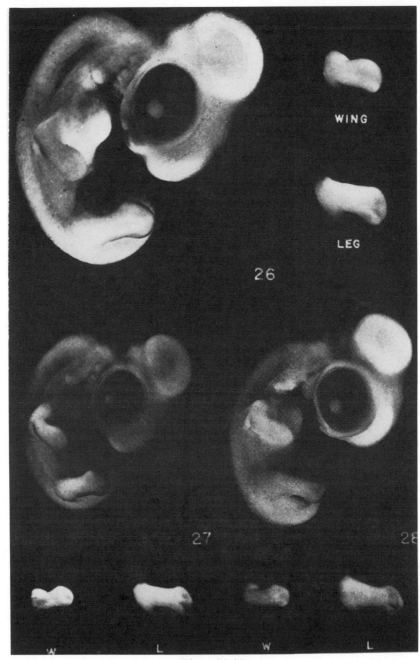

Fig. 11.11.

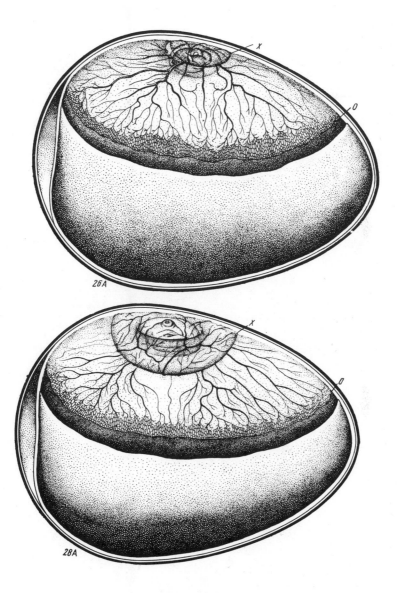

Fig. 11.12.

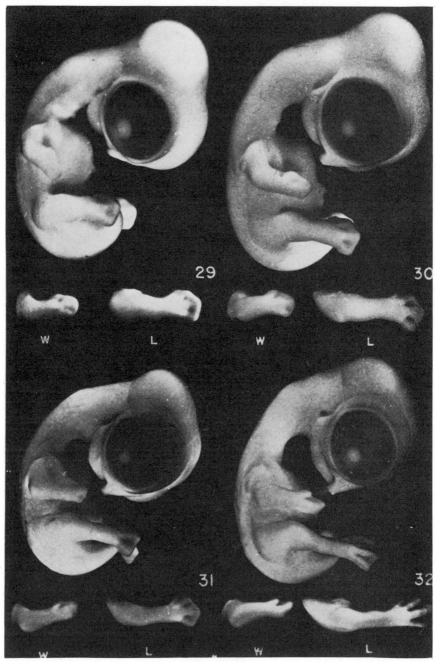

Fig.   11.13.

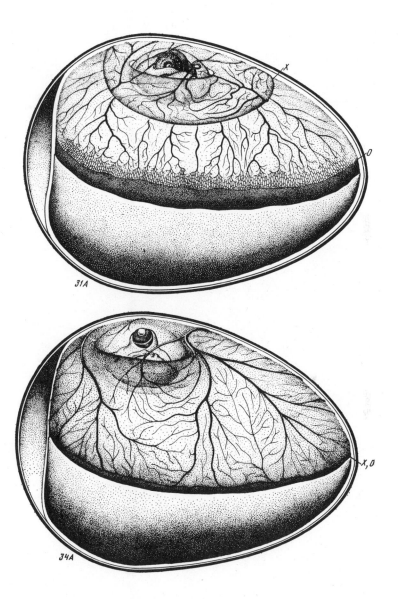

Fig. 11.14.

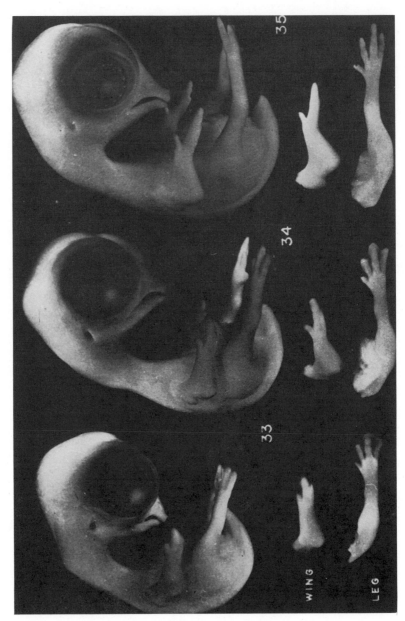

Fig. 11.15.

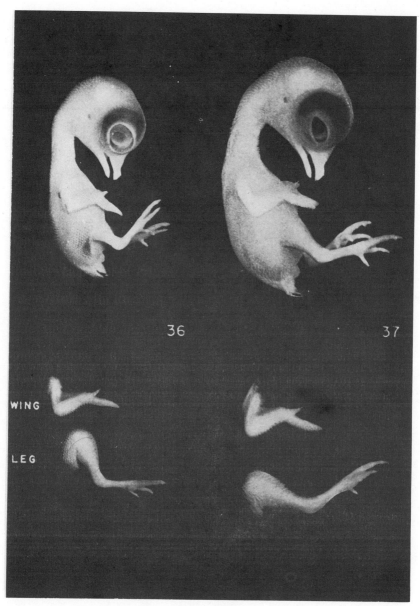

Fig. 11.16.

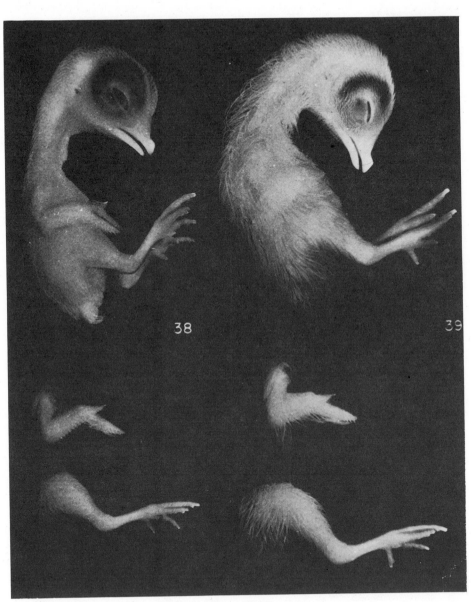

Fig.   11.17.

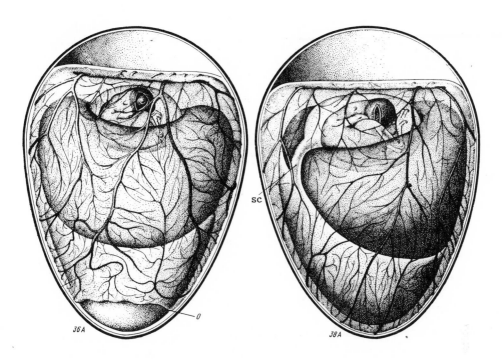

Fig. 11.18.

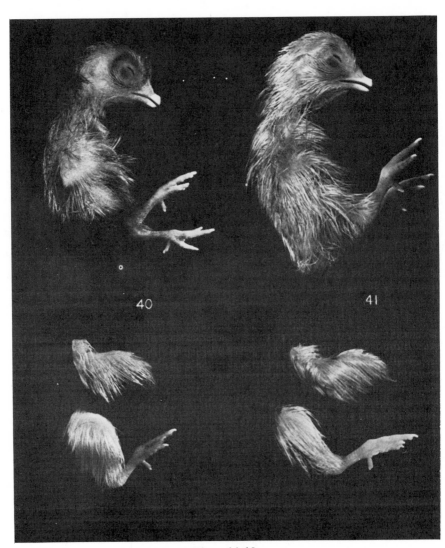

Fig.   11.19.

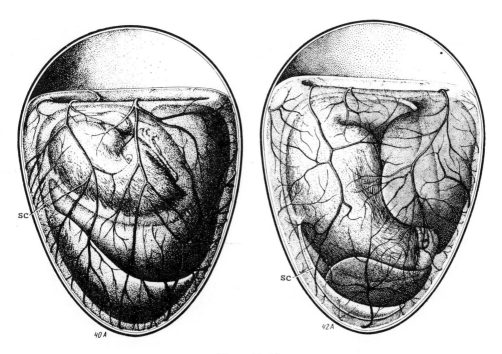

Fig. 11.20.

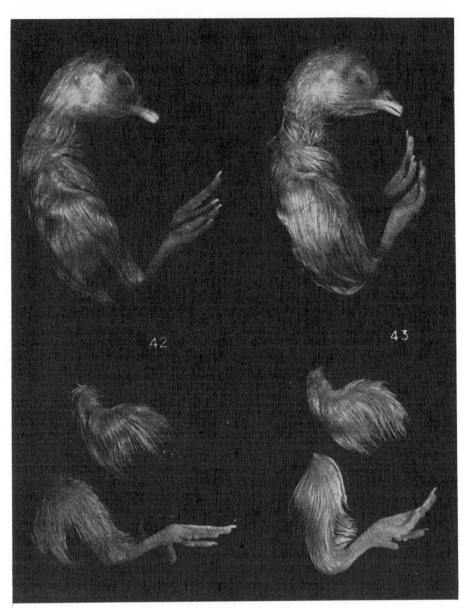

Fig. 11.21. Arrows show third toe.

Fig. 11.22.

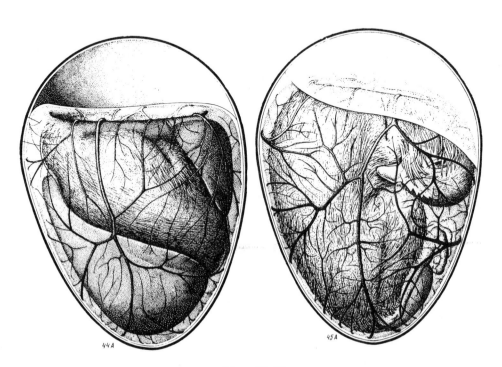

Fig.   11.23.

millimeters be appended to the number of the stage for further precision (e.g., "stage 5–0.2" would designate a notochordal blastoderm with notochord 0.2 mm in length).

Stage 6. Head fold (23–25 h). A definite fold of the blastoderm anterior to the notochord now marks the anterior end of the embryo proper. No somites have yet appeared in the mesoderm lateral to the notochord. This is a transitory stage, since the head fold and the first pair of somites develop rather closely in time. (Most embryos obtained after 24 h under the conditions of industrial incubation are at stages 6–7, minus-variants are at stage 5, and plus-variants are at stages 8–9.)

Stages 7–14 are based primarily on the numbers of pairs of somites that are clearly visible. The number of somites appears to be the simplest criterion for staging this phase of development, and it is sufficiently accurate for practical purposes. A stage is assigned to every third pair of somites which is added; embryos with in-between numbers of somites are designated by adding a + or – sign to the appropriate stage. Thus, stage 7 designates an embryo with one pair of somites; stage 7+, two pairs; stage 8–, three pairs; stage 8, four pairs, etc.

Stage 7. One pair of somites (23–26 h). This is actually the second pair of somites of the series; the first pair is not yet clearly defined. Neural folds are visible in the region of the head.

Stage 8. Four pairs of somites (26–29 h). Neural folds meet at the level of the midbrain. Blood islands are present in posterior half of blastoderm.

Stage 9. Seven pairs of somites (29–33 h). Primary optic vesicles are present. Paired primordia of the heart begin to fuse. (Arrhythmic heartbeat begins.)

Stage 10. Ten pairs of somites (33–38 h). The first pair of somites is becoming dispersed; it is not included in the counts for subsequent stages.* First indication of cranial flexure. Three primary brain vesicles are clearly visible. Optic vesicles not constricted at bases. Heart bent slightly to the right.

Stage 11. Thirteen pairs of somites (40–45 h). Slight cranial flexure. Five neuromeres of hindbrain are distinct. Anterior neuropore is closing. Optic vesicles are constricted at bases. Heart bent to the right.

Stage 12. Sixteen pairs of somites (45–49 h). Head is turning onto left side. Anterior neuropore closed. Telencephalon indicated. Primary optic vesicles and optic stalk well established. Auditory pit is deep, but wide open. Heart is slightly S-shaped. Head fold of amnion covers entire region of forebrain.

Stage 13. Nineteen pairs of somites (48–52 h). Head is partly to fully turned to the left. Cranial and cervical flexures make broad curves. Distinct enlargement of telencephalon. Slight narrowing of opening to deep auditory pit. No indication of hypophysis. Atrioventricular canal indicated by constriction. Head fold of amnion covers forebrain, midbrain, and anterior part of hindbrain. *(Most embryos obtained after 48 h of industrial incubation are at stage 13; minus variants are at stages 11–12 and plus variants at stage 15. Typical embryos at this stage are characterized by the onset of formation of blood vessels in the mesoderm of the forming yolk sac. Transverse axis of its vascular field now equals 1.5 cm.)*

Stage 14. Twenty-two pairs of somites (50–53 h). Cranial flexure: axes of forebrain and hindbrain form approximately a right angle. Cervical flexure a broad

---

*It is suggested that embryos which have gained one pair of somites beyond stage 10, but have lost somite 1 in the meantime, be designated as stage 10±; stage 10+ would then have 11 somites, not counting the rudimentary one; stage 11– = 12 somites, not counting the rudimentary one, etc.

curve. Rotation of body back as far as somites 7–9. Behind this level, a slight flexure makes its appearance, which will be referred to as "trunk flexure." Visceral arches 1 and 2 and visceral clefts 1 and 2 are distinct. Posterior arches not distinct. Primary optic vesicle begins to invaginate; lens placode is formed. Opening of auditory pit constricted. Rathke's pouch can be recognized. Ventricular loop of heart now ventral to atrioventricular canal. Amnion extends to somites 7–10.

Beyond stage 14 the number of somites becomes increasingly difficult to determine with accuracy. This is due in part to the dispersal of the mesoderm of the anteriormost somites and, in later stages, to the curvature of the tail. Total somite counts given for the subsequent stages are typical but variable enough not to be diagnostic. For these reasons, the limb buds, visceral arches, and other externally visible structures are used as identifying criteria from stage 15 onward.

Stage 15 (about 50–55 h). Lateral body folds extend to anterior end of wing level (15th–17th pair of somites). Limb primordia: prospective limb areas flat, not yet demarcated. Inconspicuous condensation of mesoderm in wing level. Somites: 24–27 pairs. Amnion extends to 7th–14th pair of somites. Cranial flexure: axes of forebrain and hindbrain form an acute angle. The ventral contours of forebrain and hindbrain are nearly parallel. Cervical flexure a broad curve. The trunk is distinct. Rotation extends to 11th–13th pair of somites. Visceral arches: arch 3 and cleft 3 are distinct; the latter is shorter than cleft 2 and usually oval in shape. Optic cup is completely formed; double contour distinct in region of iris.

Stage 16 (about 51–56 h). Lateral body folds extend to somites 17–20, between levels of wings and legs. Limbs: wing is lifted off blastoderm by infolding of lateral body fold. It is represented by a thickened ridge. Primordium of leg is still flat, represented by a condensation of mesoderm. Somites: 26–28 pairs. Amnion extends to 10th–18th pair of somites. All flexures are more accentuated than in stage 15. Rotation extends to 14th–15th pair of somites. Tail bud a short, straight cone, delimited from blastoderm. Visceral arches: cleft 3 still oval in shape. Forebrain lengthened; constrictions between brain parts are deepened. Epiphysis indistinct or not yet formed.

Stage 17 (about 52–64 h). Lateral body folds extend around the entire circumference of the body. Limb buds: both wing and leg buds lifted off blastoderm by infolding of the body folds. Both are distinct swellings of approximately equal size (Fig. 11.7). Somites: 29–32 pairs. Amnion: considerable variability, ranging from a condition in which posterior trunk and tail, from approximately 26th pair of somites, are uncovered, to complete closure except for an oval hole over 28th–36th pairs of somites. Intermediate stages with an anterior fold covering as far back as 25th pair of somites and a posterior fold covering part of the tail are common. Cranial flexure is unchanged. Cervical flexure is more sharply bent than in preceding stages, but its angle is still larger than 90°. Trunk flexure is distinct in brachial level. Rotation extends to 17th–18th pair of somites. Tail bud bent ventrad, its mesoderm unsegmented. Epiphysis a distinct knob. Indication of nasal pits. Allantois not yet formed.

Stage 18 (about 65–69 h). Limb buds enlarged; leg buds slightly larger than wing buds. L/W of wing = 6 or <6 (L = length = anterior–posterior dimension as measured along the body wall; W = width = distance from body wall to apex; see stage 20, Fig. 11.8). Somites: 30–36 pairs; extend beyond level of leg bud. Amnion: usually closed; occasionally an oval hole in lumbar region. At the cervical flexure, the axis of the medulla forms approximately a right angle to the axis of the posterior trunk. The trunk flexure has shifted to the lumbar region. The rotation extends now to the posterior part of the body; hence, the leg buds are no longer in the horizontal plane. The tail bud is turned to the right, at about an angle of 90° to the axis of the posterior trunk. Visceral arches: maxillary process absent or inconspicuous. Fourth visceral cleft indistinct or absent. Allantois: a short, thick-walled pocket, not yet vesicular.

Stage 19 (about 68–72 h). Limb buds: enlarged, symmetrical. Leg buds slightly larger and bulkier than wing buds (see Fig. 11.8). *L/W* of wing buds = 4–6. Somites: 37–40 pairs; extend into tail, but the end of the tail which is directed forward is unsegmented. In the cervical flexure the axis of the medulla forms an acute angle with the axis of the trunk. The trunk flexure has nearly or entirely disappeared due to the rotation of the entire body. The contour of the posterior part of the trunk is straight to the base of the tail. Tail bud curved, its tip pointing forward. Visceral arches: the maxillary process is a distinct swelling of approximately the same length as the mandibular process. Visceral cleft 1 is an open narrow slit at its dorsal part. It continues into a shallow furrow. Arch 2 projects slightly over the surface. Cleft 4 is a fairly distinct slit at its dorsal part and continues ventrally as a shallow groove; it does not perforate into the pharynx as a true (open) cleft but is, nevertheless, homologous to the other three clefts. Allantois: a small pocket of variable size; not yet vesicular. Eyes unpigmented. (*Most embryos at the age of 72 h of incubation are at stage 18–19, minus variants are at stage 17, and plus variants at stage 20. In typical embryos at this stage the mesodermal layer of the yolk sac (vascular field) attains 2.5 cm in cross section. Embryo is covered by the egg vitelline membrane. The latter is markedly stretched due to a significant increase of the total yolk mass at the expense of the liquid fraction of the layer transported from the albumen.*)

Stage 20 (70–72 h). Limb buds: enlarged. The wing buds are still approximately symmetrical; the leg buds are slightly asymmetrical (see Fig. 11.8). *L/W* of wing = 3–3.9; *L/W* of leg = 3–2.3. Somites: 40–43 pairs. Tip of tail still unsegmented. Cervical flexure more accentuated than at stage 19. The bend in the tail region begins to extend forward into the lumbosacral region. Contour of midtrunk a straight line. Rotation completed. Visceral arches: maxillary process distinct, equals or exceeds the mandibular process in length; arch 2 projects over surface. Arch 4 less prominent and smaller than arch 3. Cleft 4 shorter than cleft 3; a narrow slit at its dorsal part, continuing into a shallow groove. Allantois: vesicular, variable in size; on the average, of the size of the midbrain. Eye pigment: a faint grayish hue.

Stage 21 (about 3.5 days). Limbs: enlarged; both wing and leg buds are slightly asymmetrical; their proximodistal axes are directed caudad, and the apex of the bud lies posterior to the midline, bisecting the base of the bud. The posterior contours of wing and leg buds are steeper than the anterior contours; they meet the baseline at an angle of approximately 90°. *L/W* of wing = 2.3–2.7; *L/W* of leg = 2.0–2.5. Somites: 43–44 pairs; extreme tip of tail unsegmented. The posterior curvature includes the lumbosacral region. The dorsal contour of the trunk is straight or slightly bent. Visceral arches: maxillary process is definitely longer than mandibular process, extending approximately to the middle of the eye; arch 2 extends distinctly over the surface and overlaps arch 3 ventrally; arch 4 distinct; cleft 4 visible as a slit. Allantois: variable, usually larger than in stage 20; may extend to head. Eye pigmentation faint.

Stage 22 (about 3.5 days). Limbs: elongated buds, pointing caudad. The anterior and posterior contours are nearly parallel at their bases (see Fig. 11.9). *L/W* of wing = 1.5–2; *L/W* of leg = 1.3–1.8. Somites: extend to tip of tail. Flexures: little change. The dorsal contour of the trunk is a straight line or curved. Visceral arches: little change compared with stage 21; maxillary process enlarged; cleft 4 distinct as a slit. Allantois: variable in size; extends to head and may overlap the forebrain. Eye pigmentation: distinct.

Stage 23 (about 3.5–4 days). Limbs: longer than at stage 22; in particular, the proximal parts in which anterior and posterior contours run parallel are lengthened; otherwise, little change in shape. Both wing and leg buds approximately as long as they are wide. Visceral arches (see Figs. 11.9 and 11.10): maxillary process is lengthened further; visceral cleft 1 is represented by a broken line, its dorsal part is a distinct slit, and a slight protuberance (*a*) is noticeable anterior to the dorsal slit; the caudal part of arch 2 is distinctly elevated over the surface; arches 3 and 4 are still completely exposed; visceral cleft 3 is a distinct groove, and cleft 4 is reduced to a narrow oval pit at its dorsal end. Flexures: the dorsal contour from hindbrain to tail is a curved line.

Stage 24 (about 4 days). Limbs: wing and leg buds distinctly longer than wide; digital plate in wing not yet demarcated; toe plate in leg bud distinct; toes not yet demarcated. Visceral arches (see Figs. 11.9 and 11.10): visceral cleft 1 a distinct curved line; slight indication of two protuberances (*a* and *b*) on mandibular process and of three protuberances (*d, e, f*) on arch 2; part *c* of mandibular process is receding. Arch 2 longer ventrally (at *f*) and much wider than mandibular process; arch 3 reduced and partly overgrown by arch 2; arch 4 flattened; both are sunk beneath the surface. Visceral cleft 3 is an elongated groove; visceral cleft 4 reduced to a small pit. (*Most embryos at the age of 4 days of incubation are at stage 24; minus variants are at stage 20 and plus variants at stage 25. The typical embryos at this stage (see Figs. 11.9 and 11.24) are characterized by the following structural features of extraembryonic organs: amnion outlines closely the body of the embryo and there is a small amount of amniotic fluid; allantois is a small sac with equally thick walls, and from above, it comes in contact with serosa and forms chorioallantois; vitelline membrane discarded, its residue and chalazae under the bottom of forming yolk sac.*)

Stage 25 (about 4.5 days). Limbs: elbow and knee joints distinct (in dorsal or ventral view); digital plate in wing distinct, but no demarcation of digits; indication of faint grooves demarcating the third toe on leg. Visceral arches (see Figs. 11.9 and 11.10): maxillary process lengthened; it meets the wall of the nasal groove (note the notch at the point of fusion). Three protuberances on each side of first visceral cleft (*a* to *f*); in dorsal view, *a, b,* and *d* appear as round knobs, and *c* as a flat ridge; part *f* is conspicuous and projects distinctly over the surface; it will be referred to as the "collar." Dorsal part of arch 3 still visible. Visceral clefts 3 and 4 reduced to small circular pits.

Stage 26 (about 4.5–5 days). Limbs: considerably lengthened; contour of digital plate rounded; indication of faint groove between second and third digit; demarcation of the first three toes distinct. Visceral arches (see Figs. 11.10 and 11.11): contour of maxillary process a broken line; mandibular process lengthened ventrally. Protuberances *a* and *b* project over the surface; the middle protuberance (*b*) is subdivided by a shallow groove; a small knob is distinct at the dorsal edge of *c*; on arch 2, protuberances *d* and *e* are only slightly elevated over the surface; the "collar" (*f*) has broadened and overgrown visceral arches 3 and 4; a deep groove separated *f* from *c*; the two pits representing visceral clefts 3 and 4 are no longer visible.

Stage 27 (about 5 days). Limbs: contour of digital plate angular in region of first digit; grooves between first, second, and third digits indicated; grooves between toes are distinct on outer and inner surfaces of toe plate; first toe projects over the tibial part at an obtuse angle; tip of third toe not yet pointed. Visceral arches (see Figs. 11.10 and 11.11): contour of maxillary process is a curved, broken line; mandibular process has broadened ventrally (at *c*) and grown forward; protuberances *a* and *b* project over the surface; parts *d* and *e* are flat; protuberances *b* and *e* are close to fusion, but a separating line is still distinct; the collar (*f*) has broadened and continued its growth backward; it rises conspicuously above the surface; the groove between *c* and *f* has widened. Beak: barely recognizable. (*Most embryos at the age of 5 days of incubation are at stage 26 (see Figs. 11.11 and 11.25); minus variants are at stage 25 and plus variants at stage 27. Typical embryos are characterized by the following structural features of extraembryonic organs: amnion is a thin-walled vesicle filled with amniotic fluid, its wall contracting rhythmically; allantois completely covers the amniotic vesicle or exceeds it slightly in size; yolk sac markedly enlarged due to a large amount of liquid yolk fraction in it; its vascular field (upon horizontal position of the egg) comes in contact with the outer shell membrane at the air chamber upper margin; albumen condensed, its volume reduced to half.*)

Stage 28 (about 5.5 days). Limbs: second digit and third toe longer than others, which gives the digital and toe plates a pointed contour; 3 digits and 4 toes distinct; no indication of fifth toe. Visceral arches (see Figs. 11.10 and 11.11): protuberance *a* still projects over the surface; mandibular process has lengthened and grown forward; parts *b* and *e* have fused – a fine suture line is occasionally still visible; parts *b, d,* and *e* no longer project above the surface. External auditory opening is now very distinct between *a, b,* and *d.* Collar (*f*) projects distinctly over the surface; the neck between the collar and mandible has lengthened. Beak: a distinct outgrowth is visible in profile.

Stage 29 (about 6 days). Limbs: wing bent in elbow; second digit distinctly longer than the others; shallow grooves between first, second, and third digits; second to fourth toes stand out as ridges separated by distinct grooves, and with indi-

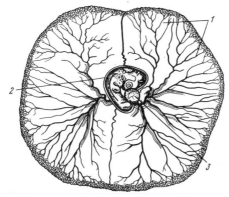

Fig. 11.24. Diagram of the embryo with extraembryonic membranes at the age of 4 days; stage 24. 1) Yolk sac vessels; 2) amnion; 3) allantois.

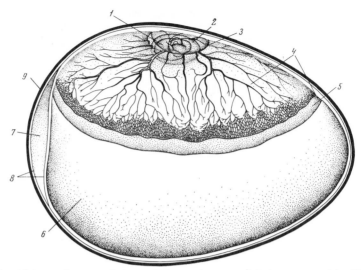

Fig. 11.25. Diagram of the embryo at the age of 5 days, stage 26. The position of the chick embryo is shown with reference to the extraembryonic organs and egg envelopes at the horizontal egg position. 1) Embryo; 2) amniotic cavity; 3) boundary of allantois (designated in the figures as X); 4) yolk-sac vascular field (designated in the figures as 0); 5) yolk-sac ectoderm; 6) albumen; 7) air chamber; 8) two-shell membranes; 9) egg shell.

cations of webs between them; distal contours of webs are straight lines, occasionally with indication of convexity; rudiment of fifth toe visible. Visceral arches: mandibular process lengthened (compare with stage 28); mandibular process and arch 2 are broadly fused. Auditory meatus distinct at dorsal end of fusion; all protuberances have flattened; neck between collar and mandibular process has lengthened; collars stand out conspicuously. Beak: more prominent than in stage 28; no egg tooth visible as yet. (*Typical embryos at the age of 6 days of incubation are at stage 29; minus variants are at stage 28 and plus variants at stage 30. Amnion contracts intensively and periodically. Its size, and hence the volume of amniotic fluid, are subject to marked variations. Allantois now covers the whole central part of yolk sac vacular field. Chorioallantois as yet without a fine net of capillaries. Yolk sac vacular field reaches the egg shell equator, and in the area of the air chamber lines its internal wall by 1/5 of its length.*)

Stage 30 (about 6.5 days). Limbs: the three major segments of wing and leg are clearly demarcated; wing bent in elbow joint; leg bent in knee joint; distinct grooves between first and second digits; contours of webs between first two digits and between all toes are slightly curved concave lines. Visceral arches: the mandibular process approaches the beak, but the gap between the two is still conspicuous; lengthening of neck between collar and mandible is very conspicuous; collar begins to flatten. Feather germs: two dorsal rows to either side of the spinal cord at the branchial level; three rows at the level of the legs; they are rather indistinct at thoracic level; none on thigh. Scleral papillae: one on either side of choroid fissure; sometimes indistinct but never more than two. Egg tooth distinct, slightly protruding. Beak more pronounced than in previous stage.

Stage 31 (about 7 days). Limbs: indication of a web between first and second digits; rudiment of fifth toe still distinct. Visceral arches: the gap between mandible and beak has narrowed to a small notch; collar inconspicuous or absent. Feather germs: on dorsal surface, continuous from brachial to lumbosacral level; approximately 7 rows at lumbosacral level; distinct feather papillae on thigh; one indistinct row on each lateral edge of the tail. Scleral papillae: usually 6, 4 on the dorsal side near the choroid tissues, and 2 on the opposite side. (*Most embryos at the age of 7 days of incubation are at stage 31. State of extraembryonic organs and egg envelopes: amnion contracts intensively; allantois covers over 1/3 of yolk sac vascular field area; albumen condensed, its weight amounting to 20% of the total egg weight.*)

Stage 32 (about 7.5 days). Limbs: all digits and four toes have lengthened conspicuously; rudiment of fifth toe has disappeared; webs between digits and toes are thin and their contours are concave; differences in size of individual digits and toes become conspicuous. Visceral arches: anterior tip of mandible has reached the beak; collar has disappeared or is faintly recognizable. Feather germs: eleven rows or more on dorsal surface at level of the legs; one row on tail distinct, second row indistinct; scapular and flight feather germs barely perceptible at optimal illumination, or absent. Scleral papillae: 6 to 8, in two groups, one group on dorsal and one on ventral side; circle not yet closed.

Stage 33 (about 7.5–8 days). Limbs: web on radial margin of arm and first digit becomes discernible; all digits and toes lengthened. Visceral arches: mandible and neck have lengthened conspicuously (compare the ventral contour of the body, from the heart region along neck to tip of mandible in this and the preceding stages). Feather germs: scapular and flight feather germs not much advanced over stage 32; in tail, three rows distinct, the middle row considerably larger than the others. Scleral papillae: 13, forming an almost complete circle, with gap for one missing papilla at a ventral point near the middle of the jaw.

Stage 34 (about 8 days). Limbs: differential growth of second digit and third toe conspicuous; contours of webs between digits and toes are concave and arched. Visceral arches: lengthening of mandible and of neck continues (see previous stage). Feather germs: on scapula, on ventral side of neck, on procoracoid, and posterior (flight) edge of wing, feather germs are visible under good illumination; feather germs next to dorsal midline, particularly at lumbosacral level, extend slightly over surface when viewed in profile; those on thigh protrude conspicuously; one row on inner side of each eye; none around umbilical cord. Scleral papillae: 13 or 14; nictitating membrane extends halfway between outer rim of eye (eyelid) and scleral papillae. (*Most embryos at the age of 8 days of incubation are at stage 34. Extraembryonic organs and egg envelopes: amnion is filled with a large amount of fluid, and its wall contracts intensively and rhythmically; allantois covers the whole yolk sac vascular field surface or a small part of its lower margin remains uncovered; in the region of obtuse egg end, allantois comes into contact with the outer shell membrane of air chamber along 1/2 of its internal wall length; a fine small-looped net of capillaries has developed in chorioallantois; yolk sac decreases in volume due to disappearance of a part of the liquid yolk fraction; its weight with yolk amounts to 60% of total egg weight; albumen maximally condensed (see Fig. 11.14).*)

Stage 35 (about 8–9 days). Limbs: webs between digits and toes become inconspicuous; a transitory protuberance on the ulnar side of the second digit is probably a remnant of the web; phalanges in toes are distinct. Visceral arches: lengthening of beak continues; compare the distance between the eye and the tip of the beak in this and the preceding stages. Feather germs: all are more conspicuous; middorsal line stands out distinctly in profile view; at least 4 rows on inner side of each eye; new appearance of feather germs near midventral line, close to sternum, and extending to both sides of umbilical cord. Nictitating membrane has grown conspicuously and approaches the outer scleral papillae; eyelids (external to nictitating membrane) have extended toward the beak and have begun to overgrow the eyeball; the circumference of the eyelids has become ellipsoidal.

Stage 36 (about 10 days). Limbs: distal segments of both wing and leg are proportionately much longer; length of third toe from its tip to the middle of its metatarsal joint = 5.4 ± 0.3 mm; tapering primordia of claws are just visible on termini of the toes and on digit 1 of the wing; protuberance on posterior side of digit 2 of wing is missing. Visceral arches: primordium of the comb appears as a prominent ridge with slightly serrated edge along the dorsal midline of the beak; a horizontal groove (the "labial groove") is clearly visible at the tip of the upper jaw, but is barely indicated on the tip of the mandible; nostril has narrowed to a slit; length of beak from anterior angle of nostril to tip of bill = 2.5 mm. Feather germs: flight feathers are conspicuous; coverts are just visible in web of wing; feather germs now cover the tibiofibular portion of the leg; at least 9–10 rows of feather germs between each upper eyelid and the dorsal midline; sternal tracts prominent, with 3–4 rows on each side of ventral midline when counted in anterior part of sternum, merging into many rows around the umbilicus. Eyelids: nictitating membrane covers anteriormost scleral papillae and approaches cornea; lower lid has grown upward to level of cornea; circumference of lids is a narrowing ellipse with its ventral edge flattened. *(Extraembryonic organs in typical embryos; amniotic cavity filled with a large amount of amniotic fluid; amnion still contracts rhythmically; allantois spread markedly over the surface of albumen concentrated under the yolk sac bottom (upon vertical position of the egg in the region of egg shell sharp end), its margins not yet closed; the weight of the yolk sac with yolk amounts to 41% of the total egg weight (see Fig. 11.18).)*

Stage 37 (about 11 days). Limbs: claws of toes are flattened laterally and curved ventrally; dorsal tips are opaque, indicating onset of cornification; tip of claw on wing is also opaque; pads on plantar surface of foot are conspicuous; transverse ridges along the superior surfaces of the metatarsus and phalanges are first indication of scales; length of third toe = 7.4 ± 0.3 mm. Visceral arches: labial groove on mandible is now clearly marked off. The comb is more prominent and clearly serrated; length of beak from anterior angle of nostril to tip of bill = 3.0 mm. Feather germs: much more numerous and in most-advanced tracts (e.g., along back and on tail) elongated into long, much-tapered cones; external auditory meatus is nearly surrounded by feather germs; circumference of eyelids is bordered by a single row of just-visible primordia; none on remainder of lids; sternal tracts contain 5–6 prominent rows when counted at anterior end of sternum. Eyelids: nictitating membrane has reached anterior edge of cornea; upper lid has reached dorsal edge of cornea; lower lid has covered 1/3 to 1/2 of cornea; circumference of lids now bounds a much-narrowed and ventrally flattened biconvex area.

Stage 38 (about 12 days). Limbs: primordia of scales are marked off over entire surface of leg; ridges have not yet grown out to overlap surface; tips of toes show a ventral center of cornification as well as a more extensive dorsal center; main plantar pad is ridged when seen in profile; length of third toe = 8.4 ± 0.3 mm (see Fig. 11.17). Visceral arches: labial groove marked off by a deep furrow at the end of each jaw; length of beak from anterior angle of nostril to tip of bill = 3.1 mm. Feather germs: coverts of web of wing are becoming conical; external auditory meatus is surrounded by feather germs; sternum is covered with feather germs except along midline; upper eyelid is covered with newly formed feather germs; lower lid is naked except for 2–3 rows at its edge. Eyelids: lower lid covers 2/3 to 3/4 of cornea; opening between lids is much reduced. (*Extraembryonic organs: amnion wall no longer contracts, and, in the region of dorsal side of embryo, the amniotic vesicle communicates with the albumen sac cavity via a wide but as yet closed serosoamniotic canal; allantois margins embrace the whole albumen mass and fuse in the region of egg shell sharp end; liquid fraction absent in yolk sac; the weight of yolk sac with yolk amounts to 34% of the total egg weight (see Fig. 11.26).*)

Stage 39 (about 13 days). Limbs: scales overlapping on superior surface of leg; major pads of phalanges covered with papillae; minor pads are smooth; length of third toe = 9.8 ± 0.3 mm. Visceral arches: mandible and maxilla cornified (opaque) back as far as level of proximal edge of egg tooth; the channel of the auditory meatus can be seen only at the posterior edge of its shallow external opening; length of beak from anterior angle of nostril to tip of bill = 3.5 mm. Feather germs: coverts of web of wing are very long tapering cones; note great increase in length of feather germs in major tracts; four to five rows of feather germs at edge of lower eyelid. Eyelids: opening between lids reduced to a thin crescent.

Stages 40–44 are based mainly on the length of the beak and on the length of the third (longest) toe, since other external features have lost their diagnostic value.

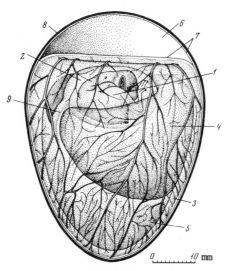

Fig. 11.26. Diagram of the embryo at the age of 12 days, stage 39. The position of the embryo is shown with reference to the extraembryonic organs and egg envelopes at the vertical egg position. 1) Embryo; 2) amniotic cavity; 3) allantois; 4) yolk sac with allantois; 5) albumen residue under allantois; 6) air chamber; 7) shell membranes; 8) egg shell; 9) serosoamniotic canal.

Of these two criteria, the length of the beak is the better, because it is more easily and accurately measured (with calipers) and shows less variability.

Stage 40 (about 14 days). Visceral arches: length of beak from anterior edge of nostril to tip of bill = 4.0 mm; the main channel of the auditory meatus is not visible in strictly lateral view of its external chamber. Limbs: length of third toe = 12.7 ± 0.5 mm; scales overlapping on inferior as well as superior surfaces of leg; dorsal and ventral loci of cornification extend to base of exposed portion of toenail; entire plantar surface of phalanges is covered with well-developed papillae. (*Extraembryonic organs: amnion markedly stretched, amniotic fluid contains albumen and the albumen mass decreased correspondingly; meconium is found in the allantoic fluid as whitish jelly inclusions; chorioallantoic capillaries have approached markedly to the outer shell membrane internal surface; weight of yolk sac with its contents amounts to 30% of the total egg weight (see Fig. 11.20). The position of chick body axis changed with respect to the long egg axis: they have approached.*)

Stage 41 (about 15 days). Beak: length from anterior angle of nostril to tip of upper bill = 4.5 mm. Third toe: length = 14.9 ± 0.8 mm.

Stage 42 (about 16 days). Beak: length from anterior angle of nostril to tip of upper bill = 4.8 mm. Third toe: length = 16.7 ± 0.8 mm. (*Extraembryonic organs: amniotic fluid contains a great amount of albumen; the albumen sac contains a small amount of albumen or none at all; yolk sac divided into three distinct lobes; its weight with yolk amounts to 26% of the total egg weight (see Fig. 11.20).*)

Stage 43 (about 17 days). Beak: length from anterior angle of nostril to tip of upper bill = 5.0 mm. "Labial grooves" are reduced to a white granular crust at the edge of each jaw; that of the lower jaw may be partially or completely sloughed off. Third toe: length = 18.6 ± 0.8 mm.

Stage 44 (about 18 days). Beak: length from anterior angle of nostril to tip of upper bill = 5.7 mm. The translucent peridermal covering of the beak is starting to peel off proximally. Third toe: length = 20.4 ± 0.8 mm. (*Extraembryonic organs: in the typical chick embryos the amnion closely outlines the body since the amount of amniotic fluid is insignificant; yolk sac is overgrown by a thin layer of abdominal wall muscles; no changes in yolk sac size; inner surface of air chamber markedly curved (Fig. 11.23).*)

Stage 45 (about 19–20 days). Beak: length is no longer diagnostic; in fact the beak is usually shorter than in stage 44, due to a loss (by sloughing off) of its entire peridermal covering; as a consequence, the beak is now shiny all over and more blunt at its tip; both labial grooves have disappeared with the periderm. Third toe: average length is essentially unchanged from that of stage 44, except in those breeds with a longer period of incubation (21 days) and a heavier build of body. For these latter, length of third toe is about 21.4 ± 0.8 mm. Extraembryonic membranes: yolk sac is half enclosed in body cavity. Chorioallantoic membrane contains less blood and is "sticky" in the living embryo. (*Extraembryonic organs in typical chicks at the age of 20 days of incubation: amnion can be ruptured by the egg tooth; allantoic vessels are still filled with blood or go to waste partially; allantoic fluid practically absent; one or two yolk sac lobes drawn in the thick abdominal cavity; air chamber attains the maximum size, curvature of its cavity increased; one margin of the outer shell membrane internal layer is at a distance of up to 2/3 of the egg shell long axis from the apex of egg obtuse end (see Fig. 11.23).*)

Stage 46 (20–21 days). Newly hatched chick. Umbilical opening is closed by the fusion of muscle and cutaneous layers at this position.

REFERENCES

1. U. K. Abbott, "Avian developmental genetics," in *Methods in Developmental Biology*, F. H. Wilt and N. Wessels, eds., Crowell (1967), pp. 13–52.
2. E. B. Abramova, Yu. G. Yurovitsky, and L. S. Milman, "Glycogen synthetase isozymes in hen tissues," *Biokhimiya* **47**, 1172–1186 (1982).
3. P. Ancel, "Sur les rapports de l'axe cephalo-caudal de l'embryon de poulet avec le grande axe de l'oeuf," *C. R. Acad. Sci. Paris* **236**, 24 (1953).
4. K. E. von Baer, *Über Entwicklungsgeschichte der Tiere, Beobachtung, und Reflexion, Theil I,* Bornträger, Königsberg (1828).
5. K. E. von Baer, *Über Entwicklungsgeschichte der Tiere, Beobachtung, und Reflexion, Theil II*, Bornträger, Königsberg (1837).
6. V. G. Bekhtina, "Some data on fertilization and initial developmental stages of the chick embryos," in *Proceedings of the Pushkin Research Laboratory of Animal Husbandry* [in Russian], Vol. 7 (1955), pp. 70–73.
7. V. G. Bekhtina, "On morphological characteristics of fertilization in domestic fowl," *Arkh. Anat. Gistol. Embriol.* **35**(1), 92–100 (1958).
8. R. Bellairs, M. Markness, and R. D. Markness, "The vitelline membrane of the hen's egg," *J. Ultrastruct. Res.* **8**, 339–359 (1963).
9. R. Bellairs, "The mechanism of somite segmentation in the chick embryo," *J. Embryol. Exp. Morphol.* **51**, 227–243 (1979).
10. R. Bellairs, "The segmentation of somites in the chick embryo," *Boll. Zool.* **47**, 245–252 (1980).
11. R. Bellairs and M. Veini, "An experimental analysis of somite segmentation in the chick embryo," *J. Embryol. Exp. Morphol.* **55**, 93–108 (1980).
12. R. G. Board, "Nonspecific antimicrobial defenses of the avian egg, embryo, and neonate," *Biol. Rev.* **49**, 15–49 (1974).
13. O. V. Borkovskaya, "Estimation of the rate of yolk formation in the hen's egg," *Fiziol. Zh. SSSR* **40**, 722–726 (1954).
14. A. K. Dondua, "Phagocytic and inflammation reactions at different phases of ontogenesis. Experiments on chick embryos," *Dokl. Akad. Nauk SSSR* **104**, 941–944 (1955).
15. A. K. Dondua, "On changes in permeability of blood vessels in the focus of inflammation at different phases of hen ontogenesis," *Nauchn. Dokl. Vyssh. Shkoly, Biol. Nauki* **1**, 56–59 (1958).
16. M. Duval, *Atlas d'Embryologie*, Paris (1889).
17. V. I. Efremov, "Changes in proliferative activity at the early developmental stages of the chick embryo," *Vestn. Leningr. Gos. Univ.* **3**, 30–43 (1968).
18. V. I. Efremov and V. I. Sergovskaya, "Proliferative activity and kinetics of cell populations of the chick blastoderm during gastrulation and early organogenesis. I. Proliferative activity and mitotic cycle patterns in the cells," *Sov. J. Dev. Biol.* **10**, 448–460 (1979).
19. V. I. Efremov, I. N. Morozova, and T. V. Sergovskaya, "Proliferative activity and kinetics of cell populations of the chick blastoderm during gastrulation and early organogenesis. II. Mitotic index, S-phase index, and composition of cell populations," *Sov. J. Dev. Biol.* **11**, 246–250 (1980).
20. H. Emmanuelson, "Cell multiplication in the chick blastoderm up to the time of laying," *Exp. Cell Res.* **39**, 386–399 (1965).

21. H. Eyal-Giladi and S. Kochav, "From cleavage to primitive streak formation; a complementary normal table and a new look at the first stages of development of the chick," *Dev. Biol.* **49**, 321–337 (1976).
22. R. M. Fraps, "Egg production and fertility in poultry," in *Progress in the Physiology of Farm Animals*, Vol. 2, J. Hammond, ed., Butterworths, London (1955).
23. V. Hamburger, *A Manual of Experimental Embryology*, University of Chicago Press, Chicago (1947).
24. V. Hamburger and H. L. Hamilton, "A series of normal stages in the development of chick embryo," *J. Morphol.* **88**, 49–92 (1951).
25. V. Hamburger and R. Oppenheim, "Prehatching motility and hatching behavior in the chick," *J. Exp. Zool.* **166**, 171–204 (1967).
26. H. L. Hamilton, *Lillie's Development of the Chick*, H. Holt and Co., New York (1952).
27. B. W. Heywang, "The time factor in egg production," *Poultry Sci.* **17**, 240 (1938).
28. A. Huettner, "The embryology of the chick," *in Fundamentals of Comparative Embryology*, Macmillan, New York (1950), pp. 142–233.
29. H. Kionka, "Die Fürchungen des Hühnereis," *Anat. Heft.* **3** (1894) [cited by Lillie (1930)].
30. G. P. Korotkova, "Antibiotic properties of the hen's egg white (with reference to the problem of embryonic immunity)," *Zh. Obshch. Biol.* **18**, 275–286 (1957).
31. E. B. Krichinskaya, "Phagocytic activity of endothelium of some organs of the chick embryo at different developmental stages," *Byull. Eksp. Biol. Med.* **40(8)**, 57–59 (1955).
32. E. B. Krichinskaya, "Phagocytic reactions to India ink and microorganism during the development of chick and duck embryos," *Nauchn. Dokl. Vysshei. Shkoly, Biol. Nauki* **1**, 48–50 (1959).
33. E. B. Krichinskaya and V. I. Efremov, "Avians," in *Methods in Developmental Biology*, V. Ya. Brodsky et al., eds. [in Russian], Nauka, Moscow (1974), pp. 20–216.
34. I. V. Kudryavtsev, Z. A. Oschchepkova, V. D. Antal, and E. A. Ptashkina, "On susceptibility of the RSV infected and aleukemie fowl to parthenogenesis," *Zh. Obshch. Biol.* **42**, 467–469 (1981).
35. I. V. Kudryavtsev, A. A. Yakovlev, L. D. Shirnova, and M. D. Pigareva, "Embryogenesis of the Japanese quail (*Coturnix coturnix japonica*) and classification of incubation waste," in *Problems of Genetics, Selection, and Immunogenetics of Animals* [in Russian], Nauka, Moscow (1972), pp. 189–204.
36. F. R. Lillie, *The Development of the Chick*, H. Holt and Co., New York (1930).
37. E. H. McNally, "The origin and structure of the vitelline membrane of the domestic fowl's egg," *Poultry Sci.* **22**, 40–43 (1943).
38. S. Meier, "Development of the chick embryo mesoblast. Formation of the embryonic axis and establishment of metameric pattern," *Dev. Biol.* **73**, 24–45 (1979).
39. L. S. Milman, E. B. Kuzmishcheva, and Yu. G. Yurovitsky, "Glycogen phosphorylase and glycogen synthetase isozymes in embryonic tissues," *Adv. Clin. Enzymol.* **2**, 175–182 (1981).

40. J. Moran and H. P. Hale, "Physics of the hen's egg. I. Membranes in the egg," *J. Exp. Biol.* **13**, 35–40 (1936).

41. E. J. Moticka, "Development of immunological competence in chicken," *Am. Zool.* **15**, 135–146 (1975).

42. B. H. Neher, M. W. Olsen, and R. W. Fraps, "Ovulation of the excised ovum of the hen," *Poultry Sci.* **29**, 554–557 (1950).

43. D. U. T. New, *The Culture of Vertebrate Embryos*, Academic Press, New York (1966).

44. G. Nicolet, "Analyse autoradiographique de localisation des différentes ébauches présomptives dans la ligne primitive de l'embryon de Poulet," *J. Embryol. Exp. Morphol.* **23**, 79–108 (1970).

45. I. E. Novik, *Biology of Reproduction and Artificial Insemination of Domestic Fowl* [in Russian], Izd. Akad. Nauk SSSR, Moscow (1964).

46. M. W. Olsen, "Maturation, fertilization, and early cleavage in the hen's egg," *J. Morphol.* **70**, 513–533 (1942).

47. M. W. Olsen, "Early-ovarian fertilization in the fowl," *Poultry Sci.* **30**, 6 (1951).

48. M. W. Olsen and E. G. Buss, "Role of genetic factors and fowl pox virus in parthenogenesis of turkey eggs," *Genetics* **56**, 727–732 (1967).

49. M. W. Olsen and R. M. Fraps, "Maturation changes in the hen's ovum," *J. Exp. Zool.* **114**, 475–487 (1950).

50. G. K. Otryganiev, V. A. Khmyrov, and G. M. Kolobov, *Incubation* [in Russian], Kolos, Moscow (1964).

51. C. S. Pagett and W. D. Ivey, "The normal embryology of the *Coturnix* quail," *Anat. Record* **137**, 1–12 (1960).

52. J. T. Patterson, "Studies on the early development of the hen's egg. History of the early cleavage and of the accessory cleavage," *J. Morphol.* **21**, 101–134 (1910).

53. R. E. Philips and D. C. Warren, "Observations concerning the mechanics of ovulation in the fowl," *J. Exp. Zool.* **76**, 117–136 (1937).

54. M. D. Pigareva, *The Breeding of Quails* [in Russian], Rosselkhozizdat, Moscow (1978).

55. M. N. Ragozina, "Dependence between the position of duck embryo inside the egg and its position during hatching," *Tr. Inst. Morfol. Zhivotn. Akad. Nauk SSSR* **14**, 286–303 (1955).

56. M. N. Ragozina, *Development of the Chick Embryo and Its Correlation with the Yolk and Egg Envelopes (with the Tables of Normal Development)* [in Russian], Izd. Akad. Nauk SSSR, Moscow (1961).

57. G. Reynaud, "Table et courbe de développement dudindon domestique *Meleagris gallopavo*," *Bull. Soc. Zool. Fr.* **97**, 95–101 (1972).

58. V. V. Rolnik, *Biology of Avian Embryonic Development* [in Russian], Nauka, Moscow (1968).

59. A. L. Romanoff, *The Avian Embryo: Structural and Functional Development*, Macmillan, New York (1960).

60. A. L. Romanoff and A. J. Romanoff, *The Avian Egg*, Wiley, New York (1949).

61. A. L. Romanoff and A. J. Romanoff, *Pathogenesis of the Avian Embryo: An Analysis of Causes of Malformations and Prenatal Death*, Wiley-Interscience, New York (1972).

62. G. C. Rosenquist, "A radioautographic study of labeled grafts in the chick blastoderm. Development from primitive streak stage to stage 12," *Contrib. Embryol. Carnegie Inst.* **38**, 71–110 (1966).
63. R. Rugh, *Experimental Embryology*, Burgess Publ. Co., Minnesota (1962).
64. J. W. Saunders, Jr., "The proximodistal sequence of origin of the parts of chick wing and the role of ectoderm," *J. Exp. Zool.* **108**, 363–404 (1948).
65. V. N. Shalumovich, "A study of the hen's egg vitelline membrane by luminescent and ultraviolet microscopy and some histochemical methods," *Dokl. Akad. Nauk SSSR* **105**, 584–586 (1955).
66. N. T. Spratt, Jr., "Location of organ-specific regions and their relationship to the development of the primitive streak in the early chick blastoderm," *J. Exp. Zool.* **89**, 69–101 (1942).
67. N. T. Spratt, Jr., "Formation of the primitive streak in the explanted chick blastoderm marked with carbon particles," *J. Exp. Zool.* **103**, 259–304 (1946).
68. N. T. Spratt, Jr., "Analysis of the organizer center in the early chick embryo. I. Localization of prospective notochord and somite cells," *J. Exp. Zool.* **128**, 121–164 (1955).
69. N. T. Spratt, Jr., "Analysis of the organizer center in the early chick embryo. II. Studies of the mechanics of notochord elongation and somite formation," *J. Exp. Zool.* **134**, 577–612 (1957).
70. L. Vakaet, "Some new data concerning the formation of the definitive endoblast in the chick embryo," *J. Embryol. Exp. Morphol.* **10**, 38–57 (1962).
71. M. Volk and H. Eyal-Giladi, "The dynamics of antigenic changes in the epiblast and hypoblast of the chick during the processes of hypoblast, primitive streak, and head process formation as revealed by immunofluorescence," *Dev. Biol.* **55**, 33–45 (1977).
72. D. C. Warren and H. N. Scott, "Ovulation in the domestic hen," *Science* **70**, 461–462 (1934).
73. A. A. Yakovlev, "Some peculiarities of the Japanese quail (*Coturnix coturnix japonica*) as a model object for biological studies," in *Problems of Genetics, Selection, and Immunogenetics of Animals* [in Russian], Nauka, Moscow (1972), pp. 169–189.
74. L. P. Yermolayeva, "Factors of gluconeogenesis control in the chick embryo liver," *Biokhimiya* **43**, 1335–1340 (1978).
75. Yu. G. Yurovitsky, L. S. Milman, and I. P. Krivopishin, "Enzymes of glycogen metabolism in the chick embryo liver," *Biokhimiya* **43**, 1602–1615 (1978).
76. A. M. Zacchei, "Lo svilippo embrionale della guaglia giapponese (*Coturnix coturnix japonica* T. et S.)," *Arch. Ital. Anat. Embriol.* **66**, 36–62 (1961).

Chapter 12

LABORATORY MAMMALS: MOUSE *(Mus musculus)*, RAT *(Rattus norvegicus)*, RABBIT *(Oryctolagus cuniculus)*, AND GOLDEN HAMSTER *(Cricetus auratus)*

A. P. Dyban, V. F. Puchkov, N. A. Samoshkina, L. I. Khozhai, N. A. Chebotar', and V. S. Baranov

Mice, rats, rabbits, and hamsters are widely used in developmental biology for embryological, teratological, genetic, cytogenetic, molecular-biological, and many other studies.

Irrespective of the specific problem, it is possible to work successfully with the embryos of these animals only if the investigator has sufficient knowledge of the biological characteristics of each species, including the peculiarities of their reproductive function. Certain guidelines for keeping these animals in the laboratory should be followed; these ensure that a high fertility rate is maintained and will lead to a decreased frequency of spontaneous embryo mortality, as well as of spontaneous inborn developmental abnormalities. It is recommended that embryos with well-known genetic characteristics be used; in other words, genetic information about corresponding strains should be available. Furthermore, it is important to pinpoint the exact date of pregnancy, since this allows the use of uniform groups of embryos corresponding to a certain developmental stage. It is impossible to discuss such a wide range of questions in a single chapter; moreover, it is not even necessary, because the biology of each species of the above-listed laboratory mammals has been discussed in numerous works, where the reader can find countless answers to these questions.

The Biology of the Laboratory Mouse by Earl L. Green [53] is still the primary handbook on laboratory mice. Reference information regarding the biology of laboratory rats, rabbits, and guinea pigs can be found in the series of books published by the American College of Laboratory Animal Medicine (ACLAM). The series includes a book on the rabbit [13], while two volumes are devoted to the rat [14, 15]. Individual chapters of these treatises deal with the genetics of rats and the rules for keeping, feeding, and reproducing them; one chapter deals with embryonic development and teratogenesis in rats [19].

Basic information on the genetics of inbred and random-bred strains and varieties of laboratory animals can be found in the two volumes written by Altman and Katz [3, 4]. The first volume deals with mice and rabbits, and the second with hamsters, guinea pigs, rabbits, and chickens.

Green's handbook [54] gives comprehensive information on the genetics of all mouse strains, while rat genetics is covered comprehensively in Robinson's book [97]. Current information on inbred and mutant strains of mice can be found in *The Mouse Newsletter*, while similar information on rats is being published in *The Rat Newsletter*. These periodicals have been published in Britain since 1977 by the Animal Laboratory Center, MRC. This laboratory also supervises the publication of *The International Index of Laboratory Animals* (edited by M. Festing).

Basic information about keeping laboratory animals can be found in the Porter and Lane-Petter handbook [94; see also 71] and in the treatise edited by Melby and Altman [80]. Useful information can be found in the books by Kovalevsky [68], Burov [29], Gambaryan and Dukelskaya [49], Zapadnyuk et al. [133], and Kozlyakov [69].

The embryonic development of the mouse has been described in a chapter in Green's handbook [53] as well as in the books by Rugh [100] and Theiler [118]. Rafferty's book [95] contains useful information regarding the technique of working with mouse embryos. Information on the embryonic development of rats can be found in a number of studies [28, 34, 43, 61, 86, 114]. Similar information about the rabbit can be found in the studies of Gregory [55], Waterman [124], and Schmidt [104].

Meanwhile, it is rather difficult to find literature which compares the stages of the embryonic development of the above-mentioned laboratory mammals. These data, however, could be particularly important if experiments are performed with several species of laboratory rodents, as required by the current practice of testing the teratogenicity of drugs.

This chapter attempts to close this gap, since we have divided the embryonic development of laboratory rodents into a number of stages which are well characterized by different morphological features. We have done this in such a way as to facilitate the identification of similar (equivalent) stages in the embryonic development of these animals.

Although this chapter has been written on the basis of material accumulated over many years in the Department of Embryology at the Institute of Experimental Medicine, Academy of Medical Sciences of the USSR, we have not restricted ourselves to our own data. In fact, we have compiled tables showing normal development in such a way as to facilitate the comparison of our tables with the stages marked by other authors in the embryonic development of laboratory rodents.

In this chapter, the reader will find information on the earlier stages of mouse development; this allows an accurate comparison of various treatments to be made with certain transformation phases of the zygote chromosome apparatus. We will be describing the methods which accurately pinpoint the exact date of pregnancy in laboratory rodents. This is essential to embryological, teratological, and molecular-biological studies, particularly at the early stages of embryogenesis.

This chapter is not a substitute for treatises dealing with the biology of laboratory mammals. However, we do hope that it will be useful to anyone wishing to use our experience with the embryos of laboratory rodents and the accompanying tables showing the normal development of mice, rats, rabbits, and hamsters which we have compiled.

12.1. MAIN GUIDELINES FOR KEEPING ADULTS AND YOUNG

12.1.1. The Mouse

The mouse, *Mus musculus*, belongs to the subfamily Murinae, family Muridae, order Rodentia. The mean weight of the adult animal is 18–35 g; the life span is 2.5–3 years; and the fertility period last 14 months (beginning at the age of 2 months). During this period a pair of mice may yield 10 litters, i.e., about 100 descendants.

Mice should be kept in air-conditioned rooms with strictly controlled temperature, humidity, and illumination. There are precise methods for calculating the optimal room size, which depends on the number of animals kept in it. Similarly, there are various types of cages designed either for keeping mice in groups (10–30 animals) or a single pregnant mouse, a single male, or between one and several females. Each type of cage should contain appropriate drinking vessels and litter.

The type of litter is important in teratological experiments, since certain materials may be contaminated by pesticides or other chemical substances, thereby affecting the accuracy of the teratological experiment.

Mice should be kept on a correct diet [63]. The diet offered to mice by the Jackson laboratory has been described (Table 12.1). When mice are kept on a briquetted diet, it is beneficial to add fresh vegetables and milk, as well as germinated oats or wheat.

Depending on genetic characteristics, mice attain sexual maturity at the age of 4–5 weeks. Sexual maturity in females is accompanied by the opening of the

TABLE 12.1. Diet Formulations at the Jackson Laboratory in Parts per Thousand by Weight [36]

Ingredients	Diet 1	Diet 2	Diet 3	Diet 4
Ground milling wheat (Pennsylvania or Ohio)	515.00	329.50	560.00	602.00
Nonfat skimmed milk, edible	200.00	120.00	200.00	200.00
50% dehulled soybean meal	112.50	67.50	112.50	25.00
Corn oil, edible grade	102.50	34.50	57.50	102.50
Dried brewer's yeast	40.00	24.00	40.00	40.00
Sodium chloride	13.75	8.75	13.75	13.75
Dicalcium phosphate	10.00	10.00	10.00	10.00
Ferric citrate	1.25	0.75	1.25	1.25
Vitamin premix	5.00	5.00	5.00	5.00
Wheat germ meal, edible		400.00		
Total	1000.00	1000.00	1000.00	1000.00

vagina and the beginning of active ovarian function (maturation of follicles, ovulation, formation of corpora lutea, etc.). Mice are polyestrous animals, with a mean duration of the estrous cycle of 4–5 days [2].

To obtain healthy offspring, the first mating should be performed with female mice at the age of at least 2–3 months. Pregnant mice are extremely sensitive to environmental effects. They should be protected from temperature fluctuations, noise, and other stress-inducing stimuli. The term of pregnancy in mice of different strains varies from 18 to 21 days, the average being 19 days. Parturition as a rule takes place at night. The size of the litter depends on the strain, the number of previous pregnancies, and the age of the female.

Newborn mice weigh between 1–2 g. The sex of newborns can be determined by the distance between the sex bulge and the anus. In male fetuses it is at least three times longer than in females. By the 4th to 6th day the young mice are covered with hair. By days 21–25 they can feed independently, although it is not recommended to wean them from their mother before they are 1 month old. The lactation period normally lasts about 4 weeks, with the peak of lactation being the 10th day after delivery. As early as 20 h after delivery the female enters estrus and the normal cycle is resumed.

12.1.2. The Rat

The rat, *Rattus norvegicus*, belongs to the family Muridae, order Rodentia. Many rat strains are known, each of which has its own peculiar characteristics (for details see [14, 15]). The length of an adult rat varies from 15 to 25 cm, its weight from 150 to 230 g. The life span is 3–4 years, and of this rats are in the phase of active reproduction for 2 years.

The general guidelines for keeping rats are similar to those for mice, although rats are less sensitive to changes of conditions. Detailed information about the size of the room and conditions such as temperature, humidity, and lighting has been described [13, 14]. These references also provide valuable information regarding the optimal design of cages.

Adult rats require 28–30 g of food per day: this amount increases during pregnancy (Table 12.2).

Rats attain maturity at the age of 8 or 9 weeks, and, like mice, they are polyestrous animals having spontaneous ovulation every 4–5 days [73]. The stages of the estrous cycle are identical to those in mice, except they are more regular. Rats have a high fertility rate. One female may deliver 4–7 litters a year, each containing 5–12 descendants. The length of pregnancy varies somewhat for different strains of rats (20–26 days), but is 22 days on average. The incidence of spontaneous embryonic mortality is lower in rats than mice. Young rats are born blind and nude, just like mice, weighing 2.5–3 g. The body temperature of newborn rats is unstable. Cutter teeth appear on the 8th day; between 9 and 12 days old young rats grow hair; and at days 14–17 they begin to see. The lactation period lasts about 1 month. By the end of this period young rats are usually weaned from the mother. At the age of 1.5 months females are kept separately from males.

12.1.3. The Rabbit

The rabbit, *Oryctolagus cuniculus*, belongs to the family Leporidae, order Lagomorpha. Rabbits of different varieties and strains are used for scientific purposes. White New Zealand rabbits are extremely fertile and show no response to such teratogens as thalidomide. They are frequently used in embryological and teratological studies. Rabbits may be kept out in the open or in a shed, but preferably in air-conditioned housing with controlled temperature, humidity, and lighting.

Diet and optimal rations for rabbits are described in the literature [68, 69, 101, 102, 117, 133].

Rabbits are sexually mature by 4–5 months. Like other rodents, rabbits are polyestrous animals. Estrus (heat) in rabbits takes place regularly every 8–9 days during summer and lasts 3–5 days. Pregnancy lasts 27–32 days. One litter contains between 5 and 12 young rabbits, the mean weight of each being 40–50 g. One female may deliver 5–6 litters a year, with a total of 35–45 young rabbits. The lactation period lasts 20–25 days. Beginning from day 20, young rabbits may be transferred to a normal diet, while at the age of 40–45 days they are taken away from their mother and placed in special cages. Rabbits remain fertile for 2–3 years.

12.1.4. The Hamster

The hamsters of the genus *Cricetus* belong to the family Cricetidae, order Rodentia. Several species of hamster are used in laboratory practice. The Chinese hamster *Cricetus griseus* possesses a unique karyotype (the diploid chromosome number is 22, all homologous pairs of autosomes as well as the sex X and Y chromosomes being easily identified without differential staining). Therefore, this species may be successfully used in cytogenetic studies; owing to difficult breeding under laboratory conditions, however, this species has not found wide application in embryology. The Djungarian hamster, *Phodopus sungorus*, is a new species of laboratory animal. Breeding this animal in captivity began only comparatively recently. Full information on keeping and breeding the Djungarian hamster can be found in Pogosianz's and Sohova's paper [93]. The Transcaucasian hamster *Mesocricetus brandti* and the ratlike hamster *Cricetulus triton* are used in microbiological and parasitological studies.

The Syrian (golden) hamster, *Cricetus auratus*, is frequently used in embryological studies. Its life span varies from 1–3 years (average 18–24 months). The active reproductive function lasts from 1.5–12 months.

Breeding golden hamsters under laboratory conditions is possible only if the rules of feeding and breeding are strictly followed [26]. The temperature in the animal house should be kept around 18–21°C, with a humidity of 40–60%. The

TABLE 12.2. Daily Nutrient Requirements for Breeding of Female Rats [13]

	Maintenance	Gestation	Lactation
Average total feed (g)	13	19	33
Gross energy per meal (kcal)	52	76	131
Metabolizable energy (kcal)	47	68	118
Essential fatty acids (mg)	Trace	20	80
Crude protein (g)	0.91	3.8	6.6
Net protein (g)	0.52	2.3	4.0

Based on data from Warner [123].

photoperiod should last at least 12 h (usually from 6 a.m. to 6 p.m.), preferably 14–16 h. According to existing standards [94], the housing area per nursing female should be not less than 450 cm^2, and 25–35 cm^2 for each adult hamster. Three types of cages should be used. Cages for the colony should be 30 × 30 × 60 cm in size; each is designed for 8–12 hamsters and should contain a nest. Cages for mating should be 25 × 15 × 50 cm, and those for the offspring 25 × 20 × 15 cm. Hamsters can be kept on the same diets usually used for other small rodents. In addition, they should be regularly fed on lettuce, carrots, fruit, brewer's yeast, milk, and meat. A complete basic diet for the golden hamster has been described [26].

The golden hamster reaches maturity by the age of 40–60 days; females mature somewhat earlier than males. Females are allowed to mate from the age of 6–8 weeks. Pregnant females are immediately set aside from males, and 14 days after mating they are put on an enriched diet. They are also provided with material (such as paper and leaves) for building a nest. Pregnancy lasts 16 days. The mean size of the offspring for each female depends on the season, keeping conditions, and the genetic characteristics of the colony. The number of young in the litter varies from 5–10. Hamsters are born blind, with closed ear openings and underdeveloped legs. The mean weight of the newborn hamster is about 2.5 g. From day 8 onward hamsters can eat solid food. Their eyes open at 14–15 days.

12.2. METHODS FOR MARKING THE ANIMALS

Animals with dated term of pregnancy should be marked. Small rodents such as mice, rats, and hamsters with pigmented hairs can be marked using the scheme shown in Fig. 12.1. Marking should be done as follows. Using a special punch or perforator, holes or marks are made in the ears, marks on the right ear corresponding to units, and marks on the left ear to tens [36, 79].

Small albino rodents can be marked according to the scheme shown in Fig. 12.1. A saturated solution of picric acid, 0.5% fuchsin, or alcoholic solution of brilliant green may be used as dyes.

Marking animals can be done using either a special metal plate attached to the ear or by using tatooing tongs to put the number of the animal on the ear tip [68].

12.3. METHODS FOR OBTAINING ANIMALS
WITH DATED PREGNANCY

Laboratory rodents with dated pregnancy may be obtained from colonies which specialize in breeding and supplying these animals. Although this seems to be convenient at first sight, it may actually lead to serious problems, especially if experiments with early terms of pregnancy are planned. Furthermore, pregnancy dating in colonies may not be as accurate as the experimenter requires. Therefore, it is usually necessary to keep a colony of males and females in the animal house attached to the laboratory, and to obtain females with dated pregnancy terms by controlled mating.

The long-term breeding of mice, rats, hamsters, and particularly rabbits in the laboratory animal house is extremely expensive and labor intensive. It is only worthwhile if there is a need to use animals which cannot be systematically bought

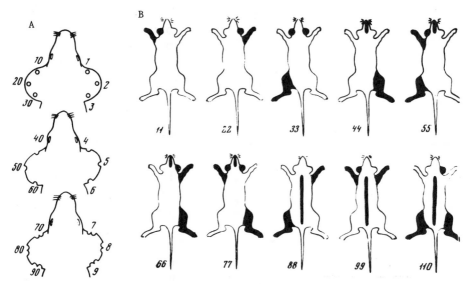

Fig. 12.1. Schematic illustration of marking pigmented mice (A) and albino mice and rats (B). A) Marks on the right ear correspond to units and on the left ear to tens [79]. B) Marks on ears, spine, and nose correspond to units, those on paws to tens.

from commercial colonies (e.g., when certain rare mutant strains of mice or those with Robertsonian chromosomal translocations are required, etc.).

If the investigator needs to keep a colony of laboratory rodents for a long period in order to produce the required number of embryos, he should strictly follow certain guidelines which have been well developed both for the inbred and random-bred (outbred) reproduction of animals [13, 14, 44]. It should be borne in mind that if these guidelines are not followed, i.e., if breeding is random, the offspring of a relatively small group of animals inevitably contains individuals (males and females) with an impaired reproductive function. This not only makes obtaining a sufficient number of embryos difficult, but also markedly complicates the interpretation of experiments, particularly those where the action of various agents upon gametogenesis or early embryogenesis is being studied. Such experiments include mutagenicity and teratogenicity testing, studies on the role of genetic or environmental factors in gametogenesis, fertilization, early embryonic development, etc. Therefore, it is recommended that mature males and females be obtained from commercial suppliers and that they be kept in the laboratory animal house only for the time required to obtain pregnant females. When animals with reduced mating capacity become numerous in the colony, they are killed and a new group of animals required for further experiments is obtained from the commercial supplier. This approach seems to be practical, since it allows experiments to be performed using a large number of embryos with the minimal expenditure of labor and money. This is particularly important in teratological or molecular-biological studies which require a very large number (several thousands) of embryos.

The results of experiments with laboratory rodents largely depend on whether sufficiently uniform groups of embryos have been used. This is not only important in experiments with early (preimplantation) embryos, but also in experiments at later embryonic stages. Many examples from experimental teratology point out that the frequency of developmental abnormalities and their nature depend on the stage at which the embryos were subjected to teratogenic agents. The stage at which sensitivity to teratogenic agents is maximal lasts a few hours for several such agents [38]. Therefore, although determining the exact length of pregnancy is not particularly vital to routine teratogenicity studies, it may be critical to the studies of the mechanisms underlying induced teratogenesis. Furthermore, studying fertilization, cleavage, blastocyst formation, and implantation can only be performed if the length of embryonic development has been determined exactly. This is also true for investigations dealing with the genetics and cytogenetics of early embryogenesis and for molecular studies with mammalian embryos.

All laboratory rodents are polyestrous animals and ovulation takes place during estrus; under natural conditions females mate prior to ovulation. In rabbits, ovulation is induced by mating, while in mice, rats, and hamsters it does not depend on it. Therefore, similar approaches can be used for obtaining dated pregnancy in mice, rats, and hamsters. A simpler technique is employed for obtaining rabbit females with dated pregnancy. Meanwhile, in all cases it is useful to employ a certain stage of the estrous cycle as a reference, i.e., to select females during estrus when they are capable of mating.

The mean duration of the estrous cycle period between consecutive ovulations in mice, rats, and hamsters is 4–5 days, while in rabbits it is 8–9 days. Friedmann et al. [48] have compared the estrous cycles of laboratory rodents. The estrous cycle in small rodents consists of five consecutive stages, including proestrus, estrus, metestrus I, metestrus II, and diestrus. Each stage of the cycle manifests itself in a certain cellular composition of the vaginal smear (Fig. 12.2). During proestrus the smear consists almost exclusively of epithelial cells, present either singly or in groups. Single cells have an oval shape, while those located in groups are mostly angularly shaped; all epithelial cells contain nuclei. During estrus the epithelial cells disappear and only nucleate cornified scales present as groups can be found. The scales have an angular shape. At metestrus I the scales form whitish clusters, well discernible on glass after making the smear (during this cycle phase animals no longer mate). At metestrus II leucocytes and individual epithelial cells appear in the mass of cornified scales, i.e., all three cell types can be found. Leucocytes are numerous by the end of this stage and mucus reappears. The scales gradually disappear. At diestrus abundant leucocytes are present, along with individual epithelial cells and a considerable amount of mucus. Correlations between estrous cycle stages in rats and changes in the external genitalia, uterus, and ovaries have been described (see Table 12.3 and [13]).

Stages in the estrous cycle are usually determined on the basis of vaginal smears. These smears are prepared by making vaginal irrigations or taking vaginal epithelium surface layers using a small spatula. Systematically screening a large number of animals using this technique, however, is rather laborious. In addition, upon repeated mechanical irritation of the vagina the estrous cycle may show deviations; these deviations involve an extended period of diestrus and a stage of false I pregnancy. Therefore, methods have been developed which rather accurately identify the estrous cycle stage in laboratory rodents, without having to take vaginal smears. With enough experience, such identification can be made by inspecting the external genitalia, observing the characteristic features of animal behavior and their locomotory activity, and inducing the so-called copulation reflex by stroking the

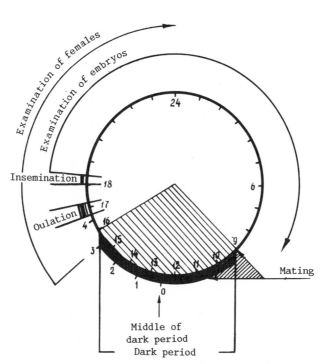

Fig. 12.2. Scheme of experiments to determine ovulation time and examine morphology of karyological transformations in the course of fertilization in mice.

pelvic region and head [13, 19]. The use of this technique for selecting female mice for mating has been described [32].

12.3.1.   Mice

If the date of pregnancy need not be very accurately pinpointed, and if the laboratory animal house contains enough mature females, kept separately from males, pregnant mice may be obtained if females are placed overnight in a cage with males. Inseminated females can then be taken in the morning, on the basis of such a criterion as the presence of vaginal plugs. Vaginal plugs in mice are retained up to 16–20 h after mating.

Usually three females are used for one male. It is recommended to select the most fertile males in advance, which produces far better results than using males taken at random from the colony. A given male should be used no more than once or twice a week. It should be borne in mind that males of certain mouse strains are capable of mating 2–3 times a week, while those of other strains are capable only once a week.

An effective procedure for presynchronizing estrus, which enables a large number of females with a given pregnancy term to be obtained on the required day, has been suggested by making use of the pheromones participating in the control of

TABLE 12.3. Changes Associated with Stages of the Estrous Cycle in the Rat

| Stage and duration | Vaginal fluids | | Superficial genitalia | Uterus | Ovary | pH of vaginal contents |
	Long and Evans [73]	Young, Boling, and Blandau [135]				
Proestrus (about 12 h)	Small, round nucleated cells only	Stage I, small round nucleated cells to 75% cornified cells. Estrus begins when 25% of cells are cornified	Vulva slightly swollen; vagina dry	Vascular engorgement; distention by fluid, lumen to 5 mm diameter	Enlargement of follicle	5.4
Estrus (about 12 h)	Cornified cells only	Stage II, from cornified cells only to cornified 25%, pavement cells 75%	Vulva swollen; vagina dry	Maximum distention in early part of this stage followed by regression to normal	Large follicles; maturation of egg	4.2
Metestrus I (about 15 h)	Late cornified stage; abundant caseous fluid	Stage III, pavement cells only	Vulva still swollen; caseous mass in vagina	Increase in vacuolar degenerational epithelium which began during estrus	Ovulation	
Metestrus II (about 6 h)	Cornified cells and leucocytes	Stage IV, pavement cells and leucocytes	No swelling; mucosa moist	Epithelium begins to regenerate	Eggs in oviduct	
Diestrus (about 57 h)	Epithelial cells and leucocytes		No swelling; mucosa moist	Epithelium regenerated; lumen to 2.5 mm in diameter	Corpora lutea formed	6.1

mouse reproduction. If cages containing 20–30 female mice are kept in a separate room, where there are no males, these mice show a retardation of the estrous cycle. If these females are then transferred to a room containing males and placed with them, the cycle is synchronously resumed in the females and most of them mate during the third night after being put with the males. The degree of the effect depends on the genetic peculiarities of a given mouse strain [127].

If females are placed overnight into a cage containing males, then the difference between the true and theoretically expected onset of pregnancy can be up to 10–16 h. This can be avoided if the females are put into a cage with males for 1–2 h [100]. If the animal house has inverted illumination (artificial light at night and darkness during the day) then, by controlling the lighting conditions, i.e., decreasing the duration of the dark period to 7–6 h, one can easily select the most convenient mating time for the experimenter. An even more accurate dating of pregnancy in mice can be achieved if the so-called delayed mating method is used, i.e., if females are put with males 1 h after ovulation and kept with them for only 20 min [70]. In this case insemination takes place after the ovulated eggs have entered the ampullary part of the oviduct.

A special study has been conducted in our laboratory in order to elucidate whether delayed mating adversely affects the pre- and postimplantation development of embryos in such females [67]. The work has been conducted using hybrid F1 CBA × C57BL mice. The animals were kept under inverted illumination conditions (the length of the dark period was 7 h) 4 weeks before and during the experiment. The first series of experiments was performed in order to identify the time of ovulation in females kept under such lighting conditions. To achieve this, females were put with males at the beginning of the dark period and then animals with vaginal plugs were selected. The animals were divided into groups; in each group the females were killed at different times, calculated from the middle of the dark period. The oviducts were removed and the accumulation of mucus with ovulated eggs was isolated from the ampullary regions using a needle. This material was transferred into a culture medium with hyaluronidase, the eggs released from the cells of cumulus were examined under a dissecting binocular microscope, and their number determined. In instances when ovulated eggs were not found in the oviducts, the ovaries were examined and the number of large follicles determined. Oocytes were released into the culture medium after puncturing the follicles, and cytological preparations were made. Thirty-nine females were examined during this series of experiments, and the data obtained (see Table 12.4) showed that mice kept under the aforementioned photoperiod conditions ovulate 3 h 45 min after the middle of the dark period; individual females ovulate within 20–30 min of each other.

The second series of experiments studied the fertility of females in the case of delayed mating. The females were placed into cages with males 1 h after ovulation, and the animals remained together for 15–30 min. The females with vaginal plugs were then selected. These animals were killed at the 18th day of pregnancy and the number of living and dead fetuses was determined, as well as the number of corpora lutea in the ovaries and the number of implantation sites. Living fetuses were examined using the standard procedure used in pathology [38] in order to reveal any developmental abnormalities.

Females mated prior to ovulation, i.e., placed into cages with males before the beginning of the dark period, served as a control. Twenty-seven pregnant mice were in the experimental group and 35 in the control group. The results of the study can be seen in Table 12.5. The two groups of mice did not differ from each

TABLE 12.4. Ovulation in CBA/C57BL Mice under Conditions of Inverted Photoperiod (Duration of Dark Period 7 h)

Time of female examination		Number of females examined			Number of ovulated eggs	
			Among these			
from beginning of dark period (h,min)	from middle of dark period (h, min)	Total	nonovulated	ovulated	in each female	mean number per female (± standard error)
6.00–6.20	2.30–2.50	3	3	–	–	–
6.30	3.00	4	4	–	–	–
6.45	3.15	5	5	–	–	–
7.00	3.30	3	3	–	–	–
7.15	3.45	7	3	4	5, 9, 5, 8	6.7 ± 2.3
7.30	4.00	10	4	6	6, 5, 5, 9, 8, 4	6.1 ± 2.2
7.45	4.15	4	–	4	7, 5, 10, 9	7.7 ± 2.4
8.00–8.30	4.30–5.00	3	–	3	6, 6, 8	6.6 ± 1.6
Total		39	22	17		

TABLE 12.5. Fertility of CBA/C57BL Mice Mated 1 h After Ovulation

Mating conditions	Number of females	Number of corpora lutea		Number of embryos that died prior to implantation		Implanted		State of fetuses at day 18 of development							
								Among implanted embryos				Among living embryos			
								Died after implantation		Living		Normal		Malformed	
		Total	per 1 female	Total	% from number of corpora lutea	Total	% from number of corpora lutea	Total	% from number of implantations	Total	%	Total	%	Total	%
Experiment One hour after ovulation	27	239	8.9 ± 1.8	16	6.7 ± 1.6	223	93.3 ± 1.6	16	7.2 ± 1.7	207	92.8	207	100	–	–
Control Several hours before ovulation	35	302	8.9 ± 1.8	23	7.6 ± 1.5	279	92.4 ± 1.5	15	5.4 ± 1.3	264	94.6	264	100	–	–

other either in the number of corpora lutea and implantation sites or in the level of pre- and postimplantation mortality. No prenatal developmental abnormalities were detected in fetuses of the experimental group.

It can thus be concluded that mating mice 1 h after ovulation adversely affects neither female fertility nor embryonic development. This technique can be employed to pinpoint the exact date of pregnancy, that is, to obtain fetuses at strictly defined developmental stages. This is particularly important when working with preimplantation embryos.

If a large number of ovulated eggs or early embryos are required, then superovulation may be induced using exogenous gonadotropic hormones. For this purpose, mice are first injected with follicle-stimulating hormone (FSH). Usually, gonadotropin from pregnant mare serum (PMS) containing FSH and traces of luteinizing hormone (LH) is used. Between 42–52 h after such an injection (usually after 48 h) a preparation of LH is given. Human chorionic gonadotropin (HCG) is often used. The doses and intervals between injecting the first and the second hormone are selected on the basis of the age and genetic characteristics of the mice. Generally it is sufficient to inject 2.5–5 IU of both FSH and HCG into one mouse. Mice usually respond by ovulating 12–13 h after the HCG injection, the exact ovulation time depending on the genetic characteristics of a given strain. Using the superovulation technique correctly enables up to 50–80 eggs to be obtained from one mouse [42, 51]. It is better to plan the experiment in such a way as to achieve ovulation during the morning or around midday, since this enables the females to be put with the males for a short period and the exact time of insemination to be pinpointed. If mice are mated 30–60 min after ovulation, then the accuracy of pregnancy dating may be increased even further.

Response to exogenous gonadotropic hormones depends not only on the genetic characteristics of the females (some strains produce far better results than others), but also on the age of the mice and the conditions under which they are kept. If immature young mice weigh 13–14 g by the time they are 21–24 days old, then they respond very well to gonadotropins. But usually at this age the mice are lighter than that and respond to gonadotropins rather poorly. In order to obtain the above-specified weight by the age of 3 weeks, particular attention should be paid to feeding; nursing females should be kept on a diet rich in fats and proteins, and nurse no more than 5 young mice. To obtain the maximum number of ovulated eggs, one should use the superovulation procedure properly, i.e., follow the guidelines listed by Gates [50, 51].

The proper use of exogenous gonadotropins results in the normal fertilization of most of the ovulated eggs *in situ*, that is, in the body of hormonally stimulated mice, as well as *in vitro*. If these embryos are left in the genital tract of the hormonally stimulated immature female, they undergo normal cleavage. However, they are transferred into the uterus too early (at the early morula stage) owing to increased motility of the oviduct, and this may result in their death. If these embryos, however, are transplanted into the oviduct or uterus of mature recipient females at synchronous stages of false pregnancy, then embryonic development continues and results in the birth of living and quite normal offspring. Thus, the technique of superovulation allows us to work with embryos at both early and late developmental stages. Meanwhile, it should be remembered that for certain experimental purposes such as microsurgery and cultivating early embryos *in vitro* eggs produced by natural ovulation should preferably be used [95].

12.3.2. Rats

Rats with dated pregnancy terms are obtained in the same way as mice, with one difference, however; with rats the presence of sperms in the vaginal smear is stronger evidence of mating than the presence of vaginal plugs. This is because vaginal plugs cannot always be found and, as they are located deeper in the vagina, they undergo resorption faster than in mice. It has been shown that 90–94% of rats with sperms in the vaginal smear are indeed pregnant, while the other 6–10% of the group had so-called sterile copulation. If, however, one uses vaginal plugs as an indicator, then pregnancy can only be diagnosed in 45–50% of the mated rats; in other words, this technique does not yield reliable results [13].

Female rats taken at random from the colony, i.e., without determining the estrous cycle stage, can be put directly with the males. This practice is used if the experimenter has a large enough colony of females and if there is no need to obtain a large number of pregnant females at a specific time. It is better, however, to use female rats in the cycle phase, when they are able to mate, that is, in estrus.

Rats in estrus can be selected by using vaginal smears. However, this is not strictly necessary, since there are other quite reliable methods of selecting females for mating [13].

Rats capable of mating are selected on the basis of behavioral characteristics, such as locomotory activity, and by measuring the pH of vaginal secretions. The simplest procedure is to select female rats for mating on the basis of copulation reflex: these females acquire a characteristic posture of lordosis after stroking their pelvic region [24]. If their spine and head are touched, this results in the characteristic shivering and trembling of the ears. Detailed information about the methods used to select rats for mating are given in the literature [13]. This publication also describes a technique for synchronizing the estrus and enabling a large number of pregnant females to be obtained by a certain date. To synchronize estrus the rats are initially given medroxyprogesterone for 6–7 days with drinking water, and then a single intramuscular injection containing 1–3 IU of the gonadotropic hormone PMS. Between 90 and 95% of the females mate 30–34 h after the PMS injection (for details see [77]). In order to improve pregnancy dating in the case of natural rat ovulation, it is recommended that males be kept with females for only 1–2 h. This can be easily achieved if the facility for keeping rats allows working with inverted lighting conditions. It may well be that "delayed mating" (see p. 361) will produce good results with rats, too. This problem, however, requires a special study since there are only a few papers dealing with temporal characteristics of spermatozoa transport through the female genital tract. It has been shown that fertilization took place 2.5 h after mating, and the first cleavage furrow appeared 20.6 h after the spermatozoon penetrated the egg [105].

If a large number of eggs or early rat embryos is required by a certain date, then the hormonal stimulation of females may be used. For this purpose the females are initially given 20–50 IU PMS, and a similar dose of HCG, 52–56 h after the first injection [134]. Only a certain proportion of the eggs may be fertilized after the superovulation of immature rats; this may be the result of inadequate mating or other factors [81, 82, 88].

12.3.3. Hamsters

Estrus in this species may be determined with certainty, not only from the cell composition of the vaginal smear, but also from a good copulation reflex. After stroking or touching the spine lightly, the female acquires a characteristic posture, i.e., she shows a characteristic bending of the lumbar region and "freezes" [65].

The beginning of estrus and the appearance of the copulation reflex usually takes place by the end of the day photoperiod. Hamsters ovulate at midnight or somewhat later. In order to obtain dated pregnancy, females at the stage of estrus are selected from the colony on the basis of a positive copulation reflex and are put into cages with males overnight. Mating is diagnosed by either the presence of vaginal plugs or spermatozoa in the vaginal smear.

12.3.4.   Rabbits

The stage of the estrus is characterized by a prominent reddening and swelling of the external genitalia and by the excited state of the animal. In contrast to mice and rats, ovulation in rabbits is induced by mating, so it is much easier and more accurate to date the pregnancy in these animals. In order to obtain dated pregnancy, a female at the stage of estrus is placed into the cage with a male. Mating usually takes place within the first hour or two, and is recorded either visually or by the presence of mucus vaginal secretions. Ovulation takes place 10–12 h after mating.

12.4. STRUCTURE AND CHARACTERISTICS OF MATURE GAMETES

12.4.1.   Eggs

All laboratory rodents ovulate before meiosis is completed, that is, at metaphase II (Fig. 12.3). The structure of rodent eggs has been described in great detail at the level of light and electron microscopy in a number of comprehensive publications [5, 16, 27, 37, 76, 85, 131].

The ovulated egg is surrounded by the zona pellucida, a homogeneous structure resulting from the secretory activity of follicle cells and oocytes. The zona pellucida plays an important part in fertilization (see Section 12.5), and if it is removed the eggs may be fertilized by heterologous spermatozoa; for example, hamster eggs can be fertilized by human sperm. This technique is used for determining the fertilizing capacity of human spermatozoa [128].

In ovulated eggs the zona pellucida is surrounded by a layer of follicle cells which form the zona granulosa. The inner layer of these follicle cells is adjacent to the zona pellucida and forms the corona radiata.

If one uses ovulated rodent eggs for such studies as determining the chromosomal complement, biochemical experiments, etc., one has to remove the follicle cells surrounding the egg. For this purpose the eggs are placed into a solution of hyaluronidase (100–300 units/ml of the culture medium). For certain experiments with ovulated eggs (such as fertilization by heterologous spermatozoa, hybridization of the egg with somatic cells, etc.), the zona pellucida has to be removed. This is achieved by pronase (0.5% w/vol solution), or by acidic Tirode solution at pH 2.5. Since the zona pellucida is retained throughout cleavage (in mice, rats, and hamsters the embryos emerge from the zona pellucida at the stage of late blastocyst), pronase or acidic Tirode solution are also used if the zona pellucida is to be removed from the cleaving embryos (for example, for the aggregation of embryos in experiments on the construction of genetic chimeras). The ovulated

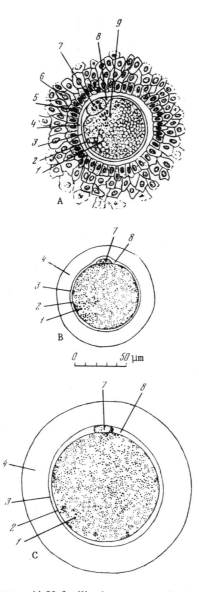

Fig. 12.3. Female gametes. A) Unfertilized mouse egg shortly after ovulation. B, C) Unfertilized rat and rabbit eggs (cells of the cumulus oophorus removed). Phase contrast, drawing. 1) Egg cytoplasm; 2) cytoplasmic membrane (vitelline membrane); 3) inner boundary of zona pellucida; 4) zona pellucida; 5) corona radiata; 6) cells of cumulus oophorus; 7) first polar body; 8) perivitelline space; 9) egg chromosomes.

eggs and early embryos of rabbits are surrounded by the mucoid membrane; this circumstance should be taken into account when working with this species.

Movement of the ovulated egg along the oviduct results in changes of the corona radiata affecting egg fertilizability and perhaps also the frequency of polyspermy.

In each species of rodent the submicroscopic structure of the egg shows characteristic features. Mouse eggs, for example, characteristically possess a large number of crystalloidlike inclusions in the cytoplasm; rabbit eggs possess characteristic ring plates [57, 110, 116, 125].

The eggs of laboratory rodents can still be fertilized for a limited time after ovulation, no longer than 10–12 h. As time goes on after ovulation, the eggs gradually develop pathological changes leading to so-called "overmaturation." This significantly increases the frequency of nondisjunction of chromosomes and leads to the impairment of early embryogenesis [39]. This circumstance should be taken into consideration when experimenting with laboratory rodents. It should be borne in mind that under normal conditions the eggs of mice, rats, and hamsters should be fertilized no later than 6–8 h after ovulation (in rabbits after 10–12 h).

12.4.2.   Spermatozoa

The microscopic and submicroscopic structure of mammalian spermatozoa has been described in detail in various books and reviews [1, 31, 35, 45, 46, 59, 72, 84, 92]. Detailed information with regard to the morphology of male germ cells and spermatogenesis in small rodents can be found in a number of publications [17, 21–23, 33, 47, 62, 89, 98, 113, 115, 120].

The spermatozoa of mice, rats, hamsters, and rabbits (Fig. 12.4) are of the flagellate type. They consist of a head, neck, and tail; the head includes the nucleus and the hyaluronidase-containing acrosome. The acrosomes of mice, rats, and hamsters have the characteristic shape of a hook bent in the posterior direction. According to our data, the length of a spermatozoon from the head to the tip of the tail is 120–125 μm in mice, 170–175 μm in rats, 190–200 μm in hamsters, and 65–70 μm in rabbits.

The acquisition of fertilizing capacity by mammalian spermatozoa requires maturation and capacitation, which take place in the reproductive tracts of the male and the female, respectively. The final stages of the maturation of sperm cells take place in the epididymis and include morphological, physiological, and biochemical changes. Specifically, such characteristics as lipid composition and ATPase activity of the spermatozoon plasma membrane undergo changes; the plasma membrane surface becomes covered with proteins and glycoproteins, forming a protective layer. It is thought that this layer stabilizes the plasma membrane.

Capacitation is a complex multistep chain of events [6] which has been studied in great detail in rabbits and hamsters, and to a lesser degree in rats and mice; it triggers the acrosome reaction which provides for gamete fusion. It has been noted (see reviews [111, 119, 129, 130]) that capacitation results in qualitative changes leading to a loss or modification of the adsorbing surface layer of the sperm plasma membrane. Capacitation is associated with the increased rate of sperm cell respiration; it leads to changes in permeability and osmotic characteristics of the spermatozoon plasma membranes.

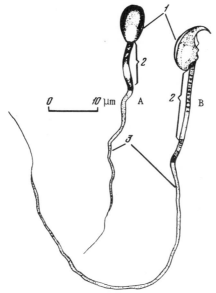

Fig. 12.4. Drawings of mammalian spermatozoa. A) Rabbit; B) mouse. 1) Sperm head; 2) neck; 3) tail.

The necessary conditions for completing sperm capacitation of various species of animals vary greatly, and this should be taken into account when experimenting with egg fertilization *in vitro*.

Usually spermatozoa retain their mobility and fertilizing capacity in the female genital tract for 6–12 h.

Cellular contact between spermatozoa and endometrium plays an important part in capacitation. This point of view is supported by Yanagimachi [130], who observed that in the hamster the epithelium of certain regions, such as the transition region from the uterus to the oviduct, can induce changes in the surface layer of the spermatozoon. The region of the reproductive system where capacitation begins is different for different species. This may be either the uterus or the oviduct, or the vagina in the case of the rabbit.

It has been shown that the genital tract of all animal species characterized by quick capacitation may induce capacitation of the sperm cells from other species. For example, hamster sperm cells undergo rapid capacitation in the rabbit oviduct. The rate of capacitation of different spermatozoa may not be the same even for a given species, and its rate depends on the male physiological condition. The hormonal status of the female, the location of sperm in its reproductive system, as well as the composition and other characteristics of the medium in the female genital tract, affect the rate of capacitation as well. There is evidence that neutrophils, eosinophils, and cumulus cells located in the female genital tract participate in capacitation.

12.5. INSEMINATION AND FERTILIZATION

Insemination, i.e., the combination of events which result in bringing together the eventual contact between spermatozoa and eggs, is similar in all four species of laboratory rodents. In the course of mating the male introduces sperm into the female genital tract. The ejaculate contains several tens of millions of spermatozoa (up to 60–80 million in mice and rats). After 15–20 min the spermatozoa enter the oviduct ampule and meet the eggs.

Fertilization in mammals includes spermatozoon penetration through the corona radiata, zona pellucida, and the perivitelline space followed by contact with the egg plasma membrane, fusion of the membranes, and penetration of spermatozoon into the egg.

A very large number of papers, including many good reviews [9, 10–12, 18, 58, 130], deal with fertilization in mice, rats, rabbits, and hamsters; they include studies using electron microscopy [8, 43, 74, 75, 132].

The penetration of the spermatozoon into the egg in mammals is mediated by the acrosome. The acrosome covers the anterior part of the sperm head and is divided into three (apical, basal, and equatorial) or two (anterior and posterior) segments (Fig. 12.5). The equatorial segment is not constant for certain mammalian

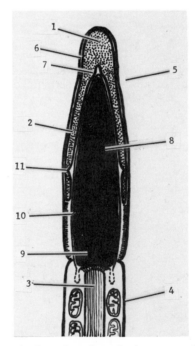

Fig. 12.5. Scheme of sagittal section through the mammalian sperm head.   1) Acrosome; 2) post-acrosomal region; 3) reduced nuclear membrane; 4) mitochondria; 5) plasma membrane; 6) outer acrosomal membrane; 7) inner acrosomal membrane; 8) nucleus; 9) nuclear membrane; 10) equatorial segment; 11) post-acrosomal lamina densa.

species, including the spermatozoa of mice and rats. The acrosome contains glycoproteins and glycolipids and, in many respects, is similar to lysosomes. The acrosome has been found to contain various enzymes such as hyaluronidase, proteinases, esterases, acid phosphatase, and others; it should be pointed out that some of them (acrosine, for example) are found in the acrosome in an inactive form.

The acrosome is an unstable structure and may undergo spontaneous degradation when the spermatozoa start to age. The acrosome undergoes characteristic changes in the course of the acrosome reaction which precedes fertilization (see below).

When capacitated spermatozoa approach the fertilization site and come into contact with cumulus cells, the acrosome reaction is triggered. In certain species, for example the rabbit, the acrosome reaction may proceed even in the absence of the egg or its cellular surrounding.

In vitro the spermatozoa undergo capacitation and the acrosome reaction proceeds after mixing with eggs in large numbers (about 3000–6000 spermatozoa per egg). Under such conditions the cumulus disperses, and spermatozoa with the intact acrosomes reach the zona pellucida and undergo acrosome reaction on its surface before penetrating through the zona pellucida.

The surface of the mature spermatozoon is covered with substances that stabilize the plasma membrane and prevent the acrosome reaction from beginning prematurely. These substances appear to be dissociated or modified in the course of sperm capacitation, which results in the sperm cell plasma membrane destabilizing, its composition undergoing characteristic changes such as a decrease in the content of cholesterol and phospholipids. These events are intimately related to the beginning of the acrosome reaction. In morphological terms the acrosome reaction is universal in all mammals (Fig. 12.6). Biologically it involves the release and activation of acrosomal enzymes. Furthermore, it plays the part of a trigger for changes of plasma membranes, enabling a fusion between the plasma membranes of the egg and the sperm, i.e., allowing the spermatozoon to penetrate the egg.

Contact between the spermatozoon and the zona pellucida results in a momentary fusion of the acrosomal membrane with the sperm plasma membrane at many places. From this moment onward the sperm head is covered by the inner acrosomal membrane (Fig. 12.6). After the sperm has penetrated the perivitelline space, its equatorial segment fuses with the plasma membrane of the egg microvilli. Rabbit oocytes possess short microvilli, while the hamster ones are longer; this explains the different times taken for the inner acrosomal membrane of the spermatozoon to fuse with the egg plasma membrane. The fusion of plasma membrane in mammals begins in the region behind the acrosome. The ultrastructural changes associated with this process have been described in a number of papers [20, 74].

Normally fertilization is monospermic in all four species of laboratory rodents; in other words, only one spermatozoon penetrates the egg cytoplasm. Its head is transformed into the male pronucleus.

Additional spermatozoa may sometimes penetrate the zona pellucida, but they remain in the perivitelline space and are not involved in fertilization. These spermatozoa may be found in the perivitelline space of the egg and of the cleaving embryo, too.

A block to polyspermy is accomplished at the level of the zona pellucida and is due to changes in the egg plasma membrane during cortical granule extrusion. The cortical granules are located in different regions of the egg. Upon fertilization they migrate to the surface layer, and their contents are released from the egg. The

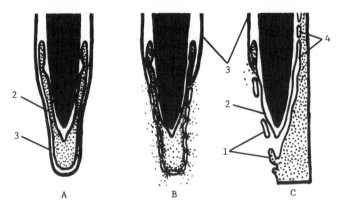

Fig. 12.6. Schematic drawings of three consecutive stages of fertilization in mammals. A) Intact acrosome near egg zona pellucida; B) acrosomal reaction; C) interaction between the sperm and egg plasma membrane after their contact. 1) Egg microvilli; 2) inner acrosomal membrane; 3) sperm plasma membrane; 4) egg plasma membrane.

cortical granules contain glycoproteins and a number of enzymes. Acid phosphatases present among the enzymes undergo activation only after the release of the cortical granule content into the perivitelline space. The release of the cortical granule content is of the merocrine secretion type; the granule membranes fuse with the egg plasma membrane, rendering it impermeable to the sperm.

Egg cortical granules belong to the class of primary lysosomes. The number of cortical granules located under the plasma membrane is different for mature eggs of different species: 32 granules are located under $100 \, \mu m^2$ of mouse plasma membrane, while the corresponding numbers for the rat and rabbit are 95 and 60, respectively. It has been suggested that some of the cortical granules are extruded even before fertilization, and that this process plays a part in separating the corona radiata from the egg. There is considerable similarity between the origin and structure of cortical granules, on the one hand, and the sperm acrosome on the other. The bringing together and fusion of membranes is the main event both in the acrosome reaction and in the secretion of cortical granule content.

When the spermatozoon enters the egg cytoplasm, its nucleus or head chromatin decomposes and a new nuclear envelope is formed only after a complete chromatin decondensation and loss of arginine-rich histones. Egg cytoplasm contains factors which lead, on the one hand, to sperm chromatic decondensation and, on the other hand, to the transformation of the spermatozoon head into the male pronucleus. These factors appear in the cytoplasm of a maturing egg, reaching their maximum level shortly before ovulation. The egg cytoplasmic factors do not possess species specificity; that is, the egg cytoplasm of one mammalian species is capable of decondensing the sperm chromatin of another. This lack of species specificity underlies a technique used to determine the fertilizing capacity of spermatozoa through heterologous *in vitro* fertilization of eggs devoid of the zona pellucida. Fertilizing hamster eggs with human sperm is one example [128].

A series of transformations both in the sperm head and in the maternal chromosomal complement takes place between the penetration of the sperm into the cytoplasm and the completion of the first cleavage division. These changes involve the formation of male and female pronuclei, which come closely together in the course of prophase and prometaphase, then paternal and maternal chromosomal complements yield a common metaphase plate.

These transformations require a considerable amount of time, in mice lasting 14–16 h; this period of early zygotic development may be divided into distinct stages. This staging is extremely important in studies of the action of various agents such as chemical mutagens and teratogens or ionizing radiation upon the zygote. In other words, the experimental results largely depend on the accurate timing of treatment relative to a certain stage of the zygote chromosomal complement transformation.

The stages of early zygote development described below have been identified from detailed karyological analysis of mouse eggs conducted in this laboratory [40, 66, 67]. In this study 1047 fertilized eggs were obtained from 158 CBA/C57BL hybrid female mice after "delayed mating" (see p. 361 and Table 12.6). The eggs were fixed by a mixture of formaldehyde and methyl alcohol stained with 0.25% lacmoid, and examined as complete preparations under a phase-contrast microscope.

Consecutive stages of karyological transformations in mouse zygotes, during the period from sperm penetration to the completion of the first cleavage divisions, are given in Fig. 12.7 and in a series of micrographs (Figs. 12.8–12.16).

It has been suggested that this period be divided into eight phases. In a number of cases distinct subphases can also be identified.

Phase I. Penetration of the Sperm into the Egg and Completion of the Second Maturation Division

The first evidence that the egg has been penetrated is the appearance of the "fertilization cone" (Fig. 12.8) at the place where the sperm head has penetrated the egg cytoplasm. Sperm capacitation in CBA/C57BL mice takes about 3 h, while the sperm takes about 45 min to pass through the membranes. The penetration stage, including maturation division completion, lasts approximately 1 h. The following subphases can be identified during this phase:

1a. The sperm head is intact or only begins to swell. The maternal chromosomal complement is located closer to the egg center and is at metaphase II.

1b. The head of the sperm has swollen but not yet throughout its entire length. The maternal chromosomal complement occupies the lateral position and is at the anaphase stage.

1c. The head of the sperm has swollen throughout its length (complete chromatin decondensation has occurred). The maternal chromosomal complement is at the telophase. Second polar body extruded.

Phase II. Formation of Pronuclei

This phase lasts about 1 h and involves the transformation of paternal and maternal chromosomal complements in pronuclei. Complete decondensation of the sperm chromatin and its subsequent contact with the cytoplasm leads to the formation of clear nucleoplasm which is surrounded by the pronucleus membrane. The male pronucleus is formed somewhat earlier than the female, and the small primary pronucleoli become visible somewhat earlier. The maternal chromosomes, after completing telophase, undergo strong condensation and are then transformed into a

TABLE 12.6. Developmental Stages (%) at Different Times after Insemination

Time after insemination (h)	Number of females	Number of eggs	Metaphase II (%)	I			II		III			IV		V		VI			VII	VIII	Two-cell embryos (%)
				Ia	Ib	Ic	IIa	IIb	IIIa	IIIb	IIIc	IVa	IVb	Va	Vb	VIa	VIb	VIc			
2	3	15	100																		
4	5	32	84.4	15.6																	
5	15	103	42.7	7.8	22.3	7.8	0.9	18.5													
6	23	145	12.4		11.7	14.5		61.4													
7	13	90	8.9		25.6	24.4		34.4	6.7												
8	11	70						42.2	57.1												
10	9	67						3.0	61.2	35.8											
12	10	74							16.2	48.7	35.1										
14	11	69							8.7	20.3	36.2	5.8	21.7	4.3					1.5		1.5
16	24	148							2.7	21.6	23.6		39.2	6.1					2.7		4.1
18	15	101							3.0	5.0			8.0	8.0	9.9	29.1	1.6	0.9	9.9	6.9	17.7
19	7	47							2.1	2.1			2.1	6.5		10.6			12.8		63.8
20	12	86							1.2	2.3			1.6	3.5		8.1					83.7
Totals:	158	1047																			

Stages of zygotic development

dense chromatin body. This is followed by chromatin decondensation and the formation of the female pronucleus with an envelope, nucleoplasm, and small primary pronuclei. The male pronucleus is larger by this time. The female pronucleus is located in the egg region, which is in contact with the second polar body, while the male is located at the opposite egg pole or closer to its center. The spermatozoon tail is similarly located in the egg cytoplasm.

Phase III. Growth of Pronuclei

This phase last 6–8 h; during this time the pronuclei mature, increase in size, and reach their maximal dimensions; DNA synthesis, corresponding to the S-phase of the cycle, begins. As the pronuclei increase in size, the number of pronucleoli changes, and the pronuclei come close to each other and migrate toward the egg center. The migration of the pronuclei involves microfilaments, i.e., the egg cytoskeleton. At first, between 3 and 10–12 small pronucleoli can be seen in the pronuclei; then the pronucleoli fuse and are transformed into large secondary pronucleoli, the number of which varies from 1–4 in each pronucleus. In the case of the mouse the number of pronucleoli in the male pronucleus is greater than in the female. In rats the number of pronucleoli in both pronuclei is markedly greater than in mouse egg pronuclei.

Phase IV. Prophase of the First Cleavage Division

Pronuclei are located close to each other and a network of chromatin fibrils becomes visible (Fig. 12.10). This process is asynchronous, i.e., chromosomes in

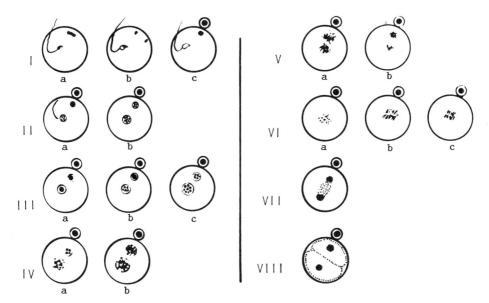

Fig. 12.7. A scheme illustrating successive karyological transformations in mouse zygotes during the period between sperm penetration and completion of the first cleavage division (see explanation in the text).

the male pronucleus can be seen earlier than in the female. Chromatin fibrils become associated with the secondary pronucleoli, and this association continues at the next stage.

Phase V. Prometaphase of the First Cleavage Division

The pronuclear membrane disappears and chromosomes associated with the remnants of pronuclei are now located in the cytoplasm. The paternal chromosomes are less coiled, while the maternal complement shows greater coiling. By the end of the stage both haploid complements come close to each other.

Phase VI. Metaphase of the First Cleavage Division

Both haploid chromosomal complements join to yield a common metaphase plate (Fig. 12.10). At first the chromosomes consist of chromatids in tight contact with each other. Subsequently a disjunction of sister chromatids becomes noticeable at the ends of the chromosomes.

Phase VII. Anaphase of the First Cleavage Division

Chromosomes segregate toward opposite poles of the spindle (Fig. 12.11).

Phase VIII. Telophase of the First Cleavage Division

Two groups of chromosomes are located near the spindle poles, chromosomal contours become less distinct, and the groups become transformed into nuclei. The appearance of the first cleavage division furrow marks the end of this stage.

The total duration of mitosis from prophase to telophase during the first cleavage division in CBA/C57BL mice is 3–3.5 h. The stage of the zygote, from egg penetration by the sperm to the end of the first cleavage division, lasts 14–16 h in these mice. The duration of this stage in different mouse strains varies depending on the genetic characteristics of both egg and sperm [51, 78, 106, 107, 126].

The scheme given in Fig. 12.12 shows the chronological sequence of karyological transformations in the course of mouse egg maturation and fertilization. Data are given for natural ovulation and superovulation induced by gonadotropic hormones. Table 12.7 compares the staging of one-cell mouse embryos as described by Kaufmann [64] and Khozhai [67].

12.6. "EQUIVALENT" STAGES OF NORMAL DEVELOPMENT

Numerous papers describe the normal development of laboratory mammals such as mice [56, 60, 99, 100, 108, 109, 118], rats [28, 30, 34, 41, 61, 86, 114], rabbits [41, 55, 83, 104, 124], and hamsters [7, 25, 26, 52, 90, 103, 121, 122].

Figs. 12.8–12.11. Successive stages of karyological transformations accompanying fertilization in the mouse (from sperm penetration to completion of the first cleavage division). Drawing numbers correspond to phase numbers. Phases I–VIII) preparations stained by lacmoid, phase contrast, magnification 12.5 × 100. Figures 1ª,1ᵇ, 1ᶜ, etc., refer to substages. (Materials of L. I. Khozai.)

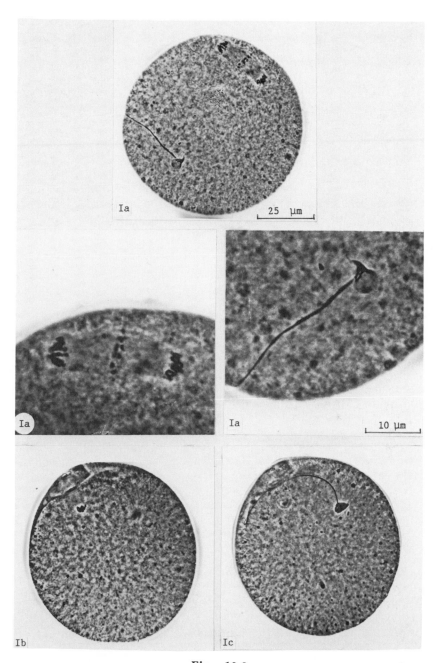

Fig. 12.8.

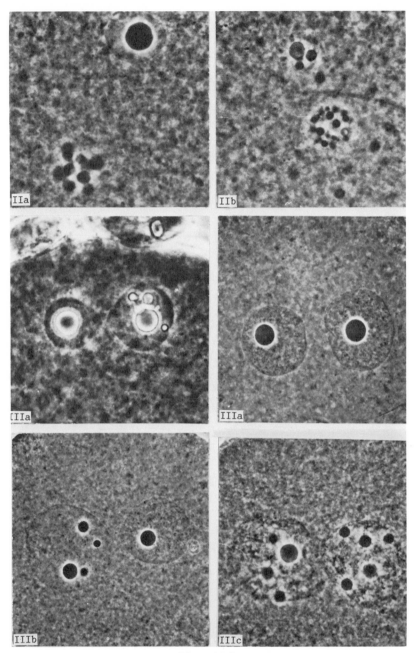

Fig. 12.9.

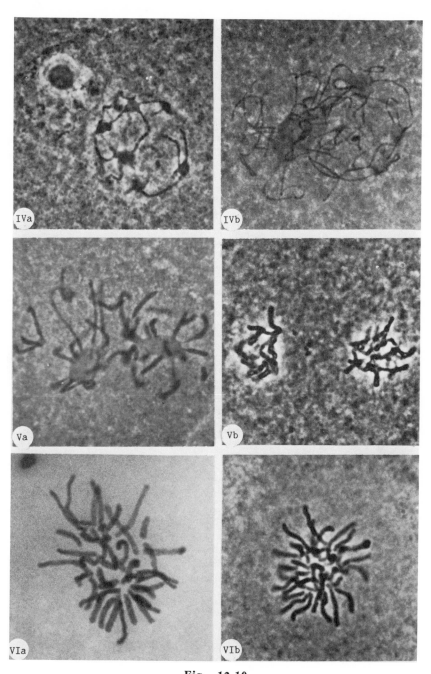

Fig. 12.10.

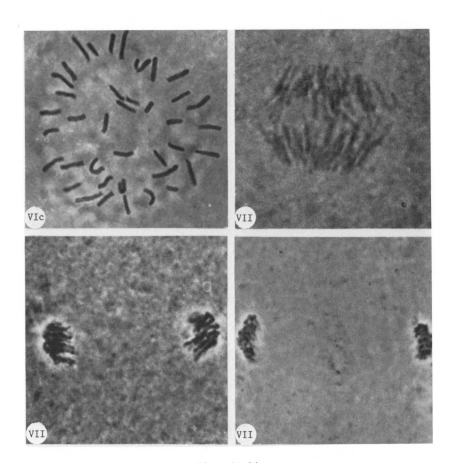

Fig. 12.11.

Tables of normal developmental stages have been compiled for rabbits, rats, and mice. Edwards [41], for example, distinguished 14 stages in the development of the albino rat and 17 stages in rabbit development. Christie [34] divided rat embryogenesis into 32 stages, while Theiler [118] described 27 stages of mouse development. In these tables the developmental stages of each animal species are described on the basis of different characters and, therefore, the stages cannot be directly compared with each other.

In order to compare the embryos of different species we have carried out a special developmental study of mice, rats, golden hamsters, and rabbits, and distinguished the developmental stages on the basis of characteristics common to different species. Our classification of stages is based on several morphological features characterizing each developmental stage and easily detectable in different species by simple examination methods. Our own observations have been complemented by written data, especially that on golden hamster and rabbit embryogenesis.

The development of the staging system was first performed using mouse embryos. We have identified 24 morphological, clearly distinct developmental stages throughout the entire embryogenesis period. Thereafter we identified the stages with similar diagnostic features in the three other species, namely rats, hamsters, and rabbits, and identified the time they began in each species.

The concept of equivalent development stages in animals and man is widely used in experimental embryology and particularly in experimental teratology [87, 91, 136]. It would be a mistake to think, however, that the embryos of different mammalian species are completely similar at these stages. Owing to a wide occurrence of heterochrony, the embryos of different species similar with respect to diagnostic characteristics and designated by us as being in one and the same developmental stage may differ in other characteristics, so that they may not necessarily be completely similar. Tables of normal development allow a more accurate identification of this heterochrony in the development of different organs and quantitative evaluation.

Theiler [118] has tried comparing stages of mouse development with the equivalent developmental stages of the human fetus, Streeter horizons [112], and the data are presented in detail in his book. Further details of Streeter horizons, including the description of 23 developmental stages, have been published [96]; this source also contains detailed references on the development of the individual system during human development.

12.7. CHARACTERISTICS OF STAGES OF NORMAL DEVELOPMENT

We present a table of consecutive developmental stages of the rat, mouse, golden hamster, and rabbit. Table 12.8 shows the diagnostic characteristics of each stage of normal development and the time they begin.

CBA and C57BL mice, Wistar albino rats, Chinchilla rabbits, and outbred golden hamsters obtained from the Rappolovo colony of the Academy of Medical Sciences of the USSR have been used for compiling tables of normal development. Photographs of the external appearance of mouse, rat, golden hamster, and rabbit embryos and their consecutive developmental stages are illustrated in Figs. 12.13–12.36; the drawing numbers correspond to stage numbers. In addition to the fig-

TABLE 12.7. Fertilization and Development of Mouse Zygote

Stages according to Khozai [67]		Stages according to Kaufmann [64]	
Stage Number	Description	Stage Number	Description
I	Penetration of egg by the sperm and completion of the second maturation division	I	Includes eggs with lightly staining pronuclei where no condensed chromatin is demonstrable
Ia	Spermatozoon head begins to swell, the egg is at metaphase II		
Ib	Spermatozoon head continues to swell, the egg is at anaphase II or telophase II		
Ic	Spermatozoon head is swollen. The second polar body is extruded		
II	Formation of pronuclei		
IIa	Primary pronucleoli appear in one of the pronuclei		
IIb	Primary pronucleoli are visible in both pronuclei		
III	Growth of pronuclei		
IIIa	One big primary pronucleolus is seen in each of the pronuclei		
IIIb	Two or more secondary pronuclei appear in each pronucleus; pronuclei increase in size		
IIIc	Two or more secondary pronucleoli appear in each pronucleus; pronuclei attain maximal size		
IV	Prophase of the first cleavage division		
IVa	Network consisting of chromatin fibers appears in one of the pronuclei. Secondary pronucleoli and the pronuclear envelope are retained.		

IIa   Includes eggs at the late pronuclear or early prometaphase stage where stainable chromatin fibers are present, though filamentous, and it is generally impossible to distinguish between individual elements

IIb   Late prometaphase includes eggs where two individual haploid groups of chromosomes can be identified. These groups are usually slightly asynchronous in their degree of chromosomal condensation

III   Early metaphase or "syngamy," where the two haploid groups have united on the spindle equator and have achieved a synchronous degree of chromosomal condensation

IV   The "chromatid" stage of metaphase where all the chromosomes are divided longitudinally into two morphologically identical chromatids

Va   Late metaphase includes eggs where the two chromatids are seen to be separating or have completely separated. The chromatids are probably still present on the spindle equator, since all preparations at this stage appear as a single group of 80 chromatids

Vb   Anaphase includes eggs where the group of 80 chromosomes has separated into two groups of 40 chromatids, and are observed to be migrating toward, or have reached, the poles of the spindle

IVb   Network of chromatin fibers appears in both pronuclei. Secondary pronucleoli are either present or disappear; the pronuclear envelope is present

V   Prometaphase of the first cleavage division

Va   Chromosomes are connected with remnants of secondary pronucleoli. The pronuclear envelope is not seen

Vb   Secondary pronucleoli disappear. Two groups of chromosomes come close to each other to yield one group

VI   Metaphase of the first cleavage division

VIa   Chromosomes are located at the spindle equator as a metaphase plate

VIb   Chromosomes consist of identical chromatids

VIc   Sister chromatids begin to separate but remain connected in the centromeric region

VII   Anaphase of the first cleavage division

VIII   Telophase of the first cleavage division

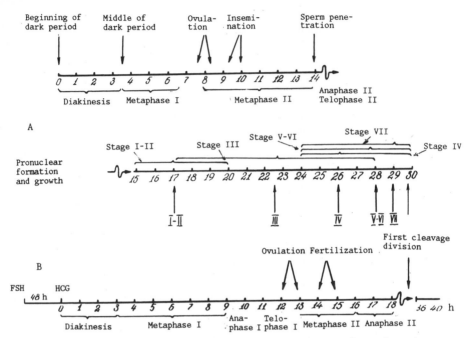

Fig. 12.12. Mouse oocyte maturation and karyological transformations in zygotes accompanying fertilization. A) After spontaneous ovulation in hybrid CBA/C57BL mice; B) after induced ovulation [42]. FSH) Follicle stimulating hormone; HCG) chorionic gonadotropin. In experiments with induced ovulation mice were given FSH followed by HCG after 48 h; immediately thereafter the animals were mated.

ures, we include in the text a more detailed description of embryo structure at different developmental stages of the mouse, rat, hamster, and rabbit; for the early stages we also describe the main changes between the adjacent stages.

In Table 12.9 the stages we have distinguished (numbered in the left column) are compared chronologically with the data given for mice by Theiler [118] and Rugh [100]. The analysis shows that, by dividing the entire mouse development into 29 stages, Theiler has limited himself to describing one or two morphological criteria for each stage of development. When comparing the embryogenesis of four different animal species, we used more general morphological characteristics as a basis for staging. Therefore, in most cases we have superimposed one, two, or three of the stages that we distinguished upon the Theiler stages. Comparing these two staging systems therefore appears to be justified and worthwhile.

Stage 1. Fertilized egg (zygote) (Fig. 12.37). The fertilized egg or zygote rapidly sheds follicular cells of the cone, loses the corona radiata, and is now surrounded on the outside only by the zona pellucida. The male and female pronuclei are formed from the spermatozoon head and the haploid egg nucleus, respectively. Well-formed pronuclei start moving toward each other and come close together near the egg center. Nuclear envelopes undergo dissolution, and both chromosomal complements join at the first cleavage spindle. The completion of the zygote stage in all four animal species takes place roughly 1 day after conception. Embryos at

TABLE 12.8. Consecutive Developmental Stages and Time of Their Onset during Rat, Mouse, Golden Hamster, and Rabbit Development

Stage	Name and short description of stage	Duration of pregnancy (days)			
		Rats	Mice	Hamster	Rabbit
1	Stage of zygote. One-cell diploid embryos are located in the oviduct ampule	1	1	1	22 h
2	Stage of 2 blastomeres. Embryos in the upper third of the oviduct	2	1.5	1.5	1 day, 2 h
3	Stage of 4 blastomeres. Embryos in the middle third of the oviduct	2.5	2.5	2.5	$1^1/_3$
4	Stage of 8 blastomeres. Embryos in the lower third of the oviduct	3	3	3	$1^3/_4$
5	Morula. 16–32 cell embryos pass to the uterus	4	3.5	3.5	2.5
6	Stage of one-layer blastocyst. Embryos (blastocyst) segregate to yield an inner cell mass and trophoectoderm	5	4.5	4	4
7	Stage of two-layer blastocyst. Embryo sheds zona pellucida and undergoes implantation (except rabbit). Inner and outer embryonic leaflets formed in inner cell mass	6	5.5	4.5	5
8	Stage of two-layer embryonic cylinder (sheath). Proamniotic cavity is formed. Ectoderm and endoderm are separated into embryonic and extraembryonic parts. Blastocoel is epibolized by distal endoderm, two-layer omphalopleura is formed	7	6	5.5	6
9	Stage of the primitive streak and three-layer embryo cylinder (shield). Primitive streak is formed. Extraembryonic and embryonic mesoderm appears. Neural plate begins to form in rats. Heart rudiment in front of oral plate; rudiment of urogenital system is formed in caudal part of mesoderm. Proamniotic cavity present in the embryo body	8	6.5	6.5	7
10	Stage of the head process and neural plate. Body of embryo contains neural plate, head process, and nerve toruli, which begin to merge in cranial region forming the forebrain rudiment. Eye invagination appears, as well as first gill arch and anterior intestinal portal	9	7.5	7	7.5

TABLE 12.8 (continued)

Stage	Name and short description of stage	Duration of pregnancy (days)			
		Rats	Mice	Hamster	Rabbit
11	Stage of deep neural groove. Neural groove is deep and open throughout its length. First 2–3 pairs of somites appear; heart rudiment appears in region located ventral to foregut; rare heart contractions begin. First aortal arch is formed. Germ cells may be identified in yolk sac epithelium	9.5	8.5	8	8
12	Formation of neural tube and 2 pairs of gill arches. Neural tube is formed; it produces 3 brain vesicles; spinal ganglia begin to appear. Stomodeum appears as well as first 2 pairs of gill arches. 10–12 somites. In rats the embryo begins to turn and its dorsal surface becomes concave. Otic placode is formed as well as IInd and IIIrd aortic arches; intermediate mesoderm undergoes condensation forming urogenital toruli. Hindgut as well as hepatic liver diverticulum appear in ectoderm. Germ cells begin to migrate from yolk sac passing the hindgut and mesentery	10	9	8.5	9
13	Stage of forelimb buds. Forelimb buds appear, lateral and medial nasal processes appear as well. Anterior neuropore is closed, optic and otic vesicles seen. Rathke's pouch comes into contact with funnel, forming rudiment of the pituitary	11	10	9	10
14	Stage of hind-limb buds. Hind-limb buds appear, maxillary processes formed, hypothalamus separates from thalamic region in mesencephalon. Genital toruli, pharyngotracheal bulge, and rudiment of dorsal pancreas appear	11.5	10.5	9.5	11
15	Stage of forelimb carpal plate and fusion of maxillary and nasolateral processes. Forelimbs separated into proximal and distal regions. Maxillary process fuses with nasolateral one. Lense and olfactory fossae appear; cerebral hemispheres begin to form	12.5	11	10	12
16	Stage of hind limb carpal plate and transformation of first gill arch into auditory meatus. Hind limbs separated into proximal and distal regions. External auditory meatus is formed, olfactory fossae become deeper, rudiments of thymus and parathyroid appear. Primary kidney begins to degenerate, germ cells migrate into undifferentiated gonads	13	12	10.5	13

TABLE 12.8 (continued)

Stage	Name and short description of stage	Duration of pregnancy (days)			
		Rats	Mice	Hamster	Rabbit
17	Stage of rounded auditory opening and first vibrissae. External auditory opening is rounded. Maxilla is formed. Mesenchyme rudiments of toes seen at first in forelimbs then in hind limbs. Mammary glands and first rudiments of vibrissae appear, first ossification sites in mandibula. Gonads may be identified	14	13	11	14
18	Stage of flipperlike extremities and formation of auricular floor. Distal parts of extremities have flipperlike pentaradial shape. Eye fissure is oval, auricular floor fold formed. Majority of trunk bones may be identified. Heart is fully divided into atria and ventricles	15.5	14	12	16
19	Stage of toe segregation and formation of hair follicles. Partial segregation of toes takes place. Hair follicles appear throughout body. Lingual papilla appear, palate processes begin to fuse	16.5	15	12.5	17
20	Stage of 5-toe extremities. Complete segregation of toes, nail rudiments appear. Omphalic hernia begins to close, kidney occupies definitive position. Palate processes fuse completely	17.5	16	13	18
21	Beginning of eyelid closure and closure of external auditory meatus. Eye fissure narrowed, auricular floor almost completely closes external auditory meatus. Omphalic hernis is completely closed	18.5	17	14	19
22	Stage of complete eyelid closure and closure of external auditory meatus. Floor of auricle completely closes external auditory metus. Eyelids completely closed	19.5	18	15	21
23	Birth of completely formed embryo. Embryonic period ends. Young animals are born. Body is extended and covered by numerous wrinkled skin folds	21	19	16	28
24	Stage of the newborn. 1-day-old rats, rabbits, mice, and hamsters are nude and wrinkled with tightly closed eyelids. Young animals of all these species nurse on mother's milk	One day after birth			

Note. Pregnancy terms are as follows: for rats, from the instant when spermatozoa are detected in the vaginal smear; for mice and hamsters, from the finding of the vaginal plug. Eight–nine hours should be added to values given in the table in order to obtain time post-fertilization; in the rabbit, fertilization takes place 10 h after verified mating. Some authors use a different system and designate the day when spermatozoa are detected in the vaginal smear as day 0. In order to pass from this (unfortunate, from our point of view) system of dating to the commonly accepted system, one should subtract 1 day from values given in our tables.

Figs. 12.13–12.36. Normal developmental stages of laboratory mammals. 13–18) Mouse; 19–24) rat; 25–30) golden hamster; 31–36) rabbit.

Figs. 12.13–12.18. Normal developmental stages of the mouse (*Mus musculus*). Drawing numbers correspond to stage numbers. Drawing 7' is an intermediate stage between stages 7 and 8. Drawing 10' is an intermediate stage between stages 10 and 11. Stages 1–6 (Fig. 12.13): photomicrography in the living state, phase contrast (data of A. P. Dyban). Stages 7–12 (Figs. 12.14 and 12.15): sections of embryos at corresponding stages (data of N. A. Samoshkina and V. S. Baranov). Stages 12–23 (Figs. 12.16–12.18): photomicrography of isolated embryos at corresponding stages (data of V. F. Puchkov and N. A. Chebotar).

Figs. 12.19–12.24. Normal developmental stages of the rat (*Rattus norvegicus*). Drawing numbers correspond to stage numbers. Stages 1–6 (Fig. 12.19): photomicrography in the living state; phase contrast (data of A. P. Dyban and N. A. Chebotar). Stages 6'–12 (Figs. 12.20 and 12.21): sections of embryos at corresponding stages (data of N. A. Samoshkina and V. S. Baranov). Stages 12–23 (Figs. 12.22–12.24): photomicrography of isolated embryos at corresponding stages (data of V. F. Puchkov and N. A. Chebotar).

Figs. 12.25–12.30. Normal developmental stages of the Golden hamster (*Cricetus auratus*). Drawing numbers correspond to stage numbers. Stages 1–4 (Fig. 12.25): photomicrography in the living state; phase contrast (data of N. A. Samoshkina and N. A. Chebotar). Stages 6–11 (Figs. 12.26 and 12.27): sections of embryos at corresponding stages (data of N. A. Samoshkina and V. S. Baranov). Stages 11'–23 (Figs. 12.28–12.30): photomicrography of isolated embryos at corresponding stages (data of V. F. Puchkov and N. A. Chebotar).

Figs. 12.31–12.36. Normal developmental stages of the rabbit (*Oryctolagus cuniculus*). Drawing numbers correspond to stage numbers. Stages 1–6 (Fig. 12.31): photomicrography in the living state; phase contrast (data of V. F. Puchkov, N. A. Samoshkina, and N. A. Chebotar). Stages 6'–9' (Fig. 12.32): sections of embryos at corresponding stages (data of N. A. Samoshkina and N. A. Chebotar). Stages 9–24 (Figs. 12.33–12.36): photomicrography of isolated embryos at corresponding stages (data of V. F. Puchkov and N. A. Chebotar).

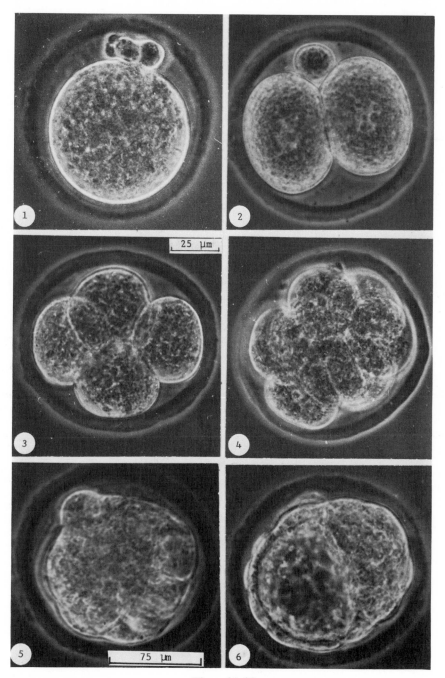

Fig. 12.13.

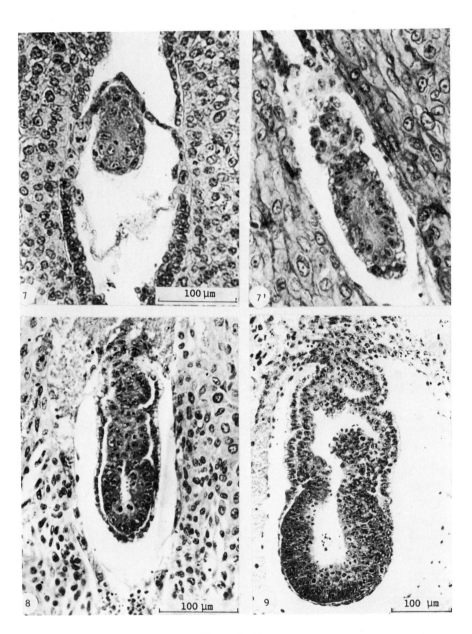

Fig.   12.14.

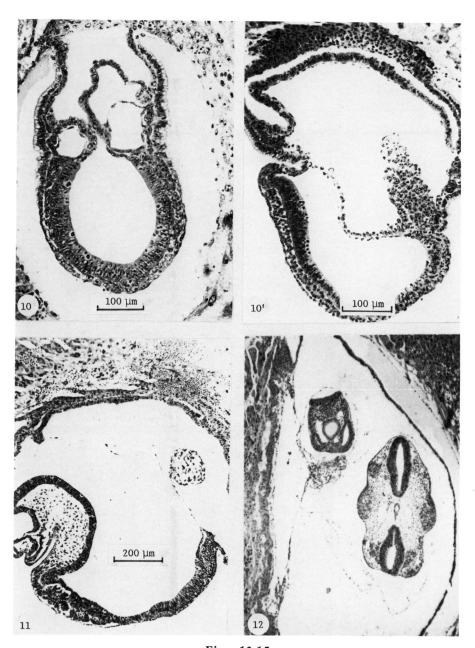

Fig. 12.15.

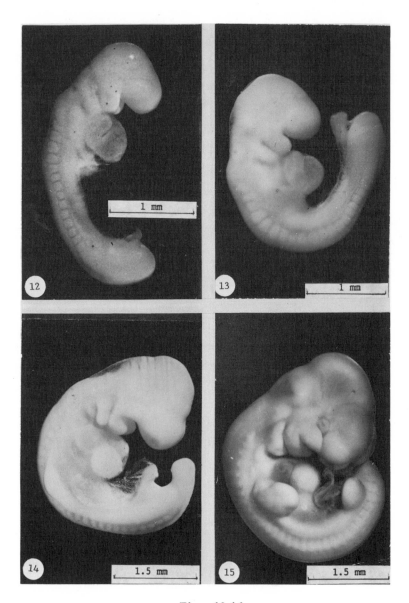

Fig. 12.16.

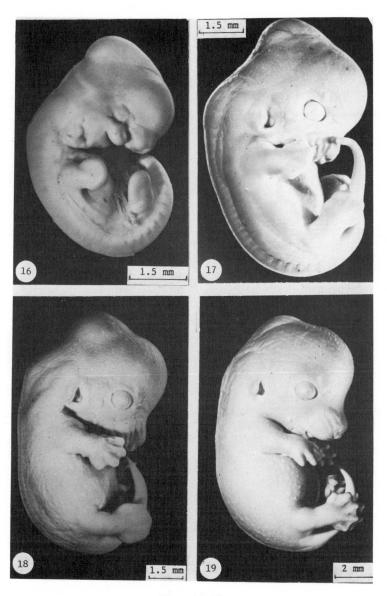

Fig. 12.17.

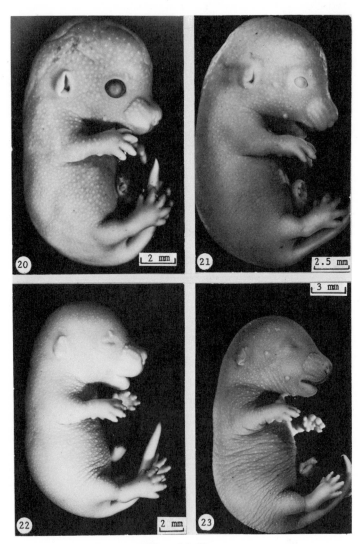

Fig.   12.18.

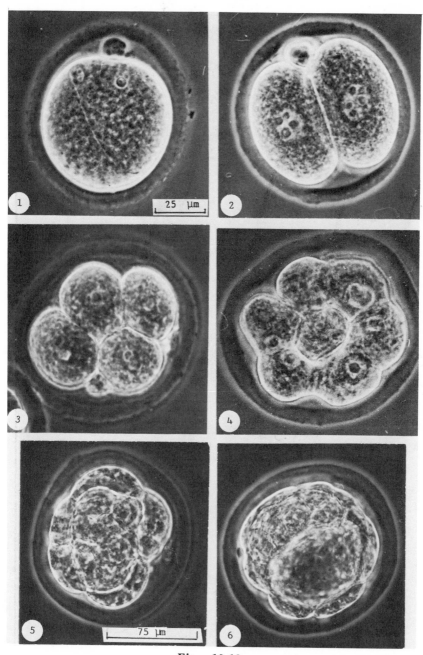

Fig.   12.19.

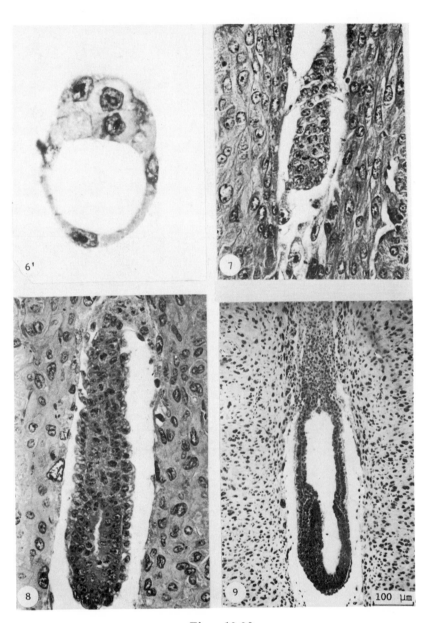

Fig.   12.20.

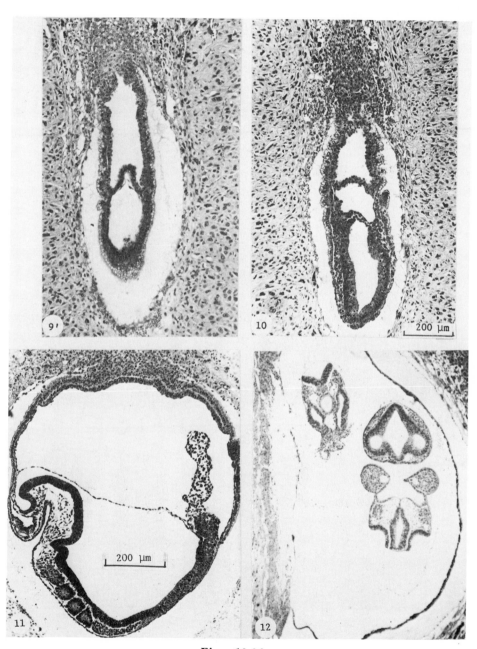

Fig. 12.21.

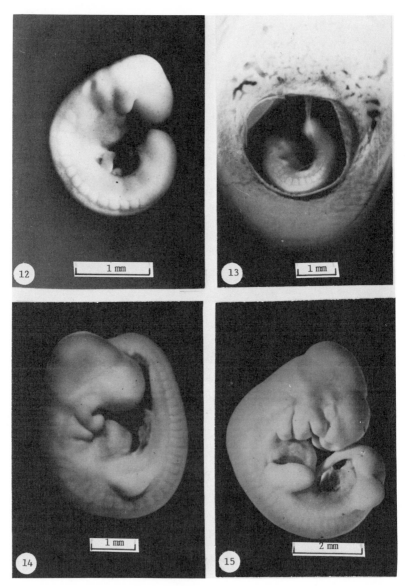

Fig.   12.22.

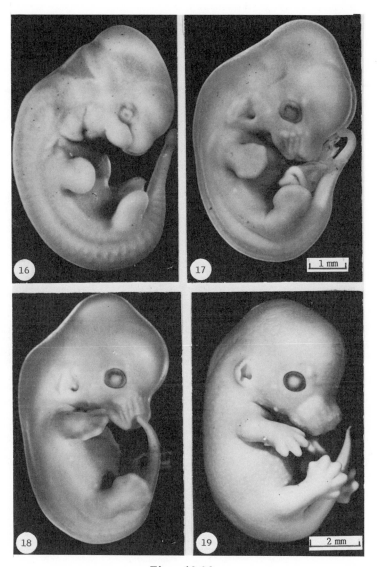

Fig. 12.23.

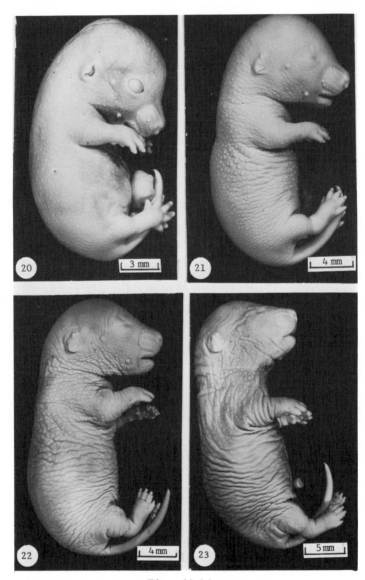

Fig.   12.24.

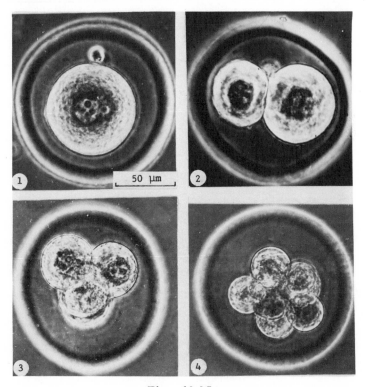

Fig.   12.25.

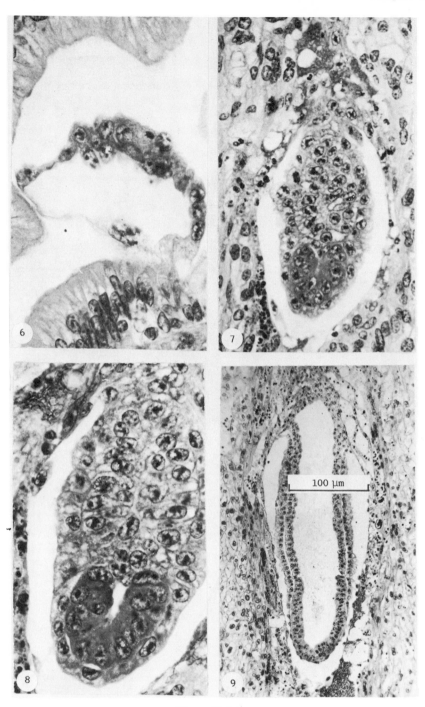

Fig. 12.26.

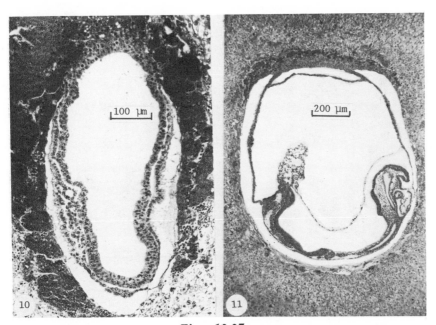

Fig.   12.27.

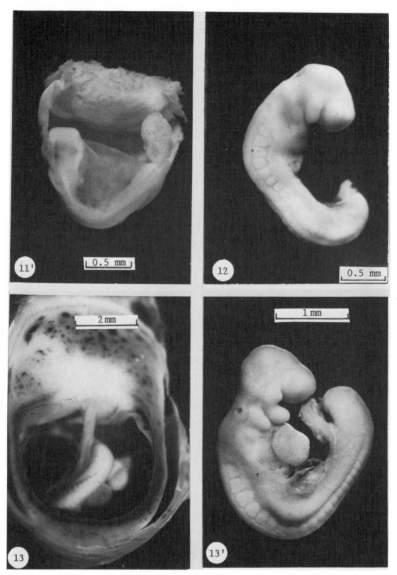

Fig.   12.28.

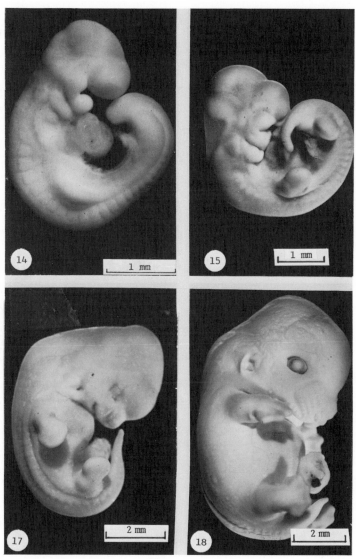

Fig. 12.29.

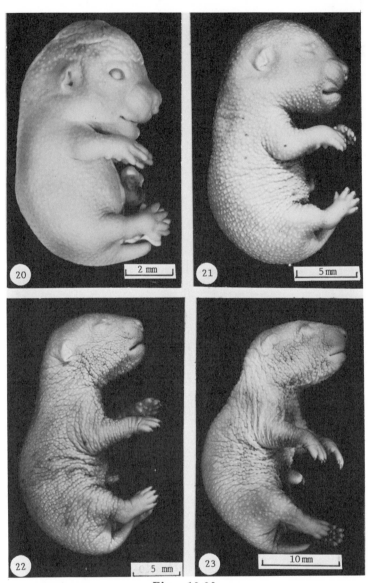

Fig. 12.30.

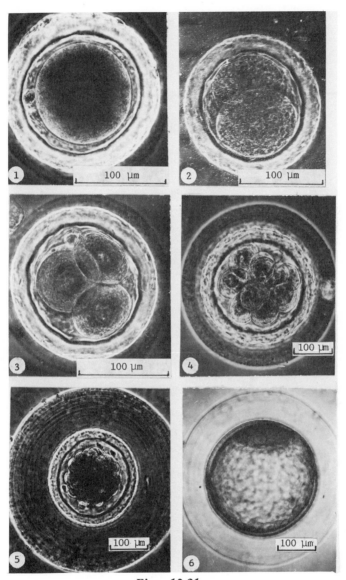

Fig. 12.31.

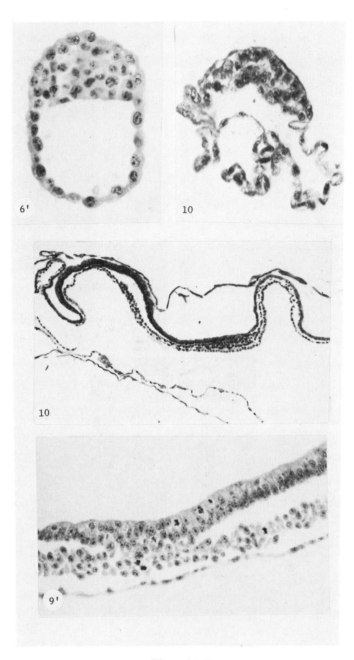

Fig.   12.32.

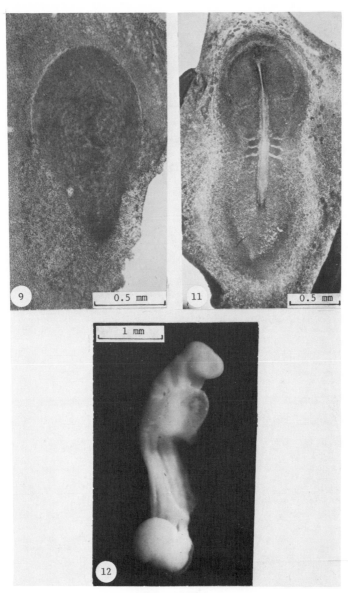

Fig. 12.33.

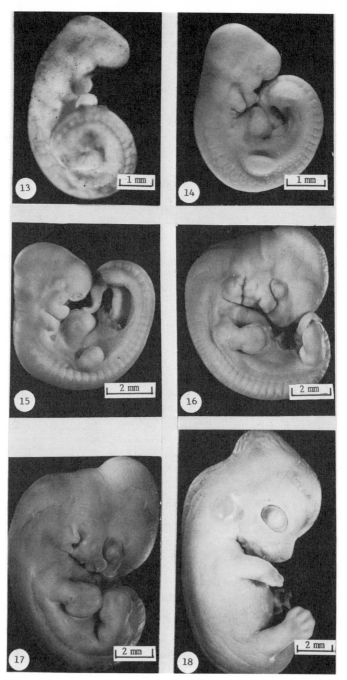

Fig.   12.34.

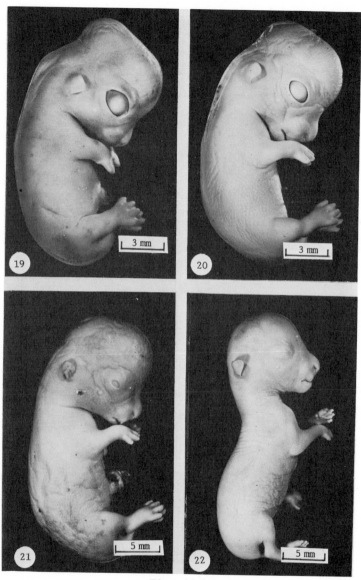

Fig. 12.35.

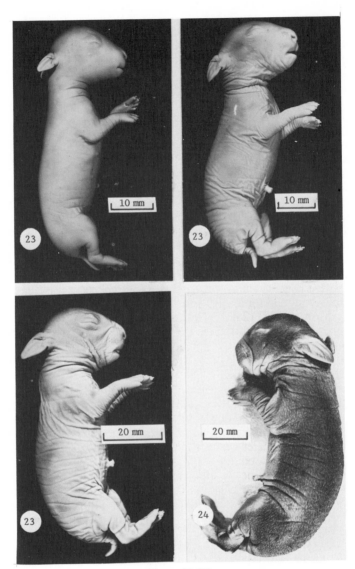

Fig.   12.36.

the zygote stage are in the oviduct ampule. In mice we divided this stage into three substages (see p. 370) on the basis of karyological changes. This is particularly interesting for experimental embryologists when precise dating of the first stages of embryonic development is required.

Stage 2. Two blastomeres (Fig. 12.38). Cleavage in laboratory mammals is completely nonuniform and asynchronous. The first cleavage division yields two cells somewhat different in size. The so-called mucin membrane, which shows intensive reaction for acid mucopolysaccharides, is formed on the surface of the zona pellucida of rabbit embryos when they pass through the oviduct. This mucous membrane is produced by the oviduct epithelial cells and is formed by gradual deposition.

Stage 3. Four blastomeres (Fig. 12.39). Second cleavage division in rats, mice, and hamsters takes place 12–24 h after the first division. In the rabbit the second division takes place 5–6 h after the first, and embryos with four blastomeres can be detected at 28–32 h after conception. The plane of the second cleavage furrow is located at right angles to the first one. One of the two blastomeres usually divided 0.5–1.5 h earlier than the other, so embryos with three blastomeres can be found during this period. Embryos with four blastomeres lie in the middle part of the oviduct.

Stage 4. Eight blastomeres (Fig. 12.40). The third cleavage division resulting in the formation of an eight-cell embryo usually takes place when the embryo moves through the lower third of the oviduct. The plane of the third cleavage division is located at right angles to the first two. The four blastomeres do not divide strictly synchronously, so intermediate states of embryos with five, six, or seven blastomeres can be found for a short period of time.

Stage 5. Morula (Fig. 12.41). The embryo consisting of 16–32 blastomeres is referred to as a morula. At this stage the embryos pass from the lower part of the oviduct into the uterus.

Stage 6. Single-layered blastocyst (Fig. 12.42). During the morula stage the embryo is transferred into the lumen of the uterus. Eccentrically located cavities appear between its cells; they become wider, fuse together, and form one common cavity. At this stage the embryo is referred to as a blastocyst. The blastocyst wall, which consists of a one-cell layer comprising trophoblast or trophectoderm on one pole, directly passes to a small group of cells, the inner cell mass, which is also called the embryoblast. Later it forms the embryo and its membranes. The embryoblast is not yet differentiated into separate parts at this stage and consists of small cells. In contrast, cells of the trophoblast are larger and have an elongated shape; they cover the embryoblast as a single layer and form the so-called Rauber layer. Embryos of rats, mice, and hamsters hardly grow at all during cleavage and blastulation. The size of the rabbit blastocyst is many times greater than that of the initial egg.

Blastocysts consisting of one cell layer are distributed along the uterus at roughly equal distances from one another. The regions of the uterus in contact with blastocysts undergo intensive morphogenetic events. The earliest response to the presence of blastocysts is shown by blood vessels. If a uterus freshly excised from a live rat is inspected immediately after the embryos have emerged from the oviducts, one can see regions with markedly dilated vessels; as a consequence, the whole uterus appears to be segmented [114]. Uterine mucosa shows increased DNA synthesis and increased cell proliferation. At this time, when the blastocysts

TABLE 12.9. Chronological Comparison of Normal Developmental Stages Distinguished by Us in Rats, Mice, Hamsters, and Rabbits, with Data of Theiler [118] and Rugh [100]

Stages by Dyban et al. [118]	Theiler [118]	Stages by Theiler	Rugh [100]	Days of pregnancy
1	1-cell stage (1–20 h)	1	1- to 2-cell stage in ampulla of oviduct	1
2, 3, 4	Beginning of cleavage (20–24 h)	2	2–16 cells, in transit through oviduct to uterus	2
	Cleavage of eggs (2 days)	3	Morula, in upper uterus	3
4, 5				
5	Advance of cleavage (3 days)	4		
			Free blastocytes in uterus. Shedding of zona pellucida	4
6	Blastocyst (4 days)	5		
7	Implantation (4.5 days)	6	Implantation beginning; inner cell mass; tropho blastic cone	4, 5
8	Formation of egg cylinder (5 days)	7	Pendant inner cell mass; endoderm; proamniotic cavity; primitive streak	5
8, 9	Differentiation of egg cylinder (6 days)	8	Implantation complete; extraembryonic parts developing; uterine reaction	6
9, 10	Advanced endometrial reaction (6.5 days)	9	Ectoplacental cone; amniotic folds, primitive streak; mesenchyme, heart, pericardium forming; head process	7
10, 11	Amnion (7 days)	10	Early neurula; neural plate, chorioamniotic stalk; embryonic lordosis; allantoic stalk beginning; pendant inner cell mass with 3 cavities; exocoelom; amniotic cavity; somites beginning to differentiate; foregut	7.5

10, 11	Neural plate, presomite stage (7.5 days)	11	Somites 1–4; visceral arch I; Reichert's membrane; pregerm cells in yolk-sac endoderm; embryonic cyst fused with ectoplacenta and also with allantoic stalk; early regression of yolk sac; heart primordia; thyroid; optic sulcus; 1st aortic arch	8
11	First somites (8 days). 1–7 somites	12	Somites 8–12; visceral arch II; disc and yolk-sac placenta; embryo begins to reverse lordosis to curve ventrally, with germ layers in proper relation; otic invagination; liver; 2 pharyngeal pouches; nephrogenic cord	9
12	Turning of embryo (8.5 days). 8–12 somites	13	Somites 21–25; yolk stalk closes; primary germ cells migrating via mesentery to final site; primitive streak gone; tail, limb, and lung buds forming; mesonephric tubules; 3 pharyngeal pouches	9.5
12, 13	Formation and closure of anterior neuropore (9 days). 13–20 somites	14	Somites 26–28; visceral arch III; lung buds; pronephros reaches cloaca; posterior limb bud and lens forming; gastrointestinal tract developing; sense organs differentiating; aortic arches I, II, and III	10
13, 14	Formation of posterior neuropore. Forelimb bud (9.5 days). 21–29 somites	15	Somites 29–42; total length 6.2 mm; thyroid; umbilical hernia; meso- and metanephros in early stages of formation; ventral pancreatic rudiment; ureteric bud and mesonephric ducts to urogenital sinus; endocardial cushion fused; epiphysis and hypophysis evaginate; subcardinals formed; olfactory fibers to the brain; pigment in retina; lens cells elongate; aortic arches IV and V	11
14	Closure of posterior neuropore; hind-limb bud and tail bud (10 days). 30–34 somites. (3.1–3.9 mm)	16		

TABLE 12.9 (continued)

Stages by Dyban et al.	Theiler [118]	Stages by Theiler	Rugh [100]	Days of pregnancy
15	Deep lens indentation (10.5 days). 35–39 somites. 3.5–4.9 mm	17	Somites 43–48; aortic arch V gone and VI reduced; vitreous humor; pancreatic rudiments; mammary welts; ribs and centrum chondrify; secondary bronchi; posterior cardinals degenerate; epithelial cords in testis; choroid fissure closed; superior vena cava enters heart	12
16, 17	Closure of lens vesicle (11 days). 40–44 somites. 5–6 mm	18	Somites 49–50; cephalization resulting in brain differentiation; spinalganglia; histological sex differentiation; active and complete circulation; atrioventricular valve; interventricular septum complete; nerves in optic stalk; otic capsule precartilaginous; esophageal submucosa thickens; neuroblasts in retina; enucleate red cells 1%	13
18, 19	Lens vesicle completely separated from surface (11.5 days). 45 somites. 6–7 mm	19	Somites 61–63; total length 9.6 mm; digital development; umbilical hernia receding; mesonephros degenerating and metanephros becoming functional; diaphragm completed; gonad primordia becoming vascular; intestinal villi; ossification of frontal and zygomatic arch; saccule and utricle separated; enucleate red cells 25%; aortic pulmonary semilunars	14
19, 20	Earliest sign of fingers (12 days). 7–9 mm	20	Somites 64–65; snout protruding from face and lifts off chest; hair follicles developing; digits clear; body contour more rounded; cartilage in humerus; centrum and ribs ossifying; nucleated red cells 5%; stratum granulosum; cerebellum fused at midline	15
20, 21	Anterior foot plate indented. Marked pinna (13 days). 9–11 mm	21	All somites formed, differentiating from anterior to posterior; fetal stage; eyelids; pinnae cover ears; umbilical hernia withdrawn	16, 17, 18

20, 21	Fingers separate distally (14 days). 11–12 mm	22	Ossification proceeding; corpus callosum formed; centra ossified; nucleated red cells down to 1%; proliferation of gastric glands; eyelids sealed; extraembryonic membranes reach maximum development; tail alone now 10 mm in length	
21	Toes separate (15 days). 12–14 mm	23		
21, 22	Reposition of umbilical hernia (16 days). 14–17 mm	24		
	Fingers and toes joined together (17 days). 17–20 mm	25		
	Long whiskers (18 days). 19.5–22 mm	26		
23	Newborn mouse (19 days)	27	Birth	19 to 20
24	Postnatal development	28		
	Weight curves	29		

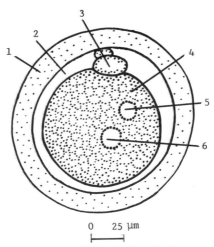

0   25 μm
├───────┤

Fig. 12.37. Stage 1. Zygote. Rat egg, drawing from phase contrast. 1) Zona pellucida; 2) perivitelline space; 3) polar body; 4) egg cytoplasm; 5) smaller female pronucleus; 6) larger male pronucleus.

lie freely in the uterine cavity, all these changes result in a local swelling of the cavity mucosa, leading to the formation of future implantation sites.

Stage 7. Two-layered blastocyst and two embryonic germ layers (Fig. 12.43). The stage of the late or two-layered blastocyst is characterized by the segregation of the cellular layer from the lower surface of the intensely proliferating inner cell mass comprising the embryoblast. This layer is a rudiment of the inner germ layer, the endoderm. The remaining cells of the inner cellular mass comprise the ectodermal rudiment.

Peripheral cells of the intensely proliferating endoderm become displaced along the inner surface of the trophoblast, forming a layer of the so-called distal or parietal endoderm, which comprises the wall of the primary yolk sac tightly apposed to the trophoblastic layer. At this stage, therefore, the blastocyst wall consists of two layers, and the blastocyst cavity becomes converted into a yolk sac cavity. The embryos shed the zona pellucida and begin to attach to the uterine wall. This attachment is generally referred to as implantation. Small mouse, rat, and hamster blastocysts undergo eccentric interstitial implantation accompanied by the transfer of the blastocyst from the lumen of the uterus into a deep diverticulum or the implantation crypt (Fig. 12.44); this is accompanied by the dying out of the epithelium present in this diverticulum followed by the blastocyst penetrating the uterine mucosa.

Implantation in rabbits takes place later, at the stage when the head protuberance and neural plate are formed (stage 10). Here implantation is of the surface type; large blastocysts remain in the uterus lumen and do not penetrate under the uterine epithelium. Prior to implantation, the mouse, rat, and hamster blastocysts become oriented so that their inner cellular masses are directed to the mesometrium side of the uterus. Blastocysts which have entered the crypt do not usually pass through to the full depth. Cells around the crypt proliferate intensively, which results in their forming distinct local thickenings; the uterus at this stage has a char-

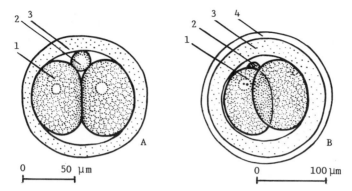

Fig. 12.38. Stage 2. Two blastomeres. A) Rat; B) rabbit. Drawings from phase contrast. 1) Blastomere; 2) polar bodies; 3) zona pellucida; 4) mucous membrane.

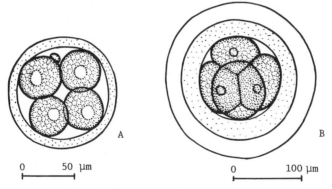

Fig. 12.39. Stage 3. Four blastomeres. A) Rat; B) rabbit. Drawings from phase contrast.

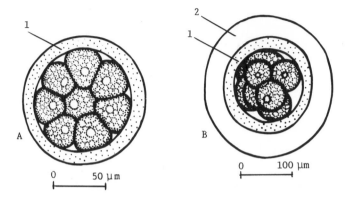

Fig. 12.40. Stage 4. Eight blastomeres. A) Rat; B) rabbit. Drawings from phase contrast. 1) Zona pellucida; 2) mucous membrane.

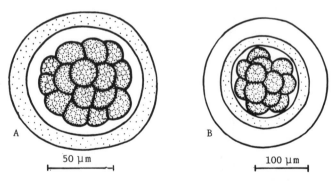

Fig. 12.41. Stage 5. Morula. A) Rat; B) rabbit. Drawings from phase contrast.

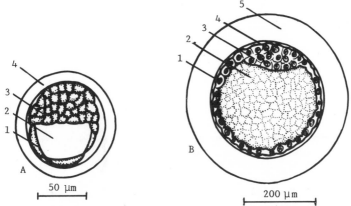

Fig. 12.42. Stage 6. Single-layer blastocyst. Drawings from phase contrast. A) Rat: 1) trophoblast; 2) blastocyst cavity; 3) inner cell mass; 4) zona pellucida. B) Rabbit: 1) trophoblast; 2) blastocoel; 3) inner cell mass; 4) zona pellucida; 5) mucous membrane.

acteristic "string-of-beads" shape. The blastocyst, located in the implantation crypt, appears as an elongated sac, its surface showing the presence of the first large cells of the trophoblast. At the places where the blastocyst contacts the crypt walls, the epithelial layer of the crypt disappears, and the blastocyst trophoblastic cells come into direct contact with the uterus decidua.

Stage 8. Two-layered egg cylinder (in the rabbit, two-layered embryonic shield). With the beginning of implantation the size of the yolk sac cavity drastically increases in mice, rats, and hamsters. This is followed by the rapid growth of the inner cell mass, and the elongated egg cylinder forms by stage 8 (Fig. 12.45). It consists of two germ layers: the inner mass of ectodermal cells and the outer layer of endodermal cells. In the rabbit, no such inversion of germ layers takes place, and the ectoderm is located externally with the endoderm inside.

The whole of the endoderm lining the yolk sac cavity may be divided into the distal part, adjacent to the trophoblast wall, and the proximal part, covering the egg cylinder. The proximal endoderm comprises a layer of cuboidal cells, the cytoplasm of which contains numerous inclusions. The endoderm of the free edge of the embryonic cylinder comprises the rudiment of the enteric endoderm and consists of flattened cells. Two parts can be clearly distinguished in the ectoderm: the ventral part is more intensely stained and possesses elongated nuclei, and the dorsal lighter region has rounded nuclei. The dorsal ectoderm produces different extraembryonic structures (the extraembryonic ectoderm), while the ventral part produces the embryonic ectoderm. The central part of the embryonic cylinder ectoderm shows a long narrow groove which is referred to as the proamniotic cavity. The dorsal extraembryonic ectoderm, in its external part turned toward the crypt cavity, forms a long outgrowth, the ectoplacental cone. Its dorsal part is loosened and the spaces between the cell strands are infiltrated by the maternal blood.

The implantation of mouse, rat, and hamster embryos is completed at stage 8 (Fig. 12.46). Decidual tissue proliferates intensively and, as a consequence, the implantation crypt lumen on the antimesometrial side narrows, its edges close, and the blastocyst is now completely submerged in the deciduoma. The decidual tissue also grows on the mesometrial side of the uterus, resulting in the formation of a subplacental part, the decidua basalis. Late decidual formations fuse together, and the primary uterine lumen at this place becomes obliterated.

In contrast to mice, rats, and golden hamsters, the rabbit blastocyst lies freely in the uterine cavity. The inner cell mass remains flattened and greatly expands, forming the so-called embryonic field or shield. It consists of two layers, ectoderm and endoderm. The trophoblast located under the embryonic field (Rauber layer) is thin or may even be completely absent. The embryonic shield is pear-shaped, its wide end corresponding to the anterior part of the embryo; the narrow end corresponds to the posterior part. From inside about half the blastocyst is covered by the extraembryonic endoderm.

Stage 9. Primitive streak and a three-layer embryonic cylinder (Figs. 12.47 and 12.48). The most characteristic features of this stage involve the formation of the primary streak and the third germ layer or mesoderm. In the rat, mouse, and hamster the posterior end of the embryonic cylinder contains an elongated and

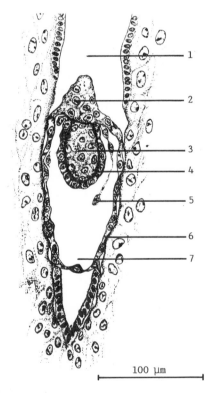

Fig. 12.43. Stage 7. Two-layered rat blastocyst. 1) Implantation crypt; 2) extraembryonic ectoderm; 3) embryonic ectoderm; 4) proximal endoderm; 5) distal endoderm; 6) trophoblast; 7) yolk cavity.

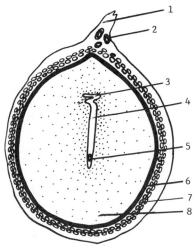

Fig. 12.44. Blastocyst implantation (rat, stage 7). Cross section, schematic drawing. 1) Mesometrium; 2) blood vessels; 3) uterus lumen; 4) implantation crypt; 5) blastocyst; 6) ring muscles; 7) longitudinal muscles; 8) decidual tissue.

thicker ectodermal region, which is called the primary streak. Mesodermal cells migrating out of the primitive streak spread laterally and below between the ectoderm and endoderm, and there form the intermediate mesoderm, in which the urogenital rudiment undergoes determination in the rat. Individual mesodermal cells quickly reach the anterior wall of the proamniotic cavity.

Sagittal sections in the middle part of the egg cylinder enable one to see the ectodermal bulge which deeply penetrates the proamniotic cavity; this is the posterior amniotic fold. Similar ectodermal protrusions may be found on the anterior and lateral walls of the embryonic cylinder. The anterior and lateral amniotic folds are poorly developed. Homogeneous and intensely stained by aniline blue, the Reichert membrane may be distinctly identified at this stage between the cells of the distal ectoderm and the trophoblast.

Rabbit embryos at this stage also have the primary streak, which is located at the posterior narrow end of the embryonic shield. Furthermore, the middle germ layer is formed in a similar way in rats, mice, and hamsters. Rabbit embryos show a characteristic difference, however, since the blastodermic vesicle has still not undergone implantation. It is quite large, its size being 4.5 × 3 mm.

Stage 10. Head process and the neural plate (Figs. 12.49 and 12.50). The embryo has a head or chordal process and the neural plate above it. The head end of the primitive streak contains a thickening, Hensen's node. The head process begins from this bulge, which is located between the ectoderm and endoderm and appears as a mesodermal wedgelike streak. It is located along the median line and directed toward the head by its pointed part. The ectoderm lying over the head process forms the neural plate, which is markedly widened at the anterior end. Its lateral regions are raised and form neural folds. The neural folds fuse in the cranial region and form the forebrain rudiment. The neural groove is located between the folds. At this stage the chordal rudiment may be distinguished in the head process

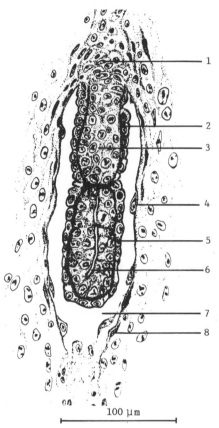

Fig. 12.45. Stage 8. Two-layered egg cylinder. Rat embryo. Drawing of section. 1) Ectoplacental cone; 2) proximal endoderm; 3) extraembryonic (dorsal) ectoderm; 4) cells of distal endoderm; 5) embryonic (ventral) ectoderm; 6) proamniotic cavity; 7) yolk cavity; 8) trophectoderm.

in the part which is located immediately under the neural groove. Small lacunar formations appear between stages 9 and 10 in the amniotic fold mesoderm. By stage 10 they increase in number and size and fuse together to yield one exocoelomic cavity. At the dorsal side the wall comprises the chorion plate, and the ventral wall is formed by the amniotic membrane. Thus, stage 10 embryos possess three proamniotic cavities: the ectoplacental, the exocoelomic, and the amniotic. The exocoelomic cavity contains the allantois rudiment, a mesodermal process beginning from the posterior end of the primary streak. The amniotic cavity of the exocoelom plays an important part in further development, while the ectoplacental cavity quickly reduces. The lateral walls of the exocoelom, which consist of extraembryonic mesoderm and extraembryonic endoderm in mice, rat, and hamsters, comprise the yolk sac and its visceral part in particular. The development of blood islets and blood vessels proceeds subsequently in the yolk sac wall.

A cavity or anterior enteric invagination comprising the foregut rudiment can be found at the anterior tip of the embryonic cylinder; just below the place where the yolk sac is located, the extraembryonic endoderm passes into embryonic endoderm. The rabbit embryo undergoes implantation at stage 10. The implantation is of the surface type, large blastocysts occupying the entire uterus lumen.

Stage 11. Stage of the deep neural groove (Figs. 12.51 and 12.52). The borders of the neural plate are thickened and rise above the remaining ectoderm: a deep groove is formed between two longitudinal neural folds. Eye fossae are present anteriorly on the inner surface of the neural fold (rudiments of the eye vesicles); in rats they appear earlier (see Table 12.7), while external thickening of the ectoderm corresponding to the otic placode is present. The head fold, which separates the head part of the embryo from the extraembryonic tissues, appears between the anterior tip of the neural fold and the site of amnion contact with the yolk sac. The rudiment of the first gill arch appears in mice, while in rats it has already formed at the preceding stage. A pair of tubular heart rudiments appears in the mesoderm near the anterior enteric invagination; they fuse to form a common rudiment. The foregut rudiment has considerably progressed in development compared with the preceding stage, and forms a deep pouch. Together with the covering ectoderm, it forms a protuberance directed toward the amniotic cavity. The posterior end endoderm has a shallow invagination, which is the rudiment of the hindgut. A deformation of the anterior and posterior enteric invaginations indicates the segregation of the body of the embryo from the extraembryonic parts. In median

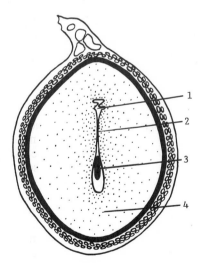

Fig. 12.46. Scheme showing uterus horn containing rat embryo (stage 8). Cross section. 1) Uterus lumen; 2) fusion of crypt borders over the embryo; 3) embryo; 4) decidual tissue.

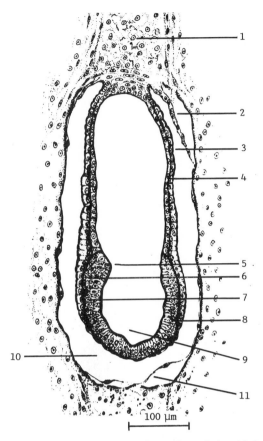

Fig. 12.47. Stage 9. Primitive streak, formation of the third germ layer. Rat embryo, drawing of section. 1) Ectoplacental cone; 2) distal endoderm; 3) proximal endoderm; 4) extraembryonic ectoderm; 5) posterior amniotic fold; 6) mesoderm; 7) primitive streak; 8) embryonic ectoderm; 9) proamniotic cavity; 10) yolk cavity; 11) Reichert's membrane.

section the embryo has a convoluted S-like shape. The stomodeum is formed in embryos at this stage as an ectodermal invagination directed toward the anterior end of the foregut. A two-layer wall consisting of the stomodeum ectoderm and the endoderm of the anterior tip of the gut forms the stomodeal plate. Two to four pairs of somites are being formed in the axial mesoderm. Rudiments of blood islets, as well as the first germ cells, are found in the yolk sac. The chorion plate is in contact with the base of the ectoplacental cone, and the ectoplacental cavity undergoes obliteration. The allantois is in contact with the chorion plate. The rudiment of the chorioallantoic placenta is formed by the fusion of chorion, ectoplacental cone, and part of the allantois.

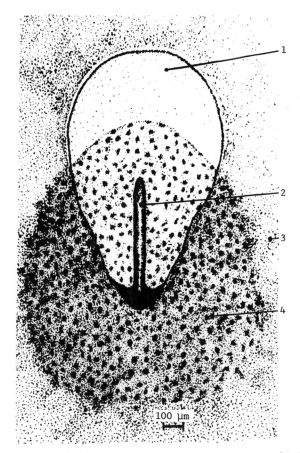

Fig. 12.48. Stage 9. Primitive streak, three-layered shield. Rabbit embryo. Drawing of whole-mount preparation. 1) Light zone; 2) primitive streak; 3) trophoblast; 4) coelomic mesoderm.

Stage 12. Formation of the neural tube and two pairs of gill arches (Figs. 12.53 and 12.54). The embryos of rats, mice, and hamsters prolapse from the dorsal position into the amniotic cavity and change from an S-shaped to a C-shaped configuration. There are 10–12 pairs of somites. The neural folds fuse to yield the neural tube possessing anterior and posterior neuropores. The rudiments of three brain vesicles, forebrain, midbrain, and hindbrain, can be seen. Neural ganglia begin to form in the cranial region. Lateral trunk folds are distinct. The caudal fold has just appeared. The second gill arch (hyoid, sublingual) is formed posteriorly from the first gill arch (mandibular).

The eye vesicle is formed; the ear placode prolapses and forms the otic invagination. The hindgut looks like a deep pouch. The tubular heart has an S-shaped
configuration, and the myocardium starts to differentiate. The posterior end of the
cardiac tube has formed a venous sinus which receives large veins, such as the umbilical and the Cuvierian ducts. The arterial stem, which continues in a short abdominal aorta, departs from the anterior end of the cardiac tube. In mice the first

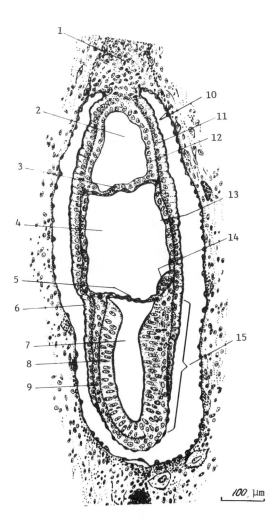

Fig. 12.49. Stage 10. Head process, neural plate. Rat embryo. Drawing of
section. 1) Ectoplacental cone; 2) ectoplacental cavity; 3) chorion plate; 4)
exocoelom; 5) amnion; 6) anterior intestinal protrusion; 7) amniotic cavity; 8)
embryonic ectoderm; 9) head process; 10) distal endoderm; 11) proximal
endoderm; 12) extraembryonic ectoderm; 13) yolk sac; 14) allantois; 15)
primitive streak.

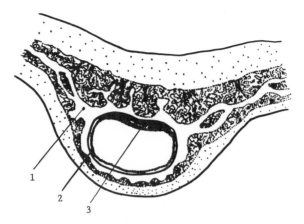

Fig. 12.50. Surface implantation of rabbit embryo (stage 10). Schematic drawing of longitudinal section of uterus. 1) Uterus cavity; 2) two-layered omphalopleura; 3) embryonic shield.

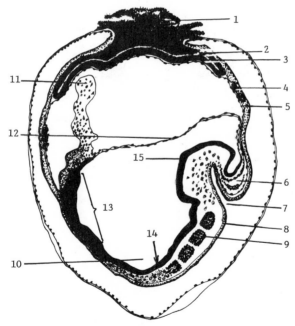

Fig. 12.51. Stage 11. Deep neural groove. Rat embryo. Drawing of section. 1) Ectoplacental cone; 2) ectoplacental cavity; 3) chorion; 4) blood islet; 5) yolk sac; 6) heart rudiment; 7) foregut; 8) endoderm; 9) somite; 10) amniotic cavity; 11) allantois; 12) amnion; 13) primitive streak; 14) ectoderm; 15) head fold.

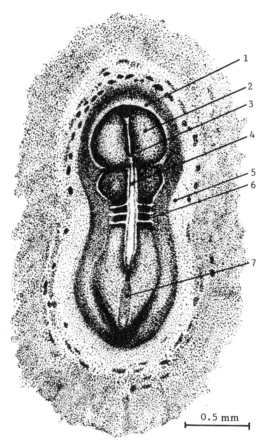

0.5 mm

Fig. 12.52. Stage 11. Deep neural groove. Rabbit embryo, whole-mount preparation. 1) Proamnion; 2) head neural fold; 3) neural groove; 4) head process; 5) light field; 6) somite; 7) primitive streak.

aortic arch is present, while in rats the second and third aortic arches are formed. Anterior and posterior enteric orifices are narrowed, and a ringlike constriction begins to form between the midgut and the yolk sac. It is now possible to speak of the beginning of umbilical cord formation. The intermediate mesoderm shows a distinct presence of the nephrogenic strand from which all three generations of kidneys develop sequentially. The foregut is extended in the region caudal to the oral plate and forms the rudiments of the pharynx. The thyroid rudiment appears in the lateral pharyngeal walls. The esophagus and the stomach are formed caudal to the throat. An excrescence, which is a common rudiment of the pancreas, liver, and yolk vesicle, is located immediately behind the stomach. The trabeculae differentiate in the liver of rat embryos. Abundant blood islets are present in the wall of the yolk sac; blood vessels begin to appear; primary germinal cells are distinguishable

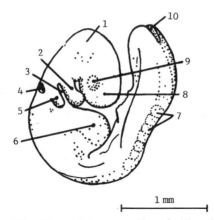

1 mm

Fig. 12.53. Stage 12. Neural tube, two pairs of gill arches. Rat embryo. Drawing of whole-mount preparation. 1) Midbrain; 2) first gill arch (mandibular); 3) first gill cleft; 4) otic fossa; 5) second gill arch (sublingual); 6) heart; 7) somites; 8) forebrain; 9) optic vesicle; 10) posterior neuropore.

in the yolk sac endoderm; in rats they begin to migrate, passing through the hind gut and the mesentery.

The rabbit embryo also segregates from the extraembryonic parts. Its cerebral region is deeply submerged into the proamnion; the anterior end is covered by the amniotic fold, which grows backward to form the amniotic membrane.

Stage 13. Forelimb buds (Fig. 12.55). The embryo bends in a ringlike fashion in such a way that the telencephalon almost touches the caudal end of the body. Up to 20 pairs of somites are present. Forelimb buds appear as elongated toruli at the level of the 8th–14th pairs of somites. Three to four pairs of gill arches are formed. The posterior neuropore closes. The brain is segmented into five regions: telencephalon, diencephalon, mesencephalon, metencephalon, and medulla oblongata. The hypothalamus may be distinguished in the diencephalon. A fingerlike invagination, the midbrain funnel, appears in the ventral wall of the mesencephalon. Rathke's pouch grows toward the funnel from the pharynx wall, forming the rudiment of the pituitary. The spinal cord differentiates into ependyma, mantle, and the edge veil. Olfactory placodes appear. The otic vesicle transforms into the rudiments of utriculus, sacculus, and the endolymphatic duct. The lens placode appears as a thickening in the ectoderm over the eye vesicle. The heart separates into ventricles and atria and now consists of four chambers; trabeculae are formed in the ventricles. First, second, and third pairs of aortic arches appear. The dorsal part of the pancreas is formed. The nephrogenic strand contains elements of the embryonic kidney. Primary germ cells move out of the yolk sac and migrate to the dorsal mesentery. Genital toruli appear in the rat, and gonocytes migrate there.

Stage 14. Hind-limb buds (Fig. 12.56). Embryos have four characteristic flexures; head, neck, trunk, and caudal. The tail rapidly grows in length in mice, rats, and rabbits. Forelimb buds elongate; formation of hind-limb buds takes place at the level of the 29th–31st somite pairs. The maxillary process appears and the ends of the mandibular arches fuse together. The heart and liver form a cardiohepatic eminence, easily noticeable from the outside. The lens and olfactory fossae are present. Rudiments of the spinal ganglia can be seen along the spinal cord toward the head end. Four pairs of aortic arches are formed. A regression of the first and second aortic arch begins. The trachea is being formed; it gives two processes corresponding to rudiments of the bronchi and pulmonary alveoli. The stomach undergoes differentiation, and the mesonephros and genital toruli are being formed in mice. Primary gonocytes migrate into the gonad rudiments.

Stage 15. Forelimb digital plate and fusion of maxillary and nasolateral processes (Fig. 12.57). Forelimb rudiments extend considerably and their distal part is flattened and widened, thereby forming the so-called digital plate. The hind-limb buds appear, each resembling a fin with a wide base. Maxillary and nasolateral processes touch each other and begin to fuse, the groove between them corre-

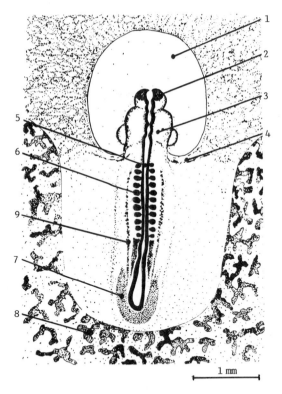

Fig. 12.54. Stage 12. Rabbit embryo. Drawing of whole-mount preparation. 1) Proamnion; 2) eye vesicle; 3) head part of the embryo; 4) yolk vein; 5) neural tube; 6) somite; 7) allantois rudiment; 8) changes in the region of the future allantoic placenta; 9) nephrogenic strand.

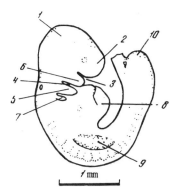

Fig. 12.55. Stage 13. Forelimb buds. Rat embryo. Drawing of whole-mount preparation. 1) Midbrain; 2) forebrain; 3) oral pit; 4) hyomandibular cleft; 5) second gill arch; 6) first gill arch; 7) third gill arch; 8) cardiohepatic eminence; 9) forelimb bud; 10) tail.

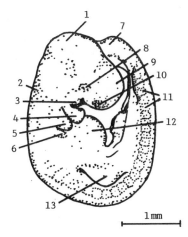

Fig. 12.56. Stage 14. Hind-limb buds. Rat embryo. Drawing of whole-mount preparation. 1) Midbrain; 2) medulla oblongata; 3) maxillar process; 4) first gill arch; 5) second gill arch; 6) third gill arch; 7) tail; 8) eye vesicle; 9) forebrain; 10) olfactory placode; 11) hind-limb bud; 12) cardiohepatic eminence; 13) forelimb bud.

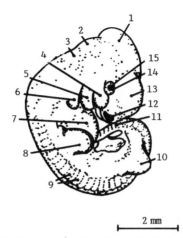

2 mm

Fig. 12.57. Stage 15. Formation of the forelimb digital plate. Rat embryo. Drawing of whole-mount preparation. 1) Mesencephalon; 2) cerebellum; 3) medulla oblongata; 4) maxillar process; 5) first gill arch; 6) second gill arch; 7) cardiohepatic eminence; 8) forelimb digital plate; 9) somites; 10) hind-limb bud; 11) tail; 12) olfactory pit; 13) telencephalon; 14) eye; 15) lens.

sponding to the rudiment of the nasolacrimal channel. The lens invagination is closed to yield the lens vesicle, and the olfactory depression becomes deeper. Rathke's pouch (somewhat later in mice) fuses with the midbrain funnel, forming the pituitary. Cerebral hemispheres are formed and the retina undergoes differentiation. Paired cranial nerves, III, IX, XI, and XII, are formed. The esophagus and the duodenum undergo intense development, and the ventral lobe of the pancreas is formed. The Wolffian ducts open to the cloaca. The gonocytes end their migration to undifferentiated gonads.

Stage 16. Hind-limb digital plates and transformation of the first gill fissure into the acoustic meatus (Fig. 12.58). The spinal part of the embryo straightens. The forebrain increases in size. The cerebellum rudiment forms as a thickened plate in the posterior brain vesicle. Vascular plexuses appear in brain ventricles I, II, and III. The first gill fissure is transformed into the external auditory meatus. The forelimbs look like flippers; the distal parts of the hindlimbs become flattened and widened, thereby forming the carpal plate. The eye-pigment membrane shows the presence of pigment, and rudiments of the upper eyelids are formed. Intestinal loops emerge outside through the omphalus and transiently form the so-called omphalic hernia. Interventricular and interatrial septa are formed in the heart. The fifth aortic arch disappears. The pulmonary arch is formed, as well as the rudiments of the hard palate. The fundus of the oral cavity is formed, and the tongue rudiment appears. Endodermal vegetations of the third gill pouch produce the thymus. The ureters become connected with the rudiment of the metanephros. Adrenal rudiments can be distinguished. Gonad differentiation begins.

Stage 17. Round auditory tube opening and the appearance of the first vibrissae (Fig. 12.59). Four rows of vibrissae are present on the maxilla. Vibrissae rudiments are formed under the eye and near the ear. The external auditory opening has a round shape; a small bulge can be seen near its posterior edge, which is the rudiment of the future auricular floor. There is a full complement of somites. The

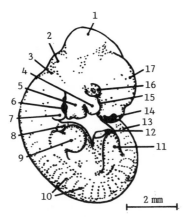

Fig. 12.58. Stage 16. Hind-limb digital plate. Rat embryo. Drawing of whole-mount preparation. 1) Mesencephalon; 2) cerebellum; 3) medulla oblongata; 4) maxillar process; 5) first gill arch; 6) first gill cleft; 7) second gill arch; 8) cardiohepatic eminence; 9) forelimb; 10) somites; 11) hind limb; 12) umbilical cord; 13) tail; 14) olfactory pit; 15) nasoocular groove; 16) eye; 17) telencephalon.

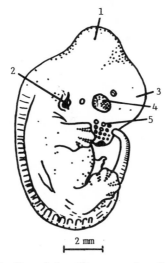

Fig. 12.59. Stage 17. Rounded auditory opening, formation of vibrissae. Rat embryo. Drawing of whole-mount preparation. 1) Mesencephalon; 2) rounded auditory opening; 3) telencephalon; 4) eye; 5) vibrissa rudiment.

spine is almost straight, and the neck flexure is smooth. The upper eyelid increases in size and the lower one is formed. Mesenchymal anlage of the toes can be seen distinctly in the digital plate of the forelimb. The primordia of the mammary glands form along the milk line. External nasal openings are formed; the tongue forms and the oral opening is apparent. Several intestinal loops are found in the omphalic hernia. Mesenchyme and cartilaginous rudiments of the cranial base bones, otic capsule, vertebral bodies and arches, ribs, and limb bones are present. The first ossification centers appear in the mandibula, and the plexus chorioideus appears in the third brain ventricle. The aorta separates from the pulmonary artery. The atrioventricular channel develops. Gonads may be identified either as ovaries or testicles. The liver has a mature configuration. Islets are differentiated in the pancreas.

Stage 18. Flipperlike limbs and the formation of auricular floor rudiment. Forelimbs and hindlimbs with wide digital plates now have the pentaradial shape of flippers. The toes begin to become distinct in the forelimb. Interdigital membranes form between the toes. The number of rows of vibrissae rudiments increases on the maxilla. The fold corresponding to the floor of the auricle can be seen distinctly near the posterior end of the external auditory opening. Hair follicles appear on the dorsal trunk surface. The rudiment of the epiphysis is formed in the midbrain. The tip of the tongue emerges through the gaping mouth. Chondrification of the vertebral bodies, cranial base, ribs, and limbs begins. The aortic arch acquires an adult configuration. The testicles begin to descend. The cloaca differentiates to yield the urogenital sinus and rectum.

Stage 19. Formation of toes. Hair follicles appear all over the body. The cervical flexure straightens; the floor of the auricle increases and partially covers the external auditory meatus. The fissure between the eyelids narrows. There is intense formation of the forelimb and hindlimb toes, and the interdigital membranes reduce. Nails are formed on the forelimb toes. Hair follicles cover the whole body surface including the head. The formation of hard palate is completed, and papillae appear on the tongue surface.

Stage 20. Pentadigital extremities. Cervical flexure disappears and the cervical curvature smoothly passes into the spinal one. Both extremities are pentadigital; nails are formed on hind-leg toes. The mouth is closed, and the tip of the tongue is no longer visible. The umbilical hernia reduces. The fissure between the eyelids narrows to half the size. The entire skeleton is cartilaginous, with ossification centers appearing in the tubular bones, maxilla, and clavicula. Fluid accumulates in nephros; renal glomeruli can be distinguished.

Stage 21. Beginning of eyelid closure and the closure of the external auditory meatus. The neck of the embryo has become markedly elongated and straight. The floor of the auricle almost completely covers the external auditory meatus; the fissure between the eyelids is narrow. Transverse wrinkles appear in the region of the neck, lateral trunk surfaces, and abdomen. Mesonephros degenerates.

Stage 22. Complete closure of the eyelids and of the external auditory meatus. Spine and neck are straight. The floor of the auricle completely covers the external auditory meatus; the eyelids are tightly closed. The muzzle is elongated and the fetus acquires the characteristic appearance of the newborn. Wrinkles appear on the head and extremities. The umbilical hernia has reduced completely. The testicles are located more laterally than the bladder; the prostate buds appear near the ureter. In rabbits the floor of the auricle has markedly increased in length.

Stage 23. Birth of the fetus. Intense increase in the total body mass of fetuses has taken place by the moment of birth. Intense ossification has occurred in all the skeletal bones. The skin is wrinkled into large folds. The mouth, eyes, and ears are closed. Parturition generally takes place at night or in the morning after the following terms: golden hamster, 16–17 days; mouse, 19–20 days; rat, 21–22 days; rabbit, 28–32 days. At birth, newborn rats weight about 2.5–3 g, mice 1–2 g, hamsters 2.5 g, rabbits 48–50 g.

Stage 24. Stage of the newborn animal. Newborn mice, rats, and golden hamsters are blind and unable to walk like most mammals living in burrows. Due to the absence of hair at birth, they have to be warmed by the mother's body; however, as early as the end of the first day of life the rudiments of primary hairs appear on the heads of young animals. The newborn young of all the species mentioned feed on their mother's milk from the first day of life.

REFERENCES

1. A. Albert, "The mammalian testis," *in Sex and Internal Secretions*, Vol. 1, W. C. Young, ed., Williams & Wilkins, Baltimore, Maryland (1961), pp. 305–365.
2. E. Allen, "The estrus cycle in the mouse," *Am. J. Anat.* **30**, 297–371 (1922).
3. P. L. Altman and D. D. Katz, *Inbred and Genetically Defined Strains of Laboratory Animals, Part I, Mouse and Rat*, FASEB, Bethesda, Maryland (1979).
4. P. L. Altman and D. D. Katz, *Inbred and Genetically Defined Strains of Laboratory Animals, Part II, Guinea Pig, Rabbit, and Chicken*, FASEB, Bethesda, Maryland (1980).
5. E. Anderson, "Comparative aspects of the ultrastructure of the female gamete," *Int. Rev. Cytol., Suppl.* **4**, 1–70 (1974).
6. C. R. Austin, "The capacitation of the mammalian sperm," *Nature (London)* **170**, 325 (1952).
7. C. R. Austin, "Ovulation, fertilization, and early cleavage in the hamster (*Mesocricetus auratus*)," *J. R. Microsc. Soc.* **75**, 1401 (1956).
8. C. R. Austin, *Ultrastructure of Fertilization*, Holt, Rinehart, and Winston, New York (1968).
9. R. E. Austin, "Fertilization," in *Reproduction in Mammals. I. Germ Cells and Fertilization*, Cambridge University Press, Cambridge (1971), p. 117.
10. C. R. Austin, "Pattern of metazoan fertilization," *Curr. Top. Dev. Biol.* **12**, 1–9 (1978).
11. C. R. Austin and A. W. H. Braden, "An investigation of polyspermy in the rat and rabbit," *Aust. J. Biol. Sci.* **6**, 674–692 (1953).
12. C. R. Austin and M. W. H. Bishop, "Fertilization in mammals," *Biol. Rev.* **32**, 296–349 (1957).
13. D. N. Baker, "Reproduction and breeding," in *The Laboratory Rat*, Vol. 1, H. J. Baker et al., eds., Academic Press, New York (1979), p. 154.
14. H. Baker, J. K. Lindsey, and S. H. Weisbroth, "Housing to control research variables," in *The Laboratory Rat*, Vol. 1, H. J. Baker et al., eds., Academic Press, New York (1979), p. 169.
15. H. J. Baker, J. K. Lindsey, and S. H. Weisbroth, in *The Laboratory Rat*, Vol. 2, H. J. Baker et al., eds., Academic Press, New York (1980).

16. T. G. Baker, "Electron microscopy of primary and secondary oocyte," in *Advances in Biosciences*, Vol. 6, G. Raspe, ed., Pergamon Press, Oxford (1971), p. 7.

17. C. Barros, "Capacitation of mammalian spermatozoa," in *Physiology and Genetics of Reproduction, Part B*, E. M. Coutinho and F. Fuchs, eds., Plenum Press, New York (1974), pp. 3–24.

18. B. D. Bavister, "Recent progress in the study of early events in mammalian fertilization," *Dev., Growth Differ.* 22, 385–402 (1980).

19. A. R. Beaudoin, "Embryology and teratology," in *The Laboratory Rat*, Vol. 2, H. J. Baker et al., eds., Academic Press, New York (1980), p. 75.

20. J. M. Bedford, "Sperm capacitation and fertilization in mammals," *Biol. Reprod. (Suppl.)*, 2, 129 (1980).

21. J. M. Bedford, "Sperm transport, capacitation, and fertilization," in *Reproductive Biology*, H. Balin and S. Glasser, eds., Excerpta Medica, Amsterdam (1972), pp. 332–392.

22. J. M. Bedford, "An electron microscopic study of sperm penetration into the rabbit egg after natural mating," *Am. J. Anat.* 133, 213–254 (1972).

23. J. M. Bedford, "Evolution of the sperm maturation and sperm storage function of the epididymis," in *Spermatozoon: Maturation, Motility, Surface Properties. Comparative Aspects*, D. W. Fawcett and J. Bedford, eds., Urban & Schwarzenberg, Baltimore, Maryland (1979), pp. 7–21.

24. R. J. Blandau, J. L. Boling, and W. C. Young, "The length of heat in the albino rat as determined by the copulatory response," *Anat. Res.* 79, 453–463 (1941).

25. C. C. Boyer, "Chronology of development for the golden hamster," *J. Morphol.* 92, 1–37 (1953).

26. C. C. Boyer, *The Golden Hamster: Its Biology and Use in Medical Research*, Hoffman, Robinson, and Magalhals, New York (1968).

27. F. Brambell, "Ovarian changes," in *Marshall's Physiology of Reproduction*, Vol. 1, Part 1, A. S. Parkes, ed., Little, Brown and Co., London (1956), p. 397.

28. P. L. Burlingame and J. A. Long, "The development of the external form of the rat, with some observations on the origin of the extraembryonic coelom and fetal membranes," *Calif. Univ. Publ. Zool.* 43, 143–183 (1939).

29. A. D. Burov, in *Laboratory Methods in Veterinary Medicine* [in Russian], Selkhosizdat, Moscow (1954), p. 205.

30. E. O. Butcher, "Development of somites in the white rat," *Am. J. Anat.* 44, 381–439 (1929).

31. R. S. Chacon and P. Talbot, "Early stages in mammalian sperm–oocyte plasma-membrane fusion," *J. Cell. Biol.* 87, 131 (1980).

32. A. Champlin, D. Dorr, and A. Gates, "Determining the stage of the estrous cycle in the mouse from the appearance of the vagina," *Biol. Reprod.* 8, 491–494 (1973).

33. M. C. Chang and R. Hunter, "Capacitation of mammalian sperm: biological and experimental aspects," in *Handbook of Physiology*, Section 7, Vol. 5, R. O. Greep, ed., Williams & Wilkins, Baltimore, Maryland (1977), p. 339.

34. G. Christie, "Developmental stages in somite and postsomite rat embryo based on external appearance, and including some features of the macroscopic development of the oral cavity," *J. Morphol.* 114, 263–286 (1964).

35. Y. Clermont, "Kinetics of spermatogenesis in mammals: seminiferous epithelium cycle and spermatogonial renewal," *Physiol. Rev.* **51**, 198–236 (1972).
36. M. M. Dickie, "Keeping records," in *Biology of the Laboratory Mouse*, E. L. Green, ed., Dover, New York (1966), p. 23.
37. M. Dvorak, J. Stasna, S. Cech, P. Travnik, and D. Horky, *The Differentiation of Rat Ova during Cleavage*, Springer-Verlag (1978), pp. 1–132.
38. A. P. Dyban, in *Methods in Developmental Biology* [in Russian], Nauka, Moscow (1974).
39. A. P. Dyban and V. S. Baranov, *Modern Problems in Oogenesis* [in Russian], Nauka, Moscow (1977).
40. A. P. Dyban and L. I. Khozhai, "Parthenogenetic development of mouse-ovulated eggs induced by ethyl alcohol," *Byull. Eksp. Biol. Med.* **139**, 487–489 (1980).
41. J. A. Edwards, "The external development of the rabbit and rat embryo," *Adv. Teratol.* **3**, 239–263 (1968).
42. R. Edwards and A. Gates, "Timing of the stages of maturation divisions, ovulation, fertilization, and the first cleavage of eggs for adult mice treated with gonadotropins," *J. Endocrinol.* **18**, 292–304 (1959).
43. R. Edwards, "Fertilization," in *Conception in the Human Female*, R. G. Edwards, ed., Academic Press, New York (1980), pp. 573–667.
44. D. S. Falconer, *Breeding Notes for Breeders of Common Laboratory Animals*, G. Poter and W. Lane-Petter, eds., Academic Press, New York (1962), p. 111.
45. D. W. Fawcett, "A comparative view of sperm ultrastructure," *Biol. Reprod.* **2** (Suppl. 2), 90–127 (1970).
46. D. W. Fawcett, "Observations on cell differentiation and organelle continuity in spermatogenesis," in *Proc. Intern. Symp. Genetics of the Spermatozoa*, Edinburgh (1972), pp. 37–68.
47. D. W. Fawcett, "The mammalian spermatozoa," *Dev. Biol.* **44**, 394–436 (1975).
48. J. C. Friedman, G. Mahovy, and H. Tuffrau, "Quelques characteristiques de la physiologie sexuelle chez les rongeurs de laboratoire," *Exp. Anim.* **1**, 111–117 (1968).
49. P. P. Gambaryan and N. C. Dukelskaya, *The Rat* [in Russian], Nauka, Moscow (1955).
50. A. H. Gates, L. L. Doyb, and R. W. Noyes, "A physiological basis for heterosis in hybrid mouse fetus," *Am. Zool.* **1**, 449–450 (1961).
51. A. H. Gates, "Maximizing yield and developmental uniformity of eggs," in *Methods in Mammalian Embryology*, J. Daniel, ed., W. H. Freeman, New York (1971), p. 64.
52. A. P. Graves, "Development of the golden hamster, *Cricetus auratus* Waterhouse, during the first 9 days," *Am. J. Anat.* **77**, 219–251 (1945).
53. E. L. Green, ed., *Biology of the Laboratory Mouse*, Dover, New York (1966).
54. M. C. Green, *Genetic Variants and Strains of the Laboratory Mouse*, Gustav Fisher Verlag, Stuttgart (1981), p. 476.
55. P. W. Gregory, "The early embryology of the rabbit," *Carnegie Inst. Wash. Contrib. Embryol.* **21**, 141 (1930).

56. H. Gruneberg, "The development of some external features in mouse embryos," *J. Hered.* **33**, 89–92 (1943).
57. B. J. Gulyas, "The rabbit zygote. II. The fate of annulate lamellae during first cleavage," *Z. Zellforsch. Mikrosk. Anat.* **133**, 187–200 (1972).
58. R. Gwatkin, "Fertilization," in *Cell Surface in Animal Embryogenesis and Development*, G. Poste and G. L. Nicholson, eds., Elsevier–North-Holland, Amsterdam, New York, London (1976).
59. E. Hafez and C. Thibault, *The Biology of Spermatozoa*, Karger, Basel (1975).
60. M. Healy, A. McLaren, and D. Michie, "Fetal growth in the mouse," *Proc. R. Soc. London* **B153**, 367 (1960).
61. B. Henneberg, *Normentafeln zur Entwicklungsgeschichte der Wanderratte (Rattus norvegicus Erleben)*, Keibel, Jena (1937).
62. W. Hilscher and H. B. Makoski, "Histologische und autoradiographische Untersuchungen zur Praspermatogenose und Spermatogenese der Ratte," *Z. Zellforsch.* **86**, 327–350 (1968).
63. W. G. Hoag and M. M. Dickie, "Nutrition," in *Biology of the Laboratory Mouse*, E. L. Green, ed., Dover, New York (1966), p. 39.
64. M. H. Kaufmann, "Timing of the first cleavage division of the mouse and the duration of its component stages: a study of living and fixed eggs," *J. Cell Sci.* **3**, 799–803 (1973).
65. G. C. Kent, "Physiology of reproduction," in *The Golden Hamster. Its Biology and Use in Medical Research*, McGraw-Hill, New York (1966), p. 113.
66. L. I. Khozhai, "Concerning characterization of stages in the course of a preimplantation development of laboratory mouse embryos," *Byull. Exp. Biol. Med.* **140**, 609–611 (1980).
67. L. I. Khozhai, "Studies of karyology in the course of early embryogenesis after fertilization or parthenogenetic development in mice," Candidate's Dissertation, Leningrad (1981), pp. 1–26.
68. K. L. Kovalevsky, *Breeding of Laboratory Animals* [in Russian], Nauka, Moscow (1951).
69. N. V. Kozlyakov, *Handbook on Feeding of Laboratory Animals and Fowl* [in Russian], Nauka, Moscow (1968).
70. M. Krishna and M. W. Generoso, "Timing of sperm penetration, pronuclear formation, pronuclear DNA synthesis, and first cleavage in naturally ovulated mouse eggs," *J. Exp. Zool.* **202**, 245–252 (1977).
71. W. Lane-Petter and A. E. G. Pearson, *The Laboratory Animal – Principles and Practice*, Academic Press, New York (1971).
72. C. Leblond, E. Steinberg, and E. Roosen-Runge, "Spermatogenesis," in *Mechanisms Concerned with Conception*, C. Hartman, ed., Pergamon Press, New York (1963), pp. 1–72.
73. J. A. Long and H. Evans, "The estrous cycle in the rat," *Mem. Univ. Calif.* (1922), p. 6.
74. F. Longo, "Fertilization: a comparative ultrastructural review," *Biol. Reprod.* **9**, 149 (1973).
75. F. Longo, "An ultrastructural analysis of spontaneous activation of hamster eggs aged *in vivo*," *Anat. Rec.* **179**, 27–56 (1974).
76. A. M. Mandl, "Preovulatory changes in the oocyte of the adult rat," *Proc. R. Soc. London* **158**, 105–141 (1963).

77. D. May and K. Simpson, "An improved method for synchronizing estrus in the rat," *J. Inst. Anim. Tech.* **22**, 133–139 (1971).
78. A. McLaren and P. Bowman, "Genetic effects on the timing of early development in the mouse," *J. Embryol. Exp. Morphol.* **30**, 491–498 (1973).
79. N. N. Medvedev, *Strains of Mice* [in Russian], Meditsina, Moscow (1964).
80. E. C. Melby and N. Altman, *Handbook of Laboratory Animal Science*, CRC Press, Cleveland, Ohio (1976).
81. B. Miller and D. Armstrong, "Superovulatory doses of pregnant mare serum gonadotropin cause delayed implantation and infertility in immature rats," *Biol. Reprod.* **25**, 253–260 (1981).
82. B. Miller and D. Armstrong, "Effects of a superovulatory dose of PMSG on ovarian function, serum estradiol, and progesterone levels and early embryo development in immature rats," *Biol. Reprod.* **25**, 261–271 (1981).
83. C. S. Minot, "The development of the rabbit," in *Normentafeln zur Entwicklungsgeschichte der Wirbeltiere*, Vol. 5, Keibel, Jena (1905), pp. 1–98.
84. H. D. M. Moore and J. M. Bedford, "Ultrastructure of the equatorial segment of hamster spermatozoa during penetration of oocytes," *J. Ultrastruct. Res.* **62**, 110–117 (1978).
85. R. M. Moor and D. G. Gran, "Intercellular coupling in mammalian oocytes," in *Development in Mammals*, M. H. Johnson, ed., Vol. 4, Elsevier–North-Holland, New York (1980), pp. 3–37.
86. J. S. Nicholas, "Experimental methods and rat embryos," in *The Rat in Laboratory Investigation*, E. J. Farris and J. Q. Griffith, Jr., eds., Hafner, New York (1962).
87. H. Nischimura and H. Yamamura, "Comparison between man and some other mammals of normal and abnormal developmental processes," in *2nd Intern. Workshop on Teratology* (1969), pp. 165–173.
88. K. Nuti, B. Sridharan, and R. Meyer, "Reproductive biology of PMSG-primed immature female rats," *Biol. Reprod.* **13**, 38–44 (1975).
89. E. F. Oakberg, "Spermatogonial stem cell renewal in the mouse," *Anat. Rec.* **169**, 515–532 (1971).
90. A. Ochs, "Die Intrauterine Entwicklung des Hamsters bis zum Beginn der Herzbildung," *Z. Wiss. Zool. Abt. A* **89**, 193 (1908).
91. E. M. Otis and R. Brent, "Equivalent ages in mouse and human embryos," *Anat. Rec.* **120**, 33–63 (1954).
92. L. Piko, "Gamete structure and sperm entry in mammals," in *Fertilization*, C. B. Metz and A. Monroy, eds., Academic Press, New York (1969), pp. 325–403.
93. H. G. Pogosianz and O. J. Sohova, "Maintaining and breeding of the Djungarian hamster under laboratory conditions," *Z. Versuchstierkd.* **9**, 292–297 (1967).
94. G. Porter and W. Lane-Petter, *Notes for Breeders of Common Laboratory Animals*, Academic Press, New York (1962).
95. A. K. Rafferty, *Methods of Experimental Embryology of the Mouse*, Johns Hopkins Press, Baltimore (1970).
96. R. Rahilly, J. Bossy, and F. Muller, "Introduction à l'étude des stades embryonnaires chez l'homme," *Bull. Assoc. Anat.* **65**, 5–99 (1981).
97. R. Robinson, *The Genetics of the Norway Rat*, Pergamon Press, Oxford (1965).

98. D. G. de Rooij, "Stem cell renewal and duration of spermatogonial cycle in the golden hamster," *Z. Zellforsch. Mikrosk. Anat.* **89**, 133–136 (1968).

99. R. Rugh, *Vertebrate Embryology; The Dynamics of Development*, Harcourt, Brace, Jovanovich, New York (1964).

100. R. Rugh, *The Mouse: Its Reproduction and Development*, Burgess Publishing Company, Minneapolis (1968).

101. P. P. Sakharov, *Laboratory Mice and Rats* [in Russian], Moscow (1933).

102. P. P. Sakharov, *Laboratory Animals* [in Russian], Biomedgiz, Moscow (1937).

103. R. Schenk, "Uber die Entwicklung des Goldhamsters," *Rontgen-Lab.* **8**, 14–25 (1955).

104. G. A. Schmidt, "Rabbit development," in *Animal Embryology, Part II* [in Russian], Nauka, Moscow (1953).

105. R. Shalgi and P. F. Kraicer, "Timing of sperm transport, sperm penetration, and cleavage in the rat," *J. Exp. Zool.* **204**, 353–360 (1978).

106. J. Shire and W. Whitten, "Genetic variation in the timing of first cleavage in mice: effect of paternal genotype," *Biol. Reprod.* **23**, 363–368 (1980).

107. J. Shire and W. Whitten, "Genetic variation in the timing of first cleavage in mice: effect of maternal genotype," *Biol. Reprod.* **23**, 369–376 (1980).

108. G. Snell and L. Stevens, "The early embryology," in *Biology of Laboratory Mouse*, E. Green, ed., Dover, New York (1966), p. 205.

109. J. Sobotta, "Die Entwicklung des Eis der Maus vom ersten Auftreten des mesoderms an bis zur Ausbildung der Embryonalenanlage und dem Auftreten der Allantois," *Arch. Mikrosk. Anat.* **78**, 271–352 (1911).

110. J. Sotelo and K. Porter, "An electron microscope study of the rat ovum," *J. Biophys. Biochem. Cytol.* **5**, 327–341 (1959).

111. R. Stambaugh, "Enzymatic and morphological events in mammalian fertilization," *Gamete Res.* **1**, 65–85 (1978).

112. G. Streeter, "Developmental horizons in human embryos," *Carneg. Inst. Wash. Contrib. Embryol,* **30**, 213–230 (1942).

113. K. K. Surkova, "Concerning cytological and cytochemical changes during spermatogenesis," *Vestn. Leningr. Gos. Univ.* **15**, 53–71 (1957).

114. P. G. Svetlov and G. F. Korsakova, "Blastocyst implantation in rats," *Dokl. Akad. Nauk SSSR* **103**, 503–506 (1955).

115. E. E. Swiersta and R. H. Foote, "Cytology and kinetics of spermatogenesis in the rabbit," *J. Reprod. Fertil.* **5**, 309–322 (1963).

116. D. Szollosi, "Modification of the endoplasmic reticulum in some mammalian oocytes," *Anat. Rec.* **158**, 59–73 (1967).

117. P. V. Terentyev, V. B. Dubinin, and G. A. Novikov, *The Rabbit* [in Russian], Nauka, Moscow (1952).

118. K. Theiler, *The House Mouse: Development & Normal Stages from Fertilization to 4 Weeks of Age*, Springer-Verlag, Berlin (1973).

119. C. Thibault, "La fécondation, sa preparation et sa réalisation," in *Fecondation Colleg. Soc. Nat. Étud. Steril. et Fécond.*, Paris (1975), p. 1.

120. T. Utakoji, "Chronology of nucleic acid synthesis in meiosis of the male Chinese hamster," *Exp. Cell Res.* **42**, 585–596 (1966).

121. J. H. Venables, "Preimplantation stages in the golden hamster (*Cricetus auratus*)," *Anat. Rec.* **94**, 105–120 (1946).

122. M. C. Ward, "The early development and implantation of the golden hamster, *Cricetus auratus*, and the associated endometrial changes," *Am. J. Anat.* **82**, 231–276 (1948).

123. R. G. Warner, "Nutrient requirements of the laboratory rat," NAS-NRC Publ. 990 (1962).

124. A. Waterman, "Normal development of the New Zealand white strain of rabbit," *Am. J. Anat.* **72**, 473–515 (1943).

125. B. Weakley, "Comparison of cytoplasmic lamellae and membranous elements in the oocytes of five mammalian species," *Z. Zellforsch.* **85**, 109–123 (1968).

126. W. K. Whitten and C. P. Dagg, "Influence of spermatozoa on the cleavage rate of mouse eggs," *J. Exp. Zool.* **148**, 173–183 (1961).

127. W. Whitten and A. Champlin, "Pheromones, estrus, ovulation, and mating," in *Methods in Mammalian Embryology*, J. Daniel, ed., W. H. Freeman, New York (1978), p. 403.

128. R. Yanagimachi, H. Yanagimachi, and B. Roger, "The use of zona-free animal ova as a test system for the assessment of the fertilizing capacity of human spermatozoa," *Biol. Reprod.* **15**, 471–476 (1976).

129. R. Yanagimachi, "Sperm–egg association in mammals," *Curr. Top. Dev. Biol.* **12**, 83 (1978).

130. R. Yanagimachi, "Mechanisms of fertilization in mammals," in *Fertilization and Embryonic Development in vitro*, L. Mastroianni and J. D. Biggers, eds., Plenum Press, New York (1981), p. 81.

131. L. Zamboni, "Ultrastructure of mammalian oocytes and ova," *Biol. Reprod. (Suppl.)* **2**, 44–63 (1970).

132. L. Zamboni, *Fine Morphology of Mammalian Fertilization*, Harper and Row, New York (1971).

133. I. P. Zapadnyuk, V. I. Zapadnyuk, and Y. A. Zakhariya, *Laboratory Animals (Their Breeding, Keeping, and Use in Experiments)* [in Russian], Kiev (1962).

134. M. X. Zarrow and R. Gollo, "Action of progesterone on PMSG-induced ovulation in the immature rat," *Endocrinology* **84**, 1274–1276 (1969).

135. W. C. Young, J. L. Boling, and R. J. Blandau, "The vaginal smear picture, sexual receptivity, and time of ovulation of the albino rat," *Anat. Rec.* **80**, 37–45 (1941).

136. V. Puchkov, "Equivalent stages of normal development of chicken, rat, and human embryos," *Dokl. Akad. Nauk SSSR* **125**, 684–687 (1959).

INDEX